ENGINEERING OF MIND

ENGINEERING OF MIND

An Introduction to the Science of Intelligent Systems

James S. Albus
Senior NIST Fellow
Intelligent Systems Division
Manufacturing Engineering Laboratory
National Institute of Standards and Technology

Alexander M. Meystel
Professor of Electrical and Computer Engineering
Drexel University
and
Guest Researcher
National Institute of Standards and Technology

A Wiley-Interscience Publication

JOHN WILEY & SONS, INC.

New York · Chichester · Weinheim · Brisbane · Singapore · Toronto

Copyright © 2001 by John Wiley & Sons, Inc. All rights reserved.

Published simultaneously in Canada.

For ordering and customer service, call 1-800-CALL-WILEY.

Library of Congress Cataloging-in-Publication Data:

Albus, James Sacra.
    Engineering of mind: an introduction to the science of intelligent systems/by James S.
 Albus and Alexander M. Meystel.
      p.   cm.   – (Wiley series on intelligent systems)
   ISBN 0-471-43854-5 (acid-free paper)
   1. Intelligent control systems. 2. Robotics. 3. Artificial intelligence. I. Meystel, A.
(Alex) II. Title. III. Series.

TJ217.5.A428   2001
629.8'92–dc21   2001024703

Printed in the United States of America

10 9 8 7 6 5 4 3 2

To Cheryl and Marina
without whose support
this book would never have been written

CONTENTS

PREFACE

WHAT IS MIND, THAT WE MIGHT ENGINEER IT?

In 1917, when Warren McCulloch entered college, he was asked by the Quaker philosopher Rufus Jones "What is thee going to be? and What is thee going to do?" McColloch replied, "I have no idea; but there is one question I would like to answer: What is a number, that a man may know it, and a man, that he may know a number?" Jones replied, "Friend, thee will be busy as long as thee lives" [McColloch61]. During his lifetime McColloch was able to answer the first part of his question and to lay the foundation for answering the second. During the first half of the twentieth century, mathematicians developed a formal theory of what a number is, and during the second, cognitive scientists have developed some plausible hypotheses for how a person might know a number. In this book, we look to a more advanced question: What is mind, that we might engineer it?

Engineering is based on science. Engineering of mind implies that there exists a scientific theory of mind upon which engineering principles and methodologies can be based. A formal theory of mind does not yet exist. Not even a widely accepted definition of mind exists. Many believe mind to be simply a vague metaphor for mental activities. Many are skeptical that a theory of mind is even possible. Many, in the tradition of Cartesian dualism, believe that the mind is of a different nature than physical reality, in the realm of the spirit and soul, and therefore not amenable to scientific study. Many believe mind to be too ill-defined to be a fit subject for scientific inquiry, much less for engineering design.

What is mind? There are many phenomena that are commonly attributed to the mind. These include imagination, thought, reason, emotion, perception, cognition, knowledge, communication, planning, wisdom, intention, motives, memory, feelings, behavior, creativity, consciousness, intelligence, intuition, and self.

There is no doubt that the mind is different from the *brain*. The brain is a physical thing. It has mass and occupies space. It is made mostly of water and hydrocarbons organized in the form of glial cells and neurons. The brain is part of the central nervous system, which is part of the body. The brain provides computing services to support perception, cognition, and control of the body. The brain is essentially a machine that computes, just as the body is a machine that acts on, and maneuvers within, the world.

The *mind* is not a physical thing. It is a process—just as life is a process. The process of mind occurs within the brain just as the process of life occurs within the body. It is tempting to say that the mind is a process that runs on the brain just as a program is a process that runs on a computer. But the analogy is flawed. First, a program is not a process. A program is simply a list of instructions. A process does not occur until a program is executed. Second, the mind is not a program. The program that runs in the brain is part of the brain itself. The brain's program is stored in the neurons and synapses of the brain. Thus, the brain is both machine and program. Third, the process of mind is not entirely, or even mostly, determined by the brain's program. Input from sensors affects the mind profoundly. The activity of mind is not determined solely by the program that is stored in the brain but also on the input received from the environment. The process of mind results from the brain executing its program while simultaneously being bombarded by signals pouring in from literally millions of sensors reporting on stimulation from the environment. The mind is a closed-loop process that senses, perceives, understands, thinks, plans, and acts in response to stimuli. The mind both acts on the world and reacts to the world.

In this book we hypothesize that the difference between mind and brain is similar to the difference between life and body. Just as life is a process that takes place within the body, so mind is a process that occurs within the brain. If our hypothesis is valid, the process of mind is not beyond the realm of scientific inquiry any more than is the process of life. In our hypothesis, *imagination* is a process of visualization and conceptualization (i.e., generating images from assumptions about state, attributes, and relationships of objects, events, situations, and classes). *Thinking* is a process of imagining what might occur if certain actions were taken and certain conditions were achieved, and analyzing the results. *Reasoning* is a process by which rules of logic are applied to representations of knowledge about the world during the process of thinking. *Emotion* is a mental state or feeling that results from a value judgment process evaluating what is good or bad, attractive or repulsive, important or trivial, loved or hated, hoped for or feared. *Feeling* is a pattern of activity that is perceived as pain, pleasure, joy, grief, hope, fear, love, hate, anxiety, or contentment. *Perception* is a process by which sensory input is transformed into knowledge about the world. *Knowledge* is information that is structured so as to be useful for thinking and reasoning. *Cognition* is a collection of processes by which knowledge is acquired and evaluated, awareness is achieved, reasoning is carried out, and judgment is exercised. *Meaning* is the set of semantic relationships that exist between the knowledge database and the external world. Meaning establishes what is intended or meant by behavioral actions, and defines what entities, events, and situations in the knowledge database refer to in the world.

Understanding occurs when the system's internal representation of external reality is adequate for generating intelligent behavior. *Planning* is a thinking process by which a system imagines the future and selects the best course of action to achieve a goal state. *Wisdom* is the ability to make decisions that are most likely to achieve high-level long-range goals. *Introspection* is a process by which a system examines its own internal state and capabilities and reasons about its own strengths and weaknesses. *Reflection* is a process by which a system rehearses, analyzes, or thinks about the meaning of situations and events. *Reflexion* is a process whereby a system considers what others think about what it is thinking. *Attention* is a mechanism by which an intelligent system directs sensors and narrows its focus to those sensory inputs and internal representations that are important to its current goals—and ignores what is unimportant. *Awareness* is a condition wherein a system has knowledge of the structure, dynamics, and meaning of the environment in which it exists. *Consciousness* is a state or condition in which an intelligent system is aware of itself, its surroundings, its situation, its intentions, and its feelings.

A scientific theory of mind does not yet exist, but an understanding of mind is developing faster than most people appreciate. Progress is rapid in many different fields. Recent results from a number of disciplines have laid the foundations for a computational theory of intelligence. In this book we outline the mainstreams of research that we believe will eventually converge in a scientific theory of mind. We also propose an avenue of research that we believe could lead to the engineering of mind. We do not presume that ours is the only road to this goal. There may be many others. However, we do argue that any road map leading to the engineering of mind must pass the milestone of highly intelligent systems.

We argue that highly intelligent behavior is the result of goals and plans interacting at many hierarchical levels with knowledge represented in a multiresolutional world model. We argue that high levels of intelligence require a rich dynamic world model that includes both a priori knowledge and information provided by sensors and a sensory processing system. We argue that intelligent decision making requires a value judgment system that can evaluate what is good and bad, important and trivial, and can estimate cost, benefit, and risk of potential future actions. We believe that our model of intelligence can enable the engineering design of intelligent systems that pursue goals, imagine the future, make plans, and react to what they see, feel, hear, smell, and taste. This will enable the development of systems that behave as if they are sentient, knowing, caring, creative individuals motivated by hope, fear, pain, pleasure, love, hate, curiosity, and a sense of priority. We believe that research on highly intelligent systems will yield important insights into elements of mind such as attention, gestalt grouping, filtering, classification, imagination, thinking, communication, intention, motivation, and subjective experience. We believe that as the systems we build grow increasingly intelligent, we will begin to see the outlines of what can only be called mind. We hypothesize that mind is a phenomenon that will emerge when intelligent systems achieve a certain level of sophistication in sensing, perception, cognition, reasoning, planning, and control of behavior.

Clearly, mind is more than intelligence. Intelligence is merely a prerequisite, a substrate in which mind may emerge. The mind goes beyond simply generating appropriate behavior. The mind distinguishes between right and wrong, good and evil. The mind is where resides a sense of justice, honor, duty, reverence, beauty, wonder, and religious experience. The mind can experience hope, fear, happiness, grief, pride, shame, and guilt. The mind can harbor a sense of the mysterious. The mind can wonder, worship, and contemplate its own place in the universe. Many aspects of mind will elude scientific understanding for a long time. Nevertheless, we predict that as more is learned about the fundamental mechanisms of intelligence, some of the deeper mysteries of mind will be revealed.

In this book we present a model of the brain as a hierarchy of massively parallel computational modules and data structures interconnected by information pathways that enable analysis of the past, estimation of the present, and prediction of the future. We hypothesize that algorithms, procedures, and data embedded within this architecture will enable the analysis of situations, the formulation of plans, the choice of behaviors, and the computation of current and expected rewards, punishments, costs, benefits, risks, priorities, and motives.

We believe that a rich internal representation of the world is indispensable to higher levels of intelligent behavior. We assert that goals, motives, and priorities are essential features of intelligent behavior, and that model based planning is the best way to find a behavioral trajectory between a starting state and a goal state. We postulate that the concept of goals, tasks, plans, and control are indispensable to understanding what the brain is for and what the mind does.

We specifically reject the behaviorist model of brain, both as a theory and a guide for intelligent systems research. Although behaviorism yields important insights into simple forms of intelligence such as can be found in insects, fish, and reptiles, it is inadequate as a tool for investigating the higher levels of intelligence where mind is likely to emerge. Behaviorism minimizes or ignores internal knowledge representation and eschews concepts such as understanding, intention, and imagination.

We believe that there is overwhelming evidence that the fundamental architecture of the brain is a hierarchy of laterally interconnected computational modules. We argue that this basic principle will be required in any system designed to approach the performance of the human brain. We claim that a multiresolutional representation of knowledge and hierarchical organization of command and control is the most efficient and effective way for individuals and groups to plan and control intelligent behavior in a complex dynamic world. We believe that the computational power to implement our model can be achieved in practical systems in the foreseeable future through hierarchical and lateral distribution of computational tasks.

We describe a reference model architecture that we believe can support an engineering methodology for the design and construction of intelligent machine systems. This architecture consists of a multiresolutional hierarchy with layers of interconnected computational nodes each containing elements of sensory processing, world modeling, value judgment, and behavior generation. At the lower levels, these elements generate goal-seeking reactive behavior. At higher levels these elements enable perception, cognition, reasoning, imagination, and planning. Within each level of our proposed architecture, the product of range and resolution in time and space is limited. At low levels, range is short and resolution is high, whereas at high levels, range is long and resolution is low. This enables high precision and quick response to be achieved at low levels over short intervals of time and space, while long-range plans and abstract concepts can be formulated at high levels over broad regions of time and space.

Our proposed reference model architecture accommodates concepts from artificial intelligence, control theory, image understanding, signal processing, and decision theory. The artificial intelligence concept of a plan is expressed as a series of subgoals, waypoints, or desired states and is shown to be equivalent to the control theory concept of a reference trajectory. A plan expressed in the form of a series of actions is equivalent to a feedforward command string. Value judgments can be expressed as cost functions, confidence factors, and measures of importance. Our reference model architecture is expressed in terms of the Real-Time Control System (RCS) that has been developed at the National Institute of Standards and Technology and elsewhere over the last 25 years. RCS provides a design methodology, software development tools, and a library of software that is free and available via the Internet. Application experience with RCS provides examples of how this reference model can be applied to problems of practical importance.

In Chapter 1 we outline evidence for the emergence of a computational theory of mind. In Chapters 2, 3, and 4 we examine the fundamental concepts of knowledge, perception, and goal-seeking behavior. Chapter 5 defines the basic building blocks of a reference model architecture for engineering intelligent systems. Chapter 6 deals with how to engineer behavior generating processes. In Chapter 7 we outline world modeling and value judgment processes and suggest data structures for designing a knowledge database. In Chapter 8 we discuss current issues in sensory processing and propose an engineering approach to designing a perception system. Chapter 9 provides an example of how our proposed reference model architecture is currently being applied to the design of intelligent unmanned military scout vehicles. In Chapter 10 we suggest what impact the ability to engineer mind might have on science, economics, politics, and the well-being of humankind.

JAMES S. ALBUS
ALEXANDER M. MEYSTEL

ENGINEERING OF MIND

1 Emergence of a Theory

A collage by A. Meystel from *A Composition* by Jacob Vin'kovetsky, and detail from the *Creation of Adam* by Michelangelo, from a fresco in the Sistine Chapel. [Courtesy of D. Vin'kovetsky and the Vatican Museums.]

What is mind? What are its fundamental properties? What are its elemental constituents? Where did it come from? How and why did it arise in nature? Can it be created artificially? If so, how?

The answers to these questions are beginning to emerge. At the dawning of the third millennium anno Domini, we are beginning to see real answers to questions that have occupied the attention of philosophers and scientists for many centuries. Since the fifth century B.C., the nature of knowledge and the mechanisms of human reason have been subjects of serious intellectual inquiry. Great thinkers such as Plato, Aristotle, Bacon, Decartes, Hume, Kant, Pascal, von Helmholtz, Darwin, James, Freud, Russell, Mach, Pierce, Piaget, and many others have thought deeply, written extensively, and provided fundamental insights into the mysteries of the mind. Today, both natural and artificial forms of intelligence are subjects of intense study. There is good evidence that we have entered a period in which the ancient mysteries of the mind are yielding to scientific inquiry.

In this century, the neurosciences have provided deep insights into the anatomical, physiological, chemical, and computational basis of cognition. Neuroanatomy has described the shape and structure of the basic computational element of the brain—the neuron, and produced extensive maps of the computational modules and interconnecting data flow pathways making up the structure of the brain. Neurophysiology is demonstrating how neurons compute a rich library of mathematical and logical functions, and how they store, retrieve, and communicate information. Neuropharmacology is discovering many of the transmitter substances that control the emotions, compute value judgments, assign priorities, generate reward and punishment, activate behavior, and mediate learning. Behavioral psychology provides information about stimulus–response behavior and instrumental conditioning. Psychophysics offers many clues about how the brain operates to perceive objects, events, time, and space, and how it reasons about relationships between the self and the external world. Cognitive psychology is exploring how the brain represents knowledge; how it perceives objects, events, situations, and relationships; how it analyzes the past and plans for the future; and how it selects and controls behavior that satisfies desires and achieves goals. Recently, new imaging techniques have begun to reveal the functional modules of the brain, much as x-rays reveal the structure of the bones. It is now possible to locate precisely brain activity that is associated with specific experiences and thought processes [Binder97].

Over the last five decades, the invention of the electronic computer has brought rapid advances in computational power, making it feasible to launch serious attempts at building intelligent systems. Artificial intelligence and robotics have produced significant results in planning, problem solving, rule-based reasoning, image analysis, and speech understanding. Research in learning automata, neural nets, fuzzy systems, and brain modeling provides insight into adaptation and learning, and understanding of the similarities and differences between neuronal and electronic computing processes. Game theory and operations research have developed methods for decision making in the face of uncertainty. Genetic algorithms and evolutionary programming have developed methods for getting computers to generate successful behavior without being explicitly programmed to do so. Autonomous vehicle research has produced advances in real-time sensory processing, world modeling, navigation, path planning, and obstacle avoidance. Intelligent vehicles and weapons systems are beginning to perform complex military tasks with precision and reliability. Research in industrial automation and process control has produced hierarchical control systems, distributed databases, and models for representing processes and products. Computer-integrated manufacturing research has achieved major advances in the representation of knowledge about object geometry, process planning, network communications, and intelligent control for a wide variety of manufacturing operations. Modern control theory has developed precise understanding of stability, adaptability, and controllability under various conditions of uncertainty and noise. Research in sonar, radar, and optical signal processing has developed methods for fusing sensory input from multiple sources and assessing the believability of noisy data. Progress is rapid in each of the foregoing fields, and there exists an enormous and rapidly growing literature in all of these areas.

From many sources, there is emerging at least the broad outline of a general theoretical framework that might best be called a *computational theory of mind*. This framework promises to transform the study of mind from a philosophical debate into an experimental science. Among the first to attempt to model the brain were Hebb, McCulloch, Pitts, and Rosenblatt. Hebb[49] proposed that the computational properties of neuronal circuits can be shaped through modification of synaptic connections between neurons. He hypothesized that the strength of synaptic connections is enhanced by simultaneous activity at pre- and postsynaptic sites. McCulloch and Pitts[43] demonstrated that a simple model of the neuron can compute mathematical and logical functions. Rosenblatt[58] built electronic neural nets called *perceptrons* in an attempt to model the brain. Perceptrons demonstrated rudimentary abilities to recognize and learn patterns. Over the past five decades there has been significant progress in neural nets and connectionist models of the brain. Widrow[95], Grossberg[89], Hopfield[82], Narendra[86], Barto[92], Sejnowski and Rosenberg[87], Kohonen[88], Miller et al.[90], Werbos[94], and many others have produced significant advances in understanding how networks of neuron models can be designed to learn, adapt, classify features, recognize patterns, and compute functions [Haykin94].

Turing[50] proved that digital computers are capable of computing any mathematical or logical function. He was among the first to hypothesize that computers might someday exhibit human levels of intelligence. Turing proposed a formal test for evaluating progress toward this goal. Weiner[48] transformed the theory of feedback control into a science of intelligent systems and coined the term *cybernetics*. Ashby[52] formulated a model of the brain containing multiple control loops. McCarthy[60], Minsky[86], Newell and

Simon[63,72], and others founded the field of artificial intelligence based on symbolic processing. McCarthy invented the LISP programming language. Newell et al.[58] developed methods for programming computers to mimic how human beings solve problems. In *Human Problem Solving*, Newell and Simon[72] attempt to explain the problem-solving powers of the human mind in computational terms. Samuel[59] wrote a computer program to play checkers and programmed it to learn. It was able to learn from experience to improve its performance until eventually it could beat human experts. Robinson[65], Raphael[76], Chang and Lee[73], and others developed methods for automatically proving theorems in the predicate calculus. Zadeh[65] showed how computations can be performed with "fuzzy" as opposed to discrete logic. Holland[75] introduced the concept of genetic algorithms.

Much has also been learned about how computations are performed and memories are stored and recalled in the brain. Sperry's studies of the split brain [Sperry62] and Penfield's analyses of memory deficits following brain lesions [Penfield and Milner58] provided early insights into brain function. During the 1960s, the experiments of Hubel and Weisel[62,68] uncovered many of the fundamental computational functions performed in the vision system. This enabled Marr[82], Rosenfeld[86], Binford[82], Winston[84], Brady and Yuille[87], Koenderink[84], Koenderink and Van Doorn[79], Riseman and Hanson[90], Van Essen[85], Ullman[96], and many others to formulate a computational theory of visual perception. In *The Cerebellum as a Neuronal Machine*, Eccles et al.[67] published a wealth of neurophysiological evidence for computational mechanisms embedded in the neuronal substrate. This led to the Marr–Albus theory of cerebellar function and the CMAC neural net model [Marr69, Albus71,75a,b]. Experiments by Sparks[86] and others [Sparks and Jay87, Sparks and Groh95] have shown how image processing is related to gaze control. Georgopoulos[95] and others have demonstrated how motor behavior is controlled by intention.

The study of language has also provided valuable clues to the structure and function of intelligence. In *Language and Mind*, Chomsky[72] suggests how the deep structures of syntax and grammar of human language can contribute to a science of mind. Miller and Chomsky[63] hypothesize that language provides a mechanism for developing inner models and plans for actions. Fodor[75,83] proposes a *language of thought*.

Selfridge[59] suggested that the brain consists of a multiplicity of agents each with a specialized competence that compete with each other for influence in controlling perception and behavior. In *The Metaphorical Brain*, Arbib[72], integrated knowledge of brain structures with control theory and behavioral science. He used automata theory and algebra to explore the concept of control schema and the relationships between symbols and actions. Arbib[92], Schank and Abelson[77], and others have proposed schema for describing plans and actions. Albus[75b,79,81,91] has suggested a model of the brain consisting of a hierarchical set of control loops for hierarchical planning and problems solving. In *How the Mind Works* Pinker[97] provides a penetrating and accessible review of the scientific and philosophical foundations of a computational theory of mind. In *Mapping the Mind*, Carter[98] summarizes the results of imaging techniques that make visible the neuronal activity that occurs in the brain during perception, thinking, planning, and behavior.

All of the works above and many others support the fundamental concept that the brain is a machine and the mind is a process that occurs in the brain. The essential nature of a machine is a physical structure with functioning parts. The essential nature of a process is motion or change in state over time. A machine may consist of a set of levers and gears, a set of switches, or an electronic circuit. A process occurs when the levers move, the gears turn, the switches open and close, or the electrons flow. A machine requires a process to make it function. A process requires a machine in which to operate. In short, a process is what a machine does. Thus, the mind is what the brain does.

What still eludes us is an understanding of the mind and brain at a level that would enable engineers to design and build intelligent systems that have the ability to perform tasks of manipulation or locomotion in the real world with anything approaching the skill and dexterity of a human being, or even that of a raccoon or squirrel. The current level of performance of artificial intelligence and robotic systems is disappointingly limited in both depth and scope. Early successes with toy problems in the laboratory have not scaled up to solve real problems in the natural world. Current laboratory robots do not have motor skills that approach those of dragonflies, spiders, birds, or squirrels; nor do they possess the commonsense understanding of 4-year-old humans.

We know how to build computers that can perform billions of arithmetic and logical computations per second. We can write software programs that can defeat the world champion in chess. Yet we cannot duplicate the capability of a 6-year-old human in understanding natural language, or even in tying a shoe. We know how to represent knowledge about complex manufactured objects such as automobiles and airplanes. We understand how to control machines that can machine parts to extreme accuracy with great reliability. Yet we have only a vague understanding of how the brain represents knowledge about the natural world, and we have not been able to endow computers with common sense. We know a great deal about brain chemistry and neurophysiology. We understand many details about how neurotransmitters cause changes in voltage that encode signals in the brain. But we are a long way from understanding how neuronal activity in the brain produces the subjective experience of emotional feelings or conscious awareness. We don't know how to build seeing or hearing systems that can perform nearly as well as those of dogs or cats in recognizing and tracking objects of interest by sight and sound. We can't build robots that duplicate the motor performance of a deer or a butterfly. We are unable to mimic the spatial and temporal reasoning abilities of a bird in controlling flight through the branches of a tree, or even that of a housefly in performing intricate maneuvers such as evading being swatted, pursuing or evading other flies, and landing upside down on the ceiling [Borst90].

Prior to the computer age, mathematics, logic, and games of skill such as chess were considered to represent the pinnacle of human intellect. Strangely, these are what have turned out to be relatively easy to duplicate in the computer. And what was considered mundane by human standards—common sense, natural language, and simple motor skills—have proven extremely difficult for computers. This is a great paradox that could not have been anticipated before actually attempting to build intelligent machines. Pinker suggests "that the discovery by cognitive science and artificial intelligence of the technical challenges overcome by our mundane mental activity is one of the great revelations of science, an awakening of the imagination comparable to learning that the universe is made up of billions of galaxies or that a drop of pond water teems with microscopic life" [Pinker97].

WHAT IS INTELLIGENCE?

For the purposes of this book, we define *intelligence* as the ability of a system to behave appropriately in an uncertain environment, where appropriate behavior is that which maximizes the likelihood of success in achieving the system's goals. This definition of intelligence addresses both biological and machine embodiments. It admits a broad spectrum of behaviors, from the simple to the complex. We deliberately do not define intelligence in binary terms (i.e., this machine is intelligent and this one is not, or this species is intelligent and this one is not) and we do not limit our definition of intelligence to behavior that is beyond our understanding. Our definition includes the entire spectrum of intellectual capabilities from that of a paramecium to that of an Einstein, from that of a thermostat to that of the most sophisticated computer system. We include the ability of a robot to spot-weld an automobile body, the ability of a bee to navigate in a field of wild flowers, a squirrel to jump from limb to limb, a duck to land in a high wind, and a swallow to catch insects in flight above a field of wild flowers. We include the ability of blue jays to battle in the bushes for a nesting site, a pride of lions to conduct a coordinated attack on a wildebeest, and a flock of geese to migrate south for the winter. We include a human's ability to bake a cake, play the violin, read a book, write a poem, fight a war, or invent a computer.

Our definition of intelligence recognizes degrees, or levels, of intelligence. These are determined by the following parameters: (1) the computational power and memory capacity of the system's brain (or computer), (2) the sophistication of the processes the system employs for sensory processing, world modeling, behavior generation, value judgment, and communication, and (3) the quality and quantity of information and values the system has stored in its memory. The measure of intelligence is success in solving problems, anticipating the future, and acting so as to maximize the likelihood of achieving goals. Success can be measured by various criteria of performance (including life or death, pain or pleasure, reliability in goal achievement, cost in time and resources, and others). Different levels of intelligence produce different probabilities of success.

Our definition of intelligence also has many dimensions. For example, the ability to understand what is visually perceived is qualitatively different from the ability to comprehend what is spoken. The ability to reason about mathematics and logic lies along a different dimension from the ability to paint or sculpt, perform music, or play sports. The ability to choose wisely involves both the ability to predict the future and the ability to assess accurately the cost or benefit of predicted future states. Along each of these dimensions, there exists a continuum. Thus the space of intelligent systems is a multidimensional continuum wherein nonintelligent systems occupy a point at the origin.

At a minimum, intelligence requires the ability to sense the environment, to make decisions, and to control action. Higher levels of intelligence may include the ability to recognize objects and events, to represent knowledge in a world model, and to reason about and plan for the future. In advanced forms, intelligence provides the capacity to predict the future, to perceive and understand what is going on in the world, to choose wisely, and to act successfully under a large variety of circumstances so as to survive, prosper, and replicate in a complex, competitive, and often hostile environment.

From the viewpoint of control theory, intelligence might be defined as a knowledgeable "helmsman of behavior." Intelligence is a phenomenon that emerges as a result of the integration of knowledge and feedback into a sensory-interactive, goal-directed control system that can make plans and generate effective purposeful action to achieve goals. From the viewpoint of psychology or biology, intelligence might be defined as a behavioral capability that gives each individual a means for maximizing the likelihood of success in achieving its goals in an uncertain and often hostile environment. Intelligence results from the integration of perception, reason, emotion, and behavior in a sensing, perceiving, knowing, feeling, caring, planning, and acting system that can formulate and achieve goals. It is from this combination of processes that we believe the phenomenon of mind can be expected to emerge.

GOALS AND SUCCESS

The notion of a goal is inherent to intelligent behavior, and hence is essential to a scientific theory of mind. A goal is something that intelligent behavior is designed to achieve or maintain. Success is achievement of the goal. Intelligent systems typically have an ultimate, or highest-level goal, and operate by decomposing that goal into subgoals. They then generate actions designed to achieve those subgoals so that the ultimate goal is eventually achieved. Ultimate goals are typically built into an intelligent system by evolution or design.

For intelligent biological creatures, the ultimate goal is propagation of the individual's genes. Intermediate goals are satisfaction of the desires, drives, and urges generated by genetically designed neural computing structures in response to sensed environmental conditions. Intermediate goals cause the organism to behave in a manner that improves the likelihood of success in achieving the ultimate goal. For intelligent machines, the highest level goals are typically defined by requirements generated by prospective users or customers. These requirements provide input to designers and software engineers who build and configure system hardware and software that are capable of achieving those goals in the real world. Users input operational goals that satisfy operational demands in the application domain.

Ultimate biological goals such as survival and gene propagation are decomposed into intermediate goals such as build nest, attract mate, and rear offspring, and finally, into actions that produce behavior. Ultimate robotic goals such as ⟨perform mission objectives⟩, or ⟨manufacture high-quality, low-cost products⟩, are decomposed into intermediate goals such as ⟨travel to location⟩, ⟨perform tasks⟩, and ⟨return safely to base⟩, or ⟨maintain inventory⟩, ⟨plan and schedule manufacturing processes⟩, ⟨inspect⟩, ⟨package⟩, and ⟨ship products⟩. These are finally decomposed into commands to actuators that produce behavior. At each level of decomposition, intelligence generates or selects plans designed to achieve higher-level goals. At each level, intelligence reacts to sensory feedback such that goals are achieved despite perturbations and unexpected events.

THE ORIGIN AND FUNCTION OF INTELLIGENCE

Natural intelligence, like the brain in which it appears and the body that it controls, is a result of the process of natural selection. This is not a new idea. It was one of Darwin's principal conclusions in *Origin of Species* [Darwin1859]. It has periodically been reiterated and elaborated by many others, e.g., Dennett[95]. Genes that produce brains that generate more successful behavior tend to survive and propagate. Genes that produce brains that generate less successful behavior tend to become extinct. Natural intelligence has thus emerged from a competitive struggle for survival and gene propagation that has taken place between trillions of brains, over millions of years. Other factors being equal, success or failure in these struggles has been determined by the intelligence of the competitors. The intelligent brain is therefore a product of natural selection.

Success for biological intelligence can be defined in terms of gene propagation. Brains producing behaviors that increase the likelihood of success in gene propagation are selected-in. Those that decrease the likelihood of successful gene propagation are selected-out. Success for machine intelligence can be defined in the marketplace of economics and ideas. Intelligent machines that are economically practical, or that are theoretically interesting, tend to survive and be replicated. Those that are impractical or theoretically vacuous die out.

For each intelligent system (natural or artificial), intelligence provides a mechanism to generate advantageous behavior. Intelligence improves an individual's ability to act effectively and choose wisely between alternative behaviors. A more intelligent animal has many advantages over less intelligent rivals in acquiring choice territory, gaining access to food, and attracting more desirable mates. The intelligent use of aggression improves an individual's position in the social dominance hierarchy. Intelligent predation improves success in capturing prey and obtaining food for offspring. Intelligent exploration improves success in hunting and establishing territory. Intelligent use of stealth gives a predator the advantage of surprise. Intelligent use of deception improves the prey's chances of escaping from danger.

Higher levels of intelligence produce capabilities in the individual for thinking ahead, planning before acting, and reasoning about the probable results of alternative actions. These abilities give to the more intelligent individual a competitive advantage over the less intelligent in the competition for survival and gene propagation. Intellectual capacities and behavioral skills that produce successful hunting and gathering of food, acquisition and defense of territory, avoidance and escape from danger, and bearing and raising offspring tend to be passed on to succeeding generations. Intellectual capabilities that produce less successful behaviors reduce the survival probability of the brains that generate them. Competition between individuals thus drives the evolution of intelligence within a species.

For groups of individuals, intelligence provides a mechanism for cooperatively generating biologically advantageous behavior. The intelligence to simply congregate in flocks, herds, schools, and packs increases the number of sensors watching for danger. The ability to recognize and communicate danger signals improves the survival probability of all members of the group. The ability to recognize and understand the signals another animal communicates regarding its emotional state and intentions reduces the likelihood of hostile encounters. Communication is most advantageous to those individuals who are the quickest and most discriminating in recognizing messages that indicate danger, anger, happiness, or sexual attraction, and most effective in responding with appropriate action. The intelligence to cooperate in mutually beneficial activities such as hunting and group defense increases the probability of gene propagation for all members of the group.

Within the group, the most intelligent individuals will tend to occupy the best territory, be the most successful in social competition, and have the best chances for their offspring surviving. The most intelligent individuals and groups within a species also tend to occupy the best territory, be the most successful, and have the best chances of passing on genes to the next generation. All else being equal, more intelligent individuals and groups win in competition with less intelligent individuals and groups.

The intellectual capacity to make and use tools, weapons, and spoken language has made humans the most successful of all predators. Humans levels of intelligence have led to the use of fire, the domestication of animals, the development of agriculture, the rise of civilization, the invention of writing, the building of cities, the practice of war, the emergence of science, and the growth of industry. These capabilities have extremely high gene propagation value for the people and societies that possess them relative to those who do not. Intelligence has thus made modern civilized humans the dominant species on the planet Earth.

Intelligence of artificial systems will probably also evolve from competitive pressures in the laboratory, the marketplace, and the battlefield. The effectiveness of different scientific theories and engineering designs will determine which experimental systems will be commercialized. The impact on productivity of future factories, construction sites, and transportation systems will determine which types of intelligent machines and systems are commercially successful. Competitive pressures of war games and actual combat will drive the evolution of intelligent machines in the military. The ability of intelligent machines to work together, to communicate information, and to acquire and use knowledge about the environment will enable entire enterprises to manufacture goods and deliver services in a competitive environment. Those systems that are successful in achieving the goals of their organizations will survive and multiply. Those that lose in market competition will be abandoned.

LEARNING, EVOLUTION, AND INSTINCT

For biological intelligence, learning is necessary to acquire knowledge from experience or from a teacher, or in the case of human intelligence, from written documents. The ability of biological creatures to communicate with language, to follow rules, to reason and plan, to think about the past, and to predict the future all require learning. Learning is crucial to the development of culture, the building of cities, the rise of civilization, the practice of medicine, the conduct of war, and pursuit of science.

For machine intelligence, knowledge can also be acquired through *learning*. For example, parameter estimation, adaptive control, and neural nets are techniques for machine learning. However, intelligent machines can also acquire knowledge through downloading of programs and data, or via the activities of human programmers.

Learning: modification of knowledge or computational mechanisms that take place within the brain of an individual system caused by experiences during its lifetime

Learning takes place within the brain of a single individual and is not passed on to the succeeding generation except through cultural exchanges such as parents teaching children how to behave, oral traditions, written documents, or an education system teaching religion, history, science, and etiquette. Learning is a process of merging what is experienced with what is already known. Learning results from both the storage of immediate experience in short-term memory and the consolidation of short-term memory into long-term memory.

Learning occurs in many places throughout the brain. Learning provides a mechanism for optimizing perception, storing knowledge about the external world, acquiring skills and values, and accumulating knowledge of how to act and think. Intelligence may be increased through the ability to learn. Learning can improve the likelihood of achieving success through improved knowledge about the world and through acquisition of skills for sensing, deciding, and acting. Intelligence may increase over the lifetime of an individual or society through learning. Intelligence may increase over successive generations through *evolution*.

Evolution: modifications of knowledge or computational mechanisms caused by mutation and natural selection that take place in a species over generations

Growth of intelligence through biological evolution is a slow and painful process. A great many individuals with random genetically based variations in behavior must live and die over many generations with a consistent statistical improvement in probability of success in gene propagation to produce learning through evolution. The intelligent capabilities that result from biological evolution are commonly termed *instinct*.

Instinct: the set of knowledge and skills that are acquired by a species through evolution

Over a series of generations, instinctive knowledge and skills that produce more successful behavior are passed along to future generations. Instinctive knowledge and skills that produce less successful behavior tend toward species extinction. Instinctive behavior can be remarkably complex. Ants, termites, spiders, dragonflies, houseflies, bees, and many other creatures that operate almost entirely on instinct can perform extraordinarily complex tasks in highly uncertain environments with extremely high probability of success. In these creatures, learning plays little or no role.

It should be noted that the distinction between learning and instinct is by no means clear and simple. As the nervous system develops under the influence of genetic factors, it forms connections and establishes parameters that govern behavior. As the nervous system experiences input from the environment, some of its neuronal connections may be modified so that perception is altered and different behavior emerges. If the behavior that emerges is the result of maturation only, it is called *instinctive behavior*. If the behavior that emerges is heavily influenced by experience, it is called *learned behavior*. The problem is that the underlying mechanisms that give rise to behavior are not easily separable into those that are modifiable by genetically controlled maturation and those that are influenced by environmental factors. The mechanisms of maturation and learning coexist in most creatures, including humans. Maturation and learning often act on the same neuronal sites during the same time frame. Thus, the effects of the two mechanisms are often impossible to distinguish.

This has fueled a long-running scientific debate about the relative importance of "nature and nurture."

In all biological creatures, even humans, most of the brain's interconnections are determined by genetic factors and most of the brain's raw computational power derives from genetically controlled growth and maturation of the neuronal substrate. For the most part, learning makes only microscopic changes in the structure defined by nature. However, in a complex system such as the human brain, these microscopic changes are capable of storing enormous amounts of information and producing profound differences in behavior. Thus, the impact of learning on human intelligence is immense.

In comparison, the impact of learning on machine intelligence is minimal. Today, most practical machines acquire their behavioral capabilities through evolutionary mechanisms (i.e., knowledge, skills, and abilities are embedded by designers in the hardware and software that make the systems function). Most software applications acquire knowledge by downloading programs or by storing data such as bank balances or airline flight reservations. Nevertheless, machine learning is an active research topic. Neural nets, genetic algorithms, and evolutionary programming are the subject of intensive research activity [Grossberg89, Widrow95, Barto92, Holland75, Fogel99]. Some process control systems use adaptive control technologies [Narendra86]. Some advanced robotic systems are capable of building dynamic world models and planning behavior that adapts to environmental conditions [Albus et al.95]. An extended discussion of learning from the perspective of intelligent systems is contained in Meystel and Albus[02]. Many machine vision systems require training to acquire system parameters and recognize targets, and some control systems can learn to optimize control parameters [Carpenter and Grossberg92, Miller et al.90]. In the future, learning will undoubtedly play a much larger role in the development of intelligent machines.

COMMUNICATION AND LANGUAGE

The relationship between language and intelligence is controversial. Many have suggested that language is the substrate upon which intelligence is constructed. Others view language as simply a behavioral mechanism by which intelligent systems communicate knowledge. In either case, language is an essential element of mind and the emergence of language in humans parallels the emergence of mind.

Language has three basic components:

1. *Vocabulary* is the set of words in the language. Words may be represented by symbols.
2. *Syntax* is the set of rules for generating strings of word symbols that form grammatically correct sentences in the language.
3. *Semantics* is the set of relationships that define meaning.

Semantics involves two kinds of relationships: (1) relationships between symbols internal to the system, and (2) relationships between internal symbols and external things (e.g., entities, events, and situations) in the world. Establishing relationships between internal representation and external reality is sometimes called *symbol grounding*.

Communication is the transmission of messages between systems or between subsystems within a system. Messages are sentences that convey semantically meaningful information. Communication requires that information be (1) encoded into messages, (2) transmitted, (3) received, (4) decoded, (5) interpreted, and (6) understood. Understanding requires that the information in the message is meaningful to both sender and receiver. Communication implies that knowledge has been encoded and decoded correctly and that the desired information has been transferred successfully from the world model of the sender to the world model of the receiver.

Communication is by no means unique to human beings. Virtually all creatures, even insects, communicate in some way and hence have some form of language. For example, many insects transmit messages announcing their identity and position. This may be done acoustically, by smell, or by some visually detectable display. The goal may be to attract a mate or to enable cooperation with other members of a social group. Species of lower intelligence, such as insects, have very little information to communicate, and hence have languages with only a few of what might be called words, with little or no grammar. In many cases, language vocabularies include motions and gestures (i.e., body or sign language) as well as acoustic signals generated by a variety of mechanisms from stamping feet, to snorting, squealing, chirping, crying, and shouting.

In any species, language evolves to support the complexity of messages that can be generated and communicated by the intelligence of that species. Depending on its complexity, a language may be capable of communicating many sophisticated messages, or only a few simple ones. More intelligent individuals have a larger vocabulary and more complex syntax and are quicker to understand and act on the meaning of messages.

Communication may be either intentional or unintentional. *Intentional communication* occurs as the result of a sender executing a task whose goal it is to alter the knowledge or behavior of the receiver. *Unintentional communication* occurs when a message is unintentionally sent or when an intended message is received and understood by someone other than the intended receiver. Preventing an enemy from receiving and understanding communication between friendly agents can often be crucial to survival.

The benefit, or value, of communication to the receiver is roughly proportional to the product of the amount of information contained in the message, multiplied by the ability of the receiver to understand and act on that information, multiplied by the importance of the action to survival and gene propagation of the receiver. The benefit of communication to the sender is the value of the receiver's action to the sender, minus the danger incurred by transmitting a message that may be intercepted by, and give advantage to, an enemy.

In social species, communication provides the basis for societal organization. Communication of alarm signals indicates the presence of danger, and in some cases, identifies its type and location. Communication of threats warning of aggression can establish a dominance hierarchy and reduce the incidence of physical harm from fights over food, territory, and sexual partners. Communication of pleas for help enables group members to solicit assistance from one another. Communication between members of a hunting pack enables them to remain in contact while spread apart, and hence to hunt more effectively by cooperating as a team in the tracking and killing of prey.

Among humans, primitive forms of communication include facial expressions, cries, gestures, body language, and pantomime. However, the human brain is capable of generating ideas of much greater complexity and subtlety than can be expressed through cries and gestures. To transmit messages commensurate with the complexity of human thought, human languages have evolved with grammatical and semantic rules capable of stringing together words from vocabularies consisting of thousands of words into sentences which express ideas and concepts with exquisitely subtle nuances of meaning. To support this process, the human vocal apparatus has evolved complex mechanisms for making a large variety of sounds. The human hearing and vision systems have developed sophisticated mechanisms for understanding spoken, signed, and written language.

Greater intelligence enhances both an individual's and a group's abilities for analyzing the environment, encoding and transmitting information, detecting messages, recognizing their significance, and acting effectively on information received. Greater intelligence produces more complex languages capable of expressing more information (i.e., more messages with more shades of meaning).

Many believe that language is a *precursor* of intelligence, a primary mechanism from which intelligence arises. We believe it more likely that language is a *product* of intelligence, an artifact resulting from the ability to perceive entities and event, to formulate and maintain an internal symbolic representation of the world, to reason about cause and effect, and to predict the future and plan behavior that anticipates situations before they occur. We prefer to think of language as a *behavior* that enables creatures to pursue goals and enhance their probability of success in survival and propagation. In this view, language is an equal among many behaviors (e.g., manipulation or locomotion) that provide survival benefits.

Language behavior is motivated by goals requiring communication. Language begins with having something to say (i.e., having symbolic knowledge to communicate). Communication requires a model of the world of discourse that is shared between the sender and receiver [Winograd72]. Language also requires computational mechanisms that are able to encode knowledge represented in the sender's world model into messages that can be transmitted, received, decoded, and interpreted as knowledge in the receiver's world model. Thus, language can emerge only in brains that already contain symbolic knowledge and are motivated by goals and enabled by mechanisms for encoding, transmitting, detecting, and interpreting knowledge. In short, intelligence is a prerequisite for language, not the other way around.

Language enables the transmission and communication of knowledge, and hence provides a means for communicating information about how to become more intelligent. Parents use language to instruct children in how to behave and what to believe. Teachers use language to instruct students in how to perform useful tasks. Intelligent persons use language to facilitate cooperation or communicate threats. Written language enables knowledge to be recorded, duplicated, and disseminated. Books and libraries enable the transmission of knowledge between generations. This is a key factor in the rise of civilization and in the development of commerce and technology. Thus language can enable the increase of intelligence in systems that are already sufficiently intelligent to understand what is being communicated. But language is not a precursor to intelligence. It is a product of intelligence. Language occurs only after two or more intelligent systems have something useful to say to each other.

A COMPUTATIONAL THEORY

The convergence of results from the neurosciences, artificial intelligence, robotics, computer-integrated manufacturing, information science, and cognitive psychology suggests that at long last we are entering a period where the study of mind can become a hard science. Theories can be formulated in terms of computational processes and tested experimentally in software and hardware.

It has often been suggested that the brain is a system of organs of computation, designed by natural selection to solve the kinds of problems that our ancestors faced in their hunting/foraging way of life, particular in understanding and outmaneuvering objects, plants, animals, and other people [Albus81, Pinker97]. This hypothesis is sometimes termed a *computational theory of mind*. Pinker suggests that this hypothesis already "has solved millennia-old problems in philosophy, kicked off the computer revolution, posed significant questions of neuroscience, and provided psychology with a magnificently fruitful research agenda." For example, computational theory of mind demystifies mental concepts and places mental states squarely within the realm of scientific discourse. Beliefs can be formulated as estimates of reality, desires as goals, thinking as computation, and imagination as simulation. *Perceptions* are data structures instantiated and maintained by computational processes operating on sensory input. *Planning* is searching for a path through state space from a starting state to a goal state. *Motivation* is a set of parameters specifying priorities. *Trying* is executing operations designed to achieve a goal. *Emotions*

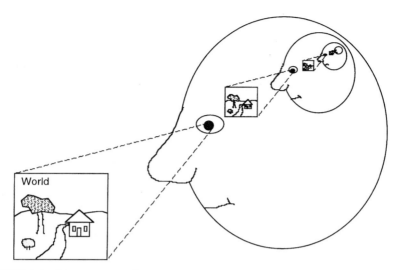

FIGURE 1.1 The concept of a homunculus (or little man), which was traditionally assumed to be required for interpreting a representation of the world in the brain. Logically, this implies a smaller little man in the head of the little man to interpret his representation, and so on ad infinitum.

are cost/benefit/risk evaluations of plans, objects, events, or situations. A "gut feeling" is the effect of emotional state variables on internal organs producing changes in heart rate, blood pressure, and flow rate in various parts of the circulatory system. A computational theory of mind completely undermines the behaviorist doctrine that mental states are not fit subjects of scientific inquiry.

Computational theory has fully eliminated that infamous philosophical bugaboo, the *homunculus*. For centuries, the concept of internal representations of knowledge in the mind was ridiculed as requiring a homunculus (a little man) sitting in the head to look at the representation, as illustrated in Figure 1.1. That, of course, would require a smaller little man inside the little man's head to look at his representation, and so on ad infinitum. Computational theory has shown that it is possible to construct a computational process (i.e., an algorithm, routine, agent, demon, or functional module) that can look at and operate on a data structure without requiring an infinite regress of agents or demons. Computational processes can operate on images to detect edges, corners, boundaries, surfaces, objects, and groups. An algorithm can compute values and attributes. A routine can recognize and classify entities and events. A functional module can aggregate entities and events into relationships and situations. Computational agents can operate on signals from sensors to detect state variables, vectors, lists, graphs, and maps. Computational demons can compute attributes of sensed objects, compare them to attributes of classes stored in memory, and recognize correlation between what is sensed and what is remembered. An infinite variety of representations can be generated and maintained within the brain with nothing more exotic than computational modules composed of neurons organized to perform mathematical, logical, and linguistic operations. There is no infinite regression. Each level of computation produces results that can be used to close control loops. There are a finite number of levels, so the process terminates [Pinker97].

Today, neuroscientists talk in terms of signals, encodings, representations, transformations, and information processing [Posner89]. Psychologists study mental images, expectations, goals, beliefs, and reasoning processes. Models of visual perception and speech understanding are expressed in terms of algorithms and computation [Flanagan91, Pylyshyn91]. In today's world, computer programs play world-class chess, handle airline reservations, manage financial transactions, control inventories, verify customer identification, dispense bank drafts, and schedule and track shipments of packages. Computational processes drive cars, fly airplanes, navigate ships, and control physical equipment such as communications networks, power grids, machine tools, steel mills, and chemical processing plants.

Yet the computational approach to intelligence has not produced the kinds of robots that have been the stuff of science fiction for a century. The robots of fiction are superhuman in strength, dexterity, agility, and reasoning. The robots of reality are relatively weak in proportion to their weight and are incredibly stupid, clumsy, and insensitive. In coping with the complexities of the natural world, artificial intelligence falls far short of the perceptual and behavioral skills of small children, animals, and even insects. Thus, the computational theory of mind remains a hypothesis with little experimental verification. Significant further progress will require that this hypothesis be subjected to experimental testing.

AN ENGINEERING APPROACH

In this book we approach the phenomenon of mind from an engineering viewpoint. Our approach is to integrate what has been learned from the neurosciences with what has been achieved in artificial intelligence and intelligent control and express the results in terms of a reference model architecture that can be used to engineer truly intelligent systems. Our first subgoal is to enable the design, engineering, and construction of intelligent systems that can rival natural intelligence in the performance of significant tasks in the real world. Such tasks include manipulating objects and tools, driving or piloting vehicles, managing organizations, controlling manufacturing plants, construction sites, and military systems. Such tasks include walking, talking, seeing, hearing, smelling, feeling, and understanding what is going on in the natural world and the social environment of everyday living. Such tasks require the ability to analyze the past, represent the present, imagine the future, and to think, plan, and act in an appropriate way under a wide variety of circumstances.

Our thesis is that mind will emerge from intelligent systems, and intelligence will emerge from the joint functioning of four fundamental processes: behavior generation, world modeling, sensory processing, and value judgment. We will show how these four processes can work together to process sensory information, to build, maintain, and use an internal representation of the external world to select goals, react to sensory input, execute tasks, and control actions. Sensory processing focuses attention, detects and groups features, computes attributes, compares observations with expectations, recognizes objects and events, and analyzes situations. World modeling constructs and maintains an internal representation of entities, events, relationships, and situations. It generates predictions, expectations, beliefs, and estimates of the probable results of future actions. Value judgment assigns value to objects, events, and situations. It computes the cost, benefit, risk, and expected payoff of future plans. It decides what is important or trivial, what is rewarding or punishing, and what degree of confidence should be assigned to entries in the world model. Behavior generation uses value judgment results to select goals, decompose tasks, generate plans, and control action.

Our approach is computational. We consider computational mechanisms that are massively parallel, full of loops, with bidirectional flow of communications, and multilevel organization of tasks, computational modules, and behavioral skills. In our proposed reference model architecture the processes of behavior generation, world modeling, sensory processing, and value judgment are linked together through a multiplicity of computational feedback control loops. Some of these loops are closed inside the intelligent system. Recursive estimation loops are closed between sensory processing and world modeling. Planning loops are closed between behavior generation and world modeling through value judgment and plan selection. Computational loops are closed within the intelligent system by feedback from sensing to acting through sensory processing, world modeling, value judgment, and behavior generation. They are closed outside the system from action to sensing through events and situations in the real world. Goals enter these computational loops from above, and sensory signals enter from below. Within these looping structures, sensed results are compared with desired goals. Rewards and punishments are computed and distributed to the appropriate places. Plans are formulated and control is exerted on the world so as to bring results into correspondence with plans and goals.

Our approach is hierarchical and multiresolutional. We believe that there is strong evidence that the fundamental processes of perception, cognition, and behavior in the brain are organized hierarchically and that knowledge is represented internally in a multiresolutional format. In our reference model, long-range high-level goals set priorities and establish context within which tasks are decomposed into subtasks at successive levels until action takes place at the lowest level. Information from sensors flow upward in a parallel hierarchy to establish and maintain knowledge about entities, events, and situations in the external world. Interaction between top-down goals and bottom-up sensory feedback is a critical element in our model of intelligence. Loops between observation, estimation, and prediction provide recursive estimation and predictive filtering. Internal links between world modeling and sensory processing generate predictions that can be compared with sensory observations. Variance between predictions and observations produces errors that can be used to update the estimated state of the world in the world model. Loops between output and input of computational modules can enable networks of neurons to function as finite-state automata, phase-locked loops, and delay lines for storing sequences of events, entities, and images. Loops between behavior generation, world modeling, simulation, and value judgment enable the simulation and evaluation of plans prior to making behavioral decisions.

Feedback loops between sensing and acting, recursive estimation loops between world modeling and sensory processing, and planning loops between behavior generation, world modeling, and value judgment are repeated many times at many different levels within an intelligent system. As a result, sensory signals are aggregated into entities, events, and situations, and goals are decomposed into behavior. At each level, world modeling maintains a knowledge database with characteristic range and resolution. At each level, plans are made and updated with different planning horizons. At each level, short-term memory traces store sensory experiences over different historical intervals. At each level, feedback control loops have a characteristic bandwidth and latency.

Our approach also involves a rich and dynamic internal representation of the external world. Our concept of a world model consists of immediate experience, short-term memory, and long-term memory. It contains both iconic images and symbolic data structures such as lists, graphs, frames, and rules. Our world model includes entities, events, and relationships that can be represented either as images, maps, and strings, or by classes, names, and pointers, or both. Our internal representations include states such as goals, intentions, desires, feelings, and moods. These states may result from, or provide input to, computational processes. In our model, understanding occurs when the internal world model corresponds to the external reality so that successful predictions can be made regarding the probable consequences of future behavior.

In subsequent chapters we argue that a model of mind consisting of a multiresolutional hierarchy of computational loops, and including a world model consisting of images, signals, and symbols, will yield deep insights into the phenomena of behavior, perception, cognition, emotion, problem solving, and learning. We suggest how this model might support a formal theory of mind. We propose a reference architecture for the design of intelligent systems based on this model. We suggest engineering guidelines for using this model to build intelligent systems for a wide variety of practical applications. We give an example of how this reference model can be applied to practical problems such as intelligent autonomous mobility. Finally, we argue that knowledge of how to engineer truly intelligent systems could have a profound impact on science, economic prosperity, military security, and human well-being.

2 Knowledge

Adam and Eve accepting fruit, from the *Tree of Knowledge of Good and Evil* by Michelangelo, from the Sistine Chapel in the Vatican. [Courtesy of the Vatican Museums.]

What does it mean to "know" something? The dictionary defines *to know* as:

1. to apprehend with the conscious mind (e.g., to know what is going on), or
2. to be acquainted with by experience (e.g., to know that the stove is hot), or
3. to have acquired a skill (e.g., to know how to water ski), or
4. to be informed about (e.g., to know what time it is), or
5. to have committed to memory (e.g., to know the Gettysburg Address), or
6. to be convinced of (e.g., to know that Jesus loves me), or
7. to be aware of (e.g., to know that you are lost)

Knowledge is all of these things, and much more. Knowledge includes all kinds of information about the world and ourselves. Knowledge is information about things, places, events, states, situations, classes, relationships, causes, effects, tasks, skills, goals, motives, values, plans, behaviors, perceptions, opinions, prejudices, feelings, experiences, tastes, rules of thumb, laws of behavior, and rules of etiquette. Knowledge includes memories of past experience and generalization from application of logical rules to the combination of many past experiences with a priori knowledge. Knowledge includes models of how the world works and knowledge of procedures for using models to generate expectations about what is likely to result from alternative choices of future behavior. Knowledge includes knowing how to do such things as tie a shoe, brush our teeth, read and write, and recognize our friends. Knowledge includes knowing how to cope with everyday unexpected events and how to compensate for failures. Skills consist of knowledge of how to do tasks. Without task knowledge we could never make it though a day at work, at school, or at home.

Most of us know how to ride a bike, drive a car, and use a telephone. Some of us know how to scuba dive, do calculus, breed orchids, or read ancient Egyptian hieroglyphics. All of us know how to recognize objects and places, family and friends, events and relationships. We know what familiar things are called and how to classify things that we have never seen before. We know how to speak and write and understand what is spoken and written. We know what we think and feel. We know what others think and feel about us. We know about time and space, places and things, events and situations. We know where we are and where things are around us. We know where things are in our homes. We know how to find our way home from distant places. We know what we have seen and heard. We know how things act and respond to our actions. We know what is probable and what is impossible. We even know what we don't know. We know if we have never before been somewhere, or done something, or seen someone.

We know what is good and bad. We know what is right and wrong. We know how to behave under a wide variety of circumstances. We know what we like and don't like. We know what hurts and what feels good. We know how things smell, taste, and feel. We know when something smells sweet or rotten. We know what things are worth. We know how much cost and risk we are willing to incur to achieve our goals or to avoid failure. We know what it means to win or lose. We know the difference between reward and punishment, praise and criticism. We know when we are hungry, thirsty, or suffocating. We know when we feel happy or sad. We know when we feel lonely and when we are sexually aroused. We know when we need a hug.

Many kinds of knowledge can be expressed by rules. There are rules of the road. There are rules of physics and chemistry. There are rules of cause and effect. There are rules that can be expressed in equations as laws of physics or theorems in mathematics or logic. There are rules for medical diagnosis and spectrum analysis. There are rules of causality and pragmatics. There are rules of syntax and semantics.

Some kinds of knowledge can be embedded in rules for how to behave. Ask a child what s/he knows and s/he might respond: "Always say 'Please' and 'Thank you,'" "Wash your hands after using the bathroom," "Look both ways before you cross the street," "Listen to your teacher," "Be nice to others," "Don't put your elbows on the table," "Brush your teeth after every meal," "Don't spit," "Keep your hands to yourself," "Wait your turn," "Don't talk with your mouth full," "Keep your fingers out of your nose." These are examples of rules of behavior that typically are learned from parents and teachers. Such rules most often are learned through physical experiences, that is, by being in situations where one is rewarded for saying "Please" and "Thank you" and reprimanded or punished for talking with one's mouth full. Much of the knowledge represented in these rules is acquired through spoken or written instructions where knowledge is encoded in linguistic strings of symbols and decoded through complex processes of speech recognition or image processing and language understanding.

Much of what we know we learn through personal experience. For example, we can learn rules of gravity from experience—by falling down, or seeing how things fall when they are dropped or thrown. However, we also have built-in gravity sensors that tell us which way is up. Much is learned through communications from others. Knowledge can be acquired by memorizing rules, historical facts, or formulas. Much of what we know we learn from reading books or attending lectures. Knowledge that enables skills such as reading, writing, and arithmetic is acquired through long hours, days, weeks, and years of practice. Knowledge of mathematics and physics enables us to understand complex spatial-temporal relationships such as that gravity causes the planets to move in elliptical paths around the sun, or that gravitational fields warp space so that rays of light from distant stars are bent.

This leads to a number of important scientific questions, such as: How does such a rich variety of different kinds of knowledge become embedded in the neural circuitry of the brain? How do the processes of genetic encoding, neurological development, and learning generate and maintain the incredible knowledge database that resides in the brain? How does this knowledge database support the processes of mind?

Certainly, some knowledge is acquired over generations from the hard school of natural selection. Instinct is a set of behavioral skills installed in the brain by evolution. Instinctive behavior is produced by neuronal computational modules that have evolved to produce useful behavior in response to environmental conditions. Instinctive knowledge is encoded in the genes so that the developing brain will have the functional modules and communication architecture needed to generate basic behavioral patterns necessary for sustaining life.

Chomsky and Pinker argue that the human ability to understand and speak language is due largely to embedded knowledge of syntax and grammar that develops in the human brain because of our genetic makeup. In *The Language Instinct*, Pinker[94] argues that human language abilities derive from how our brains are wired by our genes. At the very least, embedded knowledge provides the computational infrastructure in which the experiences of reward and punishment enable what is commonly termed *learning*. Embedded knowledge provides the basic mechanisms that generate goals for behavior, convey feedback for control, and compute reward and punishment signals for decision making.

Knowledge can be represented by signals, symbols, images, or relationships. Signals include voltages on cell membranes, impulse trains on axons, and levels of transmitter chemicals in synapses. Symbols include characters, strings, words, numbers, vectors, matrices, equations, and formulas. Symbolic knowledge can be encoded into language and communicated by vocalization, writing, or sign language. Images include scenes projected on the focal plane of a retina or in the visual cortex. Images can be formed in a camera or on a screen. Images may be generated by projection of natural scenes, photographs, paintings, icons, charts, graphs, drawings, and maps.

Knowledge about situations and events can be represented as relationships between symbolic or iconic representations. Spatial and temporal relationships can be represented by pointers that link symbolic structures such as lists, graphs, classes, or frames with images or maps. Labeled regions or features on charts and graphs can express functional relationships. Drawings are often used to capture knowledge about geometric shape and spatial relationships. For example, in manufacturing and construction, mechanical drawings are commonly used to describe the desired shape and size of parts or the desired relationship between parts in assembled products such as autos and appliances or in structures such as bridges and buildings. Maps are frequently used to represent knowledge about the location of roads, rivers, bodies of water, terrain, ground cover, buildings, bridges, cities, and armies. A good recent survey of representation in vision can be found in Edelman[99b].

Representation in the brain of knowledge about the world lies at the core of the phenomenon of mind. Arrays of signals from the retina produce immediate visual experience. Arrays of signals from tactile sensors produce the sensation of tactile feeling. Signals from the ears produce awareness of the acoustic environment. Signals from chemical detectors in the nose and tongue produce the sensations of smell and taste. Signals from thermal sensors tell us what is hot or cold. Signals from structural damage sensors produce the sensation of physical pain. All of these signals are interpreted by the mind in the context of what it knows or believes about how the world works and how things behave. Knowledge enables the mind to analyze the past, to evaluate the present, and plan for the future. Knowledge of what is going on in the world enables computational processes of the mind to generate behavior that is most likely to succeed in accomplishing system goals.

EPISTEMOLOGY

Study of how knowledge is represented in the mind is among the oldest of the scientific disciplines. The question of how the mind can know about and act on the world has occupied the attention of the world's foremost philosophers and scientists for at least 2500 years [Beakley and Ludlow92]. Epistemology is the branch of philosophy dealing with the study of knowledge, including its nature, origin, foundations, limits, and validity. *Epistemology* means "theory of knowledge" (Greek *episteme*, knowledge; *logos*, theory). Epistemology began in the Western world with the ancient Greeks.

Plato (428–348 B.C.) [see Jowett1892] was among the first to distinguish clearly between the world of ideas that exists in the mind and the world of reality that exists in nature. Plato's famous parable of the cave is an analogy of how a representation of the external world of reality projected onto the internal world of the mind is like the shadows of actors in the outside world projected onto the wall of a cave. A viewer of the shadows in the cave perceives only a truncated version of the external reality causing the shadows. Aristotle (384–322 B.C.) [see Ross28] developed a hierarchical representation of knowledge about objects and their components in the world. He formulated a theory of entity formation and classification. He realized that the whole can have attributes that are more than simply a sum of its parts.

Soon after the time of Aristotle, the Greeks were conquered by the Romans, and for almost 2000 years, serious interest in epistemology languished. During the rise of the Roman Empire, the keenest intellects were engaged with more practical issues such as military science, politics, law, and civil engineering. The Romans wasted little time on the philosophical aspects of knowledge, and simply adopted most of the concepts of the ancient Greek authorities.

With the rise of the Christianity and the fall of Rome, Western philosophy turned to theology and to metaphysics, the branch of philosophy dealing with first principles. Philosophical inquiry into knowledge about the natural world was deemed unnecessary since anything worth knowing could be obtained from Holy Scripture or other authoritative writings. The material world was considered base and inconsequential compared with matters of heaven and the means of salvation necessary to get there. Hence there was little interest in knowledge about the physical world, and few questioned established conclusions. Many of Aristotle's ideas were adopted by the Church and elevated to sacred truth. During the European Dark Ages, any attempt to question authority was suspect as a manifestation of disbelief, and possibly heresy. Virtually all scientific investigation of the material world ceased (except for largely clandestine efforts by alchemists to turn lead into gold). Eastern philosophies focused primarily on spiritual meditation and stoic acceptance of the status quo, with little interest in critical analysis or reasoned inquiry into physics or chemistry.

Not until the advent of the Renaissance and the Reformation did philosophers such as Francis Bacon (1561–1626) began to question the ancient authorities and suggest that knowledge must begin with experience and proceed by induction to general principles. It was the pursuit of this attitude toward knowledge that led René Descartes (1596–1650) [see Haldane and Ross11] to found the philosophical school of rationalism. Descartes wanted to assume nothing, to take nothing on faith, and to base his entire metaphysical explanation of man and nature on the simplest possible set of axioms. He arrived at a single axiom: "I think, therefore I am." In this statement, he formulated a division of the world into the domain of thinking and the domain of being.

Descartes's philosophy is one in which God and the human mind belong to one order of reality (i.e., of the spirit and spiritual), while the body and the rest of nature belong to another order of reality (i.e., of the material and corruptible). This is called *dualism*. Descartes fashioned his philosophy to be fully compatible with the sixteenth-century Catholic Church doctrine of eternal life of the human spirit and decay of the physical body after death. Dualism has proved amazingly durable. It has strongly influenced both science and religion for 300 years and still has respectable adherents.

David Hume (1711–1776) [see Selby-Bigge1888] argued that ideas belong to our mind and are either imposed upon the reality or are created as a result of our perception. He concluded that it was impossible to prove the existence of a real world. This is a philosophical precursor to Godel's theorem of incompleteness. Hume claimed that perception can never produce any evidence outside itself to be verified. Hence there is no justification for assuming the reality of either a material or a spiritual world. Reality beyond immediate experience just cannot be proved. Using contemporary language of semiotics, Hume argued that symbol grounding is impossible.

Immanuel Kant (1724–1804) [see Muller1881] played a major role in clarifying the relationship between the cognitive representation of space and time in the mind, and the attributes of objects, events, and agents that we observe in the world. Kant argued that while the real world is broader than the appearances of its components, the human mind forms a representation of reality by organizing the thoughts about the world through categories created by the mind. This does not refute Hume's pessimistic conjecture, but it creates an important opening for continuing analysis. Thus, even though absolute symbol grounding might be impossible, Kant argues that we can approach it asymptotically by making our categories more and more adequate to represent external reality. This is suggestive of modern recursive estimation theory that begins with a categorical hypothesis that is tested against experience until its adequacy (i.e., probability of correctness) grows high enough that the hypothesis is deemed confirmed and the categorization is verified. Under these circumstances, symbol grounding is possible, or at least approachable in the form of a "best estimate."

Charles Darwin (1809–1882) first suggested that like the body, the mind and intelligent behavior are products of evolution driven by natural selection [Darwin1859]. Darwin's analysis led the psychologist William James to subject the mind to experimental scientific investigation. James[1890] attributed to the mind the ability to represent goals mentally, to choose means to reach them, and to act to achieve them. He also explored the concepts of intention, desire, hope, expectation, belief, and the concept of free will. James focused much of his attention on the phenomenon of consciousness and put a great deal of faith in the power of introspection as an experimental method for studying the mind.

Shortly after Darwin, Sigmund Freud (1856–1939) pointed out that there is much that goes on in the brain below the level of consciousness. Freud[17] focused on dreams and the unconscious, and largely discounted introspection by the conscious mind as a reliable source of information. He saw dreams, not rational discourse, as revealing our real hopes, desires, and wishes.

During the first half of the twentieth century, Watson[13,28] and Skinner[53] established the school of *behaviorism*, which largely denied the existence of internal mental states of mind such as intention or belief and considered the notion of free will to be an illusion. Behaviorism focused entirely on stimulus–response behavior as all that can be observed scientifically, and therefore all that can reliably be studied. Behaviorism largely dominated psychology for 50 years and survives today despite the fact that Piaget[52] undermined its central premise with his observations of the progressive development of mental capabilities in children. Piaget maintained that the development of cognitive behavior in children is a manifestation of the development of internal representations of knowledge in the brain. Chomsky[88] goes further with evidence that the human brain has genetically programmed facilities for the acquisition of language. Nevertheless, the influence of behaviorism has persisted and is still evident in behaviorist approaches to robotics research.

During the second half of the twentieth century, the study of intention, goal seeking, knowledge representation, and mental imagery has again become a respectable topic of research. In the 1940s, Bertrand Russell[48] explored mental representations of belief, common sense, reason, and logic. Since the

1950s, the mainstream of artificial intelligence research has focused on mental processes of planning, problem solving, game playing, and language translation. This approach uses the tools and concepts of production rules, symbolic reasoning, propositional and predicate calculus, list processing, and heuristic search.

Today, epistemology is very much alive. The invention of the computer has made it possible to subject ideas that were the source of ancient philosophical debates to experimental analysis. Modern computer technology is rapidly making it feasible to build machines that can mimic, simulate, and emulate the functional properties of at least parts of the brain. Much of the current debate revolves around the issues of:

1. Reductionism versus holism. Can the gestalt experience be understood by reduction to the neuronal level?
2. Functionalism versus structuralism. Can the function of the brain be understood without modeling the neural substrate?
3. Computation versus connectionism. Are the fundamental operations in the brain best modeled by mathematical logic and symbolic rules, or by connectionist circuitry?

Almost all modern theories of the mind are heavily influenced by the invention of the computer and the development of a theory of computing. The modern concept of a computational process, and the relationship between a process and the machine on which it runs, provides an entirely new way of looking at the epistomological mind–brain problem. If the brain is a machine, the mind is a process that runs in the brain. Our hypothesis is that the brain is a machine, driven by physical, chemical, and electrical forces; and that the mind is a process that runs on the neural circuitry of the brain. The mind is a product of the software and firmware embedded in synaptic connections activated by transmitter chemicals and hormones that determine what the brain does. Among the computations performed in the brain are recursive estimation, correlation, prediction, attention, clustering, segmentation, computation of attributes, classification, recognition of entities, detection of events, formation of relationships, and analysis of situations.

These computational capabilities enable the brain to generate images, to imagine entities and events, and to assign them properties. Infants can imagine objects that are hidden from view. Children can create imaginary friends. People of all ages can imagine ghosts, angels, gods, spirits, demons, and leprechauns. Legends, myths, and opinions that are reinforced by repetition without contradiction may become political or religious dogma embedded in the knowledge database of the mind. Once embedded, this knowledge becomes the basis for behavior. People act on what they believe in their minds to be true.

It is, of course, clear that the computing mechanisms in the brain are not at all like those in a von Neumann digital computer. Individual neurons and computing nodes in the brain much more resemble analog than digital computers. The brain is a massively parallel machine, and although serial processing threads clearly play in important role in neuronal computation, the brain's most obvious characteristic is its distributed connectionist architecture. Each neuron computes a function of many, sometimes thousands of input variables and there are about a hundred billion (10^{11}) neurons in the brain. Each neuron has an estimated average of 1000 synapses, with each synapse simultaneously computing a product about 100 times per second. Thus the total computational power of the brain may be on the order of 10^{16} operations per second. If each synapse has the capacity to store a single byte of information, the total memory capacity may be on the order of 10^{14} bytes of storage [Kurzweil99, Moravec98].

However, neurons are noisy and unreliable both as computing machines and memory devices. It has been estimated that about 10,000 neurons die every day after the age of 10. The brain compensates for the stochastic nature of neuronal computation through massive redundancy. Computed parameters such as the planned direction for reaching, or the fear of falling from high places, are computed by large populations of neurons. Thus the contribution of any single neuron has very little effect. This suggests that the computational power required to approximate the performance of the human brain may be two to three orders of magnitude less than estimated above, on the order of 10^{13} integer operations per second with memory capacity around 10^{11} bytes (100 gigabytes). Given that computer power is increasing by at least an order of magnitude per decade, the computational power of the human brain may be within the reach of a supercomputer within one decade and within the reach of a modest network of 10 desktop computers before the year 2020. A disk memory 100 gigabites is already available in desktop computer systems.

WHERE DOES KNOWLEDGE COME FROM?

The brain and mind within it are stimulated by two types of signals: one internal, the other external. Internal signals arise from deep limbic structures such as the hypothalamus, which generates basic drives such as hunger, thirst, and sexual arousal. Other limbic structures give rise to signals that signify hope, fear, joy, despair, love, hate, pleasure, pain, aggression, and submission. These structures are built into the brain by the genes to provide the basic goals, drives, motivation, and decision mechanisms that are necessary to survive and propagate in a complex world. External signals originate from sensory neurons that react to energy input from the environment. These provide the ability to sense what is happening in the world so that the brain can build and maintain a dynamic model of the world and how it works. This information is acquired through learning. Together, these two types of data account for the phenomena of immediate experience.

Immediate Experience

Knowledge of objects, events, relationships, and situations in the world around us comes from sights, sounds, smells, tastes, and feelings of immediate experience. Knowledge of immediate experience is dynamic, changing with every variation in sensory input. Vision is one of our most important sources of immediate experience. When we look at the world we see a rich mosaic of colors, shadows, shapes, and textures. We see motion. We see patches of light dancing in the shadows. We see birds and insects flying through the air. We see leaves and papers blowing in the wind. We see animals, people, and machines moving through the environment. And we see depth. We use stereo, parallax, image flow, and other cues, such as texture, shading, size, and occlusion relationships, to estimate how far away things are. We know where things are by sight.

We also use vision to deduce what things are. We can recognize a particular animal as a cat or dog even though we have never seen that animal before. We can recognize objects, persons, places, and pets that we know, even though we have never seen them in exactly the same place, in the same light, from the same perspective, and in the same position before. We use our visual knowledge to build an internal model of spatial and temporal relationships in the external world; and we compare attributes of what we currently see with what we know from previous experience. Usually, we can either recognize the object or situation that we are currently witnessing as something we are familiar with, or can say with confidence that we have never seen it before.

Knowledge is also conveyed by sounds that generate signals in our auditory sensors. We process auditory signals to estimate where we are and what is going on in the world around us. We can recognize sounds, guess what or who is making them, and estimate from where they are coming. We can recognize voices, and sounds made by wind, rain, insects, birds, and animals. We understand what words mean and can communicate through spoken language. We can remember tunes and songs, and recognize pieces of music. We sometimes "cannot get a tune out of our minds."

We also have internal sensors that tell us where we are in the world and what is happening in our bodies. Perhaps the most primitive of all senses is the sense of gravity that provides knowledge of what direction is "up." Our knowledge of "up" profoundly influences how we represent directionality in space. Virtually all living creatures, including some single-celled organisms, have a sense of gravity. We also have inertial sensors that measure both linear and rotational accelerations. We have sensors in our muscles and joints that let us know approximately where our limbs are positioned relative to our body. We have sensors in our skin that tells us when we feel hot or cold. We have blood chemistry sensors and sensors in our stomach to let us know when we are hungry or thirsty. We have internal clocks that let us know when we should be sleepy or wide awake. We have sensors that measure chemical imbalance in our digestive tract to make us feel nauseous. When we touch things in the world, we feel shapes and texture. We know what feels wet or dry, hard or soft. We know how things smell and taste. We know what hurts and where it itches.

Prior Experience

Clearly, we possess knowledge other than immediate experience. Much of what we know is the result of prior experience. We know what we can recall from recent experiences (short-term memory) as well as what we can remember from what we have learned from the totality of our experiences since birth (long-term memory). We use what we already know to interpret immediate experience. This enables us to recognize and interpret smells and tastes and sights and sounds. It enables us to build symbolic descriptions of what we experience. It enables us to reason about the past and to predict the future. It enables us to understand situations and events, to comprehend stories, to reenact events in our minds, and to communicate thoughts to others.

What we know determines what we expect. What we expect colors what we perceive. Knowledge of internal variables (such as drive, urges, needs, desires, and mood) influences how we assign importance to what we experience from external sources. Knowledge of what is important affects where we look, what we listen for, what we hope for, what we remember, and what we fear. Knowledge of what is urgent can affect what we hear and feel. For example, athletes in competition often block out the roar of the crowd. Soldiers in combat sometimes feel no pain despite horrible wounds.

How all this is accomplished is still largely a mystery. Although it is widely recognized that knowledge is crucial to behavior, perception, and cognition, there is little agreement on how knowledge is acquired or stored in the brain. It is clear that humans have and use a great deal of knowledge about the world and are aware of much that goes on in the world around them. But exactly how sensory input is processed to establish and maintain this internal store of knowledge, and how the brain represents and retrieves this knowledge, is largely unknown. Recently, brain imaging techniques such as positron emission tomography (PET) and functional magnetic resonance imaging (fMRI) have begun to shed light on how experiences are processed, remembered, and recalled. Attributes of objects and events are apparently stored in regions that are very close to, or overlapping with, areas that are active during processing of those attributes [Martin et al.95].

REPRESENTATION IN THE BRAIN

The neuron is the fundamental element of computation and representation in the brain. Each neuron is a tiny computer that receives input through synaptic receptor sites on its dendrites and cell body and computes an output that is distributed to other neuron inputs via an axon that may branch many times. Each axon branch terminates on a synaptic receptor site on a dendrite or cell body of another neuron, or an actuator cell such as a muscle or gland. The dendrites and cell body of the receiving neuron integrate the weighted input from each receptor site, over time and sum over all the receptor sites, as shown in Figure 2.1. The result is that to a first approximation, each neuron computes a temporal integral on the dot product between its input vector and a weight vector that represents the strengths of its synaptic receptor sites. A more exact analysis of the various types of neurons suggests that individual neurons may compute complex nonlinear functions, combining both logical and arithmetic operations. The result of each neuron's computation produces a cell body membrane potential that can represent, or be represented by, a real scalar variable. Some neurons compute positive scalar variables. Others compute negative scalar variables.

The output of each neuron is conveyed by an axon to the various locations in the brain where its information is needed. At the point on the neuron where the output axon attaches to the cell body, there is a section of membrane (the axon hillock) that transforms the cell body potential into a string of impulses for transmission over the axon. The frequency of these impulses may vary between zero and 500 impulses per second. The signal carried by each axon is thus represented by the impulse rate, or its reciprocal, the interimpulse interval. Impulses propagate down each axon at a velocity that depends on the diameter and insulation properties of the axon. Propagation velocities vary from less than 1 meter (m) per second in small noninsulated axons to 120 m per second in large well-insulated axons.

An axon may branch many times on its way to a variety of destinations. At the end of each branch, there is a termination consisting of a synaptic bouton. Inside each synaptic bouton, there are many tiny vesicles filled with transmitter chemicals, as shown in Figure 2.2. The arrival of each impulse triggers the release of a set of vesicles filled with transmitter chemicals into the gap between the axon bouton and a dendritic receptor site on the next neuron. The transmitter chemical diffuses across the gap and activates ionic electric current flow across the receptor dendritic membrane. The value of the current flow is a product of the frequency of the impulse rate on the incoming axon and the strength of the synaptic connection on the receiving neuron. The dendrites and cell body of the receiving neuron integrate the flow of current to produce a voltage. By this process, the information encoded by the membrane potential (i.e., the state) in one neuron is conveyed to, and affects, the membrane potentials (i.e., states) of other neurons. The effect of axon activity on synaptic receptor sites may be modulated by the concentration levels of hormones and chemicals in the fluid surrounding the neurons, as well as by waves of electrical potential caused by simultaneous activation of many surrounding neurons [Grossman67, Peele61, Churchland and Sejnowski92].

Neurons in various systems of computational modules secrete different neurotransmitters. Each chemical works on different types of neurons in different ways. Some 50 chemicals have been identified as neurotransmitters in the

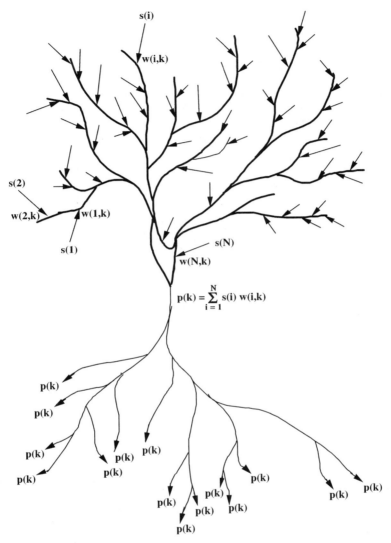

FIGURE 2.1   A neuron computes the scalar value $p(k)$ as the inner product of the input vector $s(1), s(2), \ldots, s(i), \ldots, s(N)$ and the weight vector $w(1,k), w(2,k), \ldots, w(i,k), \ldots, w(N,k)$.

brain. Among the more important are:

- *Dopamine*: controls arousal levels in many parts of the brain and is involved in motivation. Low levels are associated with Parkinsonian symptoms such as the inability to move voluntarily. High levels are implicated in schizophrenia and may cause hallucinations.

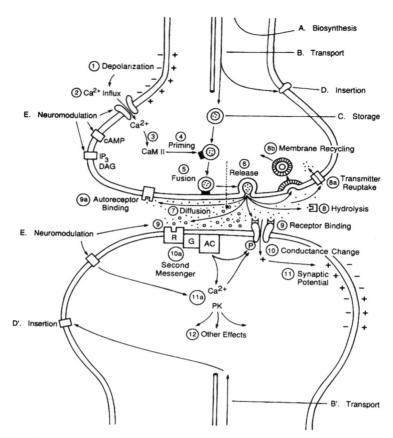

FIGURE 2.2 Synaptic junction between an axon bouton and a receptor site on a dendrite or cell body. Depolarization (1) from an action potential on the axon causes calcium ion intake (2) that primes a vesicle (4) in which a transmitter substance is stored. This causes the vesicle to fuse with the axon bouton membrane (5) and release its transmitter (6) into the junction, where it diffuses (7) and activates receptor sites (9). This produces a conductance change (10) and change in synaptic potential (11) in the dendrite or cell body. The transmitter is hydrolyzed (8) or recycled into the axon (8a, 8b). [Reprinted from *Neurobiology*, Third Ed., by Gordon Sheperd, 1988. ©1983, 1988, 1994, used by permission of Oxford University Press, Inc.]

- *Serotonin*: has a profound effect on mood and anxiety. High levels are associated with serenity and optimism. Serotonin also affects sleep, pain, appetite, and blood pressure.

- *Acetylcholine*: affects activity in brain areas associated with attention, learning, and memory. People with Alzheimer's disease typically have low levels of acetylcholine.

- *Noradrenaline*: an excitatory chemical that induces physical and mental arousal and establishes levels of alertness or aggressiveness. Production in the locus coeruleus, one of the brain's pleasure centers, brightens mood.

- *Glutamate*: an excitatory neurotransmitter involved in learning and long-term memory.

- *Endorphins*: opiatelike chemicals that reduce pain and stress and promote a sensation of calm and well-being. A variety of drugs similar to endorphins signal reward resulting from recognition of successful behavior. (Endorphin mimics such as cocaine and heroin allow drug users to bypass the requirement of successful behavior and experience the rewarding feeling of success by injecting drugs into the bloodstream artificially.)

Other drugs produce punishing effects that lead to repulsive or evasive behavior or to depression. Pain signals are specifically designed to punish behavior that leads to bodily harm. Emotional pain signals also affect behavior in profound ways, as do emotional feelings of joy or elation.

The chemical environments in various parts of the brain depend on diffusion rates that may change concentrations in individual neurons over fractions of a second, or throughout entire regions of the brain over longer periods of seconds or minutes or longer. Concentrations of transmitters in synaptic clefts can alter electrical potentials in dendrites in a matter of milliseconds. Concentrations of hormones and other chemicals in the bloodstream may alter moods or feelings (modes of operation) in the brain over periods of hours, days, or even weeks. The brain is a complex network of computing structures with many parameters that change over time on a variety of different time scales.

Feedback from output to input of a single neuron can preserve state information or perform integration or differentiation, depending on the sign of the feedback loop and nature of the information being conveyed. Feedback loops that travel through several neurons can act as delay lines or as recirculating memory structures. These types of looping structures can store dynamic strings of information temporarily and are well suited for detecting temporal patterns or forming phase-locked loops and filters. Short-term memory in the brain is thought to involve reverberating loops consisting of chains of neural elements [Zipser91].

Most theories of learning assume changes in synaptic weights that occur over time as a result of neuronal activity [Dudai89]. Long-term memory in the brain is widely thought to result from structural changes in the microstructure of the synapses. It is well known that synaptic sites can vary in strength (or "weight"). It is clear is that a network of long-term connections develops between neuronal modules as a result of experience as the brain grows and matures. Synaptic modifications may persist over periods of months or years. Many of these connections endure for a lifetime. We do know that knowledge is represented in the brain by electrical potentials in neuron dendrites and cell bodies, by impulse rates in axons, by the strengths of synaptic connections, by the network of connections between neurons, and by electrical and chemical environment surrounding the neurons [Churchland and Sejnowski92]. Whether these are the complete set of mechanisms by which knowledge is represented remains to be discovered.

There is much that remains unknown about the details of learning. No one knows exactly what causes the various kinds and rates of learning in different parts of the brain. Memories of experiences can be stored almost immediately during a single learning experience. Knowledge of motor skills may require repeated practice and repetition over long periods of time, up to weeks or years. Some knowledge, such as language skills, can be learned easily during a brief interval during early childhood, and with difficulty later in life.

Images in the Brain

How neurons in the brain represent images is relatively well understood. Arrays of photodetectors in the retina of the eyes, or arrays of pressure sensors on the surface of the skin, respond to stimuli. The position of the stimulus can be determined by observing which neurons in the array are active. At each point in space, the intensity of the stimulus is represented by the value of the neural potential or the rate of impulses on axons at the corresponding point in the array. Other attributes of the stimulus, such as its spatial or temporal derivatives may also be represented at each point by an array of neurons that compute these attributes from the intensity image. At each point in time, the spatial pattern of stimulation produces a corresponding pattern of activity on an array neurons and axons.

It is in the vision system where our understanding is most complete. The visual field is focused by a lens on a two-dimensional array of photosensitive cells in the retina of the eye. Photons impinging on each retinal photosensor produce an electrical potential that is proportional to the intensity of the incoming light (number of photons per second per square millimeter) integrated over the area of the photosensor and over a time interval determined by the cell membrane capacitance and resistance. The retinal neural image is thus an array of voltages produced by an array of photosensors. The output from each photosensor represents a pixel, and the array of pixel outputs from the photosensors produces an intensity or color attribute image.

There are rod photodectors in the retina that detect white light intensity and three types of cones that measure input energy in three different color bands. The retina also contains horizontal cells that compute spatial derivatives (or spatial frequency, or Gabor functions, or wavelets) at each point in the image. Output from the horizontal cells thus represents a spatial-derivative-attribute image that is registered with the intensity-attribute image. The retina also contains amacrine cells that compute temporal derivatives of the intensity or color image at each pixel. The output from the amacrine cells thus represents a temporal-derivative-attribute image. Figure 2.3 is a drawing of the retina showing the rods and cones that detect incoming light and a variety of interneurons that compute spatial and temporal derivatives of intensity and color.

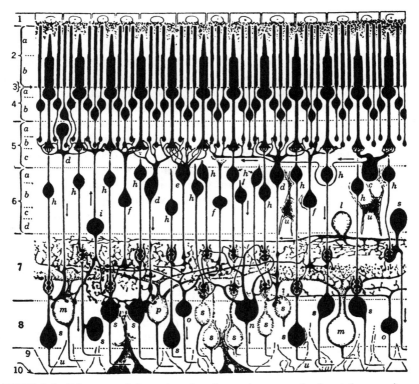

FIGURE 2.3 Diagram of the retina showing rods, cones, horizontal, amacrine, and retinal ganglion cells. [From Polyak57.]

In the lateral geniculate nucleus, visual signals from corresponding points in the visual fields of the two eyes are brought into registration. In the visual cortex, a variety of attributes are represented at each point in the visual field, with a variety of scales and resolutions. Direction and magnitude of spatial and temporal derivatives, binocular disparity, estimated range, color, texture, shading, image flow rate and direction, and names of entities that occupy regions in the image may all be represented at each point in the visual field. There are layers of cells in the visual cortex that are tuned to different stereo disparities. There is thus a three-dimensional visual mapping from the shell of space near the horoptor to the visual cortex. The radial distance of the horoptor shell from the eyes is determined by the vergence. The vergence value is represented by the neural commands to the eye muscles.

The brain also has a set of translational and rotational accelerometers in the vestibular system. Signals from these sensors can be integrated to provide estimates of position, orientation, and linear and rotational velocity with respect to the external world. Control algorithms in the vestibular-ocular reflex loop can then inertially stabilize the eyeballs in their sockets and generate an inertial reference frame for egosphere image representations. Figure 2.4 is a block diagram of 32 distinct visual processing areas known to exist in the brain of the macaque monkey. Each visual processing region contains a topographically organized map of the visual field; thus each processing area represents one or more attribute or entity images. Two of these areas, V1 and V2, are very large, each occupying about 10% of the entire cortical sheet. V1 contains approximately 250 times as many neurons (about 500 million) as there are inputs from the lateral geniculate (about 2 million). This suggests that V1 could contain many (perhaps 100 or more) attribute image representations of the visual field at a variety of resolutions. Altogether, there are about 1.3 billion neurons in the visual cortex, or about 600 times more neurons than there are input axons from the lateral geniculate. This implies that the brain dedicates many different neural computers to extract many different attributes simultaneously for each pixel. It suggests that many attribute images are being computed simultaneously in parallel. On average, each visual area has approximately 10 distinct types of inputs and 10 outputs. The total number of reported interconnecting pathways between the 32 visual processing areas is over 300 [Van Essen and Deyoe95].

It is well known that many neural computations are performed by computational maps in the brain. Neurons, as elements of the map, may be thought of as an array of processors or filters, each tuned slightly differently and operating in parallel. Arrays of signals are transformed within a few milliseconds into a distribution of neural activity within the map. The values of the stimulus parameters are represented as locations of peaks of activity within the map [Knudsen et al.87, Sparks and Groh95]. Attribute and entity images can be registered by virtue of the fact that neurons implementing various types of attribute and entity computers occupy the same pixel area in the cortex. Registered attribute images may differ in resolution, depending on the relative density and receptive field size of the various types of computing neurons. Attribute and entity images may also be registered by means of topographical mappings that exist between different visual processing areas.

Other sensory modalities may also produce maps or images that represent the spatial distribution of excitation over arrays of sensory neurons. For example, arrays of tactile (pressure, temperature, vibration, and pain) sensors exist on the surface of the skin. These arrays are quite dense (i.e., high resolution) on some parts of the body, such as fingertips, lips, and tongue, and less dense on other parts of the body. Figure 2.5 shows the topological mapping that exists between tactile sensors in the skin and regions in the brain that process tactile sensory input. This mapping is well known and often explored. The entire surface of the body is mapped onto the surface of the tactile sensory cortex. The variation in density of tactile sensors on the skin produces different sized regions on the surface of the cortex dedicated to processing the tactile information. About half of the sensory-motor neurons are dedicated to the fingers and mouth. The face and hand use an additional one-fourth, leaving only one-fourth of the sensory-motor cortex for the entire rest of the body.

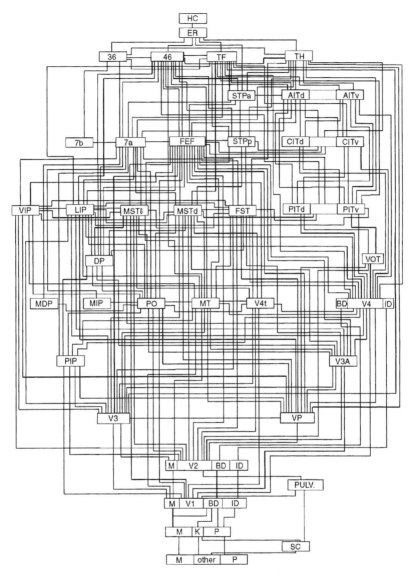

FIGURE 2.4 Hierarchy of visual areas in the macaque monkey based on patterns of anatomical connections. Each of the boxes represents a computational module located in a specific region of the visual processing system in the brain. Each computational module contains a map of the visual field. Each of the lines represents a specific neural pathway that conveys visual information from one computational module to another. [From Van Essen and Deyoe95.]

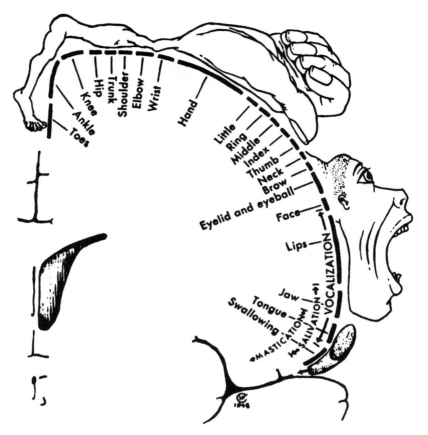

FIGURE 2.5 Topographical mapping on the surface of the sensory-motor cortex. This view is a front view of a cut through one side of the human brain along the central sulcus, perpendicular to the side view shown in Figure 2.10. The relative size of the body parts indicate the relative number of cortical neurons dedicated to sensing and control of that part of the body. [From Penfield and Rasmussen50.]

Because the number of processing neurons is roughly proportional to the area on the surface of the cortex, a distorted image of the body is produced on the surface of the brain, as shown in Figure 2.5. There is a similar mapping from the adjacent motor cortex to muscle groups that produce forces and motions.

In the auditory system, the cochlea generates a map that resembles a sonogram, in which the component frequencies of a sound are arrayed along one dimension and temporal delay along the other dimension. The superior colliculus contains a map of auditory space such that the azimuth of the incoming sound is arrayed along the horizontal dimension and the elevation is arrayed along the vertical. This is maintained in registration with a map of the visual egosphere. In the barn owl, the eyes and ears are both fixed in the head so that the auditory and visual map can easily be maintained in registration on the surface of the superior colliculus. In the monkey, the auditory map (which is fixed with respect to the head) scrolls dynamically across the surface of the superior colliculus so as to remain in registration with the visual field (which is fixed with respect to the eyes) [Jay and Sparks84,87, Sparks86, Sparks and Mays90].

There exists good evidence for cognitive maps of space and time that allow animals to reason about spatial and temporal relationships [Schmajuk98]. Place learning has been demonstrated in a number of situations: for example, in the "water maze," where rats are required to escape from a pool filled with opaque water by navigating to a submerged invisible platform. The animals demonstrate the ability to learn the position of the platform from landmarks outside the pool [Morris81].

Cognitive maps were first proposed by Tolman[32] to explain place learning in rats, including their ability to take shortcuts. The cognitive map is represented in a Euclidean coordinate system in the external world, not tied to the surface of the body. An alternative view sugggested by Hull is that navigation is achieved by following a list of stimulus–response steps. O'Keefe and Nadel[78] proposed that there are two independent neural systems that support both stimulus–response learning for route navigation and a place learning system for map-based navigation. Single unit recordings in the hipocampus of freely moving rats has revealed *place cells* that respond to entities in small portions of the rat's environment. Firing of place cells can be manipulated by changing the rat's environment. For example, rotating the cues in the environment can cause the place cell response field to rotate [Muller et al.91].

Humans apparently carry an internal map of the relationship between themselves and the external world in the parietal lobe [Carter98]. The fact that the hippocampus remains involved in memories of space can be seen from PET scans of London taxi drivers as they imagine the routes that would have to be taken to get to and from various points in the city [Maguire et al.97].

Symbols in the Brain

How the brain represents and manipulates symbolic information is much less well understood. There is no doubt that the brain stores and uses symbolic information. Most higher animals have some capabilities for reasoning, expectation, and communication that suggest symbolic computing capabilities. Humans have symbolic reasoning and language facilities that set them far apart from the rest of the animal kingdom. The ability of the human mind to formulate and manipulate symbols is awesome. All of language, mathematics, science, art, music, dance, and theater is based on manipulation of symbols in one form or another. Industry, business, finance, commerce, and military science could not exist without the use of symbols. Yet, there is little understanding of how the neuronal substrate of the brain represents symbols or performs the operations of symbol manipulation.

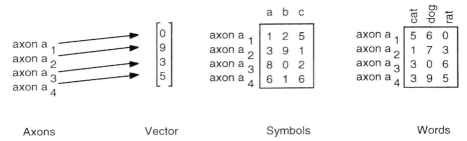

FIGURE 2.6   At a point in time, signals on an ordered set of axons can represent a vector, a symbol from an alphabet, or a word from a vocabulary. On the left an ordered set of axons carry a set of signals that define a vector. At different times, signals on an ordered set of axons can represent different symbols from an alphabet, or different words from a vocabulary. The set of all possible signals on a set of axons can define a vector space, a symbol set, or a vocabulary.

Nevertheless, at least at the lower levels of the neuronal hierarchy, there are some clues as to how the brain handles symbolic information. It is well known that the impulse rate (or interimpulse interval) on each axon is a time-dependent scalar variable that can represent a signal or the value of an attribute or state variable. An ordered set of axons each representing a scalar time-dependent variable can thus represent a time-dependent vector, such as a state, an attribute, or a geometric description of an entity. For example, Figure 2.6 shows how an ordered set of axons can represent a vector, symbols from an alphabet, or words from a vocabulary. The set of all possible signals on a set of axons can define a vector space, a state space, or a vocabulary.

Over an interval of time, a string of impulses on each axon in an ordered set can represent a string of vectors, a string of points, a trajectory through state space, a string of characters in a word, a multidigit number, or a string of words in a sentence, as shown in Figure 2.7.

Neurons in the brain are typically arranged in clusters or modules with axon pathways connecting to other modules. The brain can therefore be modeled as a collection of computing modules connected by a set of signal pathways. Within each module, synaptic strengths, the network of interconnections, and the electrical and chemical environment in the brain define the functional transformation between input and output for a collection of neurons. This is illustrated in Figure 2.8a, where a set of input axons defines an input vector $\mathbf{S} = (s_1, s_2, s_3, \ldots, s_N)$. Each output axon represents a single-valued function computed on the set of inputs $\mathbf{S}$. $p_i = h_i(\mathbf{S})$ is the scalar value computed by the ith output neuron. A set of output axons defines an output vector $\mathbf{P} = (p_1, p_2, p_3, \ldots, p_M)$. The entire module computes the vector function $\mathbf{P} = \mathbf{H}(\mathbf{S})$. A set of neurons in a module may thus compute matrix operations such as coordinate transformations. A neuronal module can also compute an IF/THEN rule. If the input vector $\mathbf{S}$ represents an IF predicate, the output vector $\mathbf{P}$ can represent a THEN consequent.

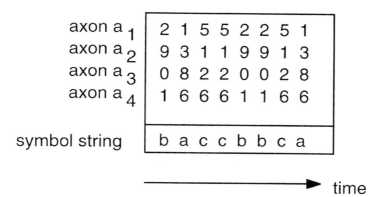

axon a$_1$	2 1 5 5 2 2 5 1
axon a$_2$	9 3 1 1 9 9 1 3
axon a$_3$	0 8 2 2 0 0 2 8
axon a$_4$	1 6 6 6 1 1 6 6

| symbol string | b a c c b b c a |

→ time

FIGURE 2.7   Over an interval in time, signals on an ordered set of axons can represent a string of symbols.

Neuronal modules can implement many types of table look-up memory, classification, recognition, relational pointers, and function calls, as described in Figure 2.8b. If the input axons represent an address, the output axons may represent the contents of that address. If the input vector **S** represents a class or object name, the output vector can represent class or object attributes. Alternatively, if the input vector **S** represents a set of object attributes, the output vector **P** can represent the name of the class to which the object belongs. If the input vector represents the name of a function plus a set of parameters, the output **P** can represent a value returned by the function call. Finally, if the input **S** is a pointer (i.e., an address), the output can be the contents of a data structure located at that address. This enables neuronal modules to represent relationships and perform list processing operations.

Functions computed within neural modules can be single-valued arithmetic or logical functions such as addition, subtraction, multiplication, division, AND, OR, XOR, NOR, NAND, or the inner product between an input vector carried by an ordered set of axons and a synaptic weight vector. A group of neuronal modules can compute a spatial cross-correlation function between image segments. If there is feedback between input and output, a neuronal module can act as a finite-state automaton. A neural module with feedback can therefore generate and recognize strings or compute correlation between temporal patterns. Thus a set of neural networks can be interconnected to perform recursive estimation or implement phase-locked-loops, delay lines, and tracking filters.

These representational and computational capabilities provide mechanisms for various types of learning and memory. Simple storage and retrieval can implement long-term memory. This enables the learning of object and class names, attributes, relationships, schema, plans, and task skills, including skills of manipulation, locomotion, and language. Recirculating loops and finite-

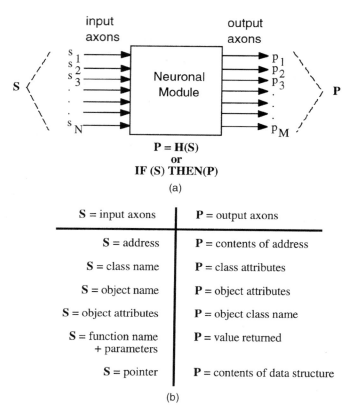

$$P = H(S)$$
$$\text{or}$$
$$\text{IF (S) THEN(P)}$$

(a)

S = input axons	**P** = output axons
S = address	**P** = contents of address
S = class name	**P** = class attributes
S = object name	**P** = object attributes
S = object attributes	**P** = object class name
S = function name + parameters	**P** = value returned
S = pointer	**P** = contents of data structure

(b)

FIGURE 2.8 Computation in neuronal modules. Axons carrying signals into a neuronal module consists of a vector that can represent a symbol, an address, an IF predicate, a name, or a list of attributes. Axons leaving a neuronal module can represent a function of the input, a THEN consequent, the attributes of the input name, or the name of an entity or event associated with the input attributes.

state automata can implement short-term memory. This enables recursive estimation and the acquisition of direct and inverse models of complex dynamical systems. Dynamic models enable prediction, simulation, planning, and feedforward control.

Thus it is possible to speculate and imagine how the neuronal modules in the brain might represent and use symbolic knowledge. We can hypothesize neuronal modules with computational and memory mechanisms that are adequate to support most, if not all, of the behavior that is observed. However, the amount of empirical evidence for how the brain actually does the job of representing symbolic information is sparse. What we do know is that sets of neurons have been observed in the motor cortex that define a motion vector that indicates the direction and magnitude of planned movement by the

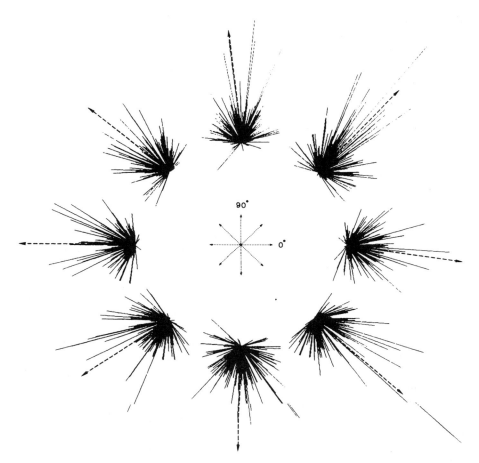

FIGURE 2.9   Population vector analysis applied to eight movement directions in two-dimensional space. Each cluster represents the same population of neurons. The length of the vectors represents the firing rate of the neurons in the population. The direction of the vectors indicates the perferred direction of each of the neurons in the population. The movement directions of the hand in hand coordinates are shown in the diagram at the center. The resultant population vector (dashed arrow) points in or near the direction of movement. [Reprinted from Georgopoulos et al.83.]

hands in hand coordinates. Single neurons in these sets are broadly tuned so that each neuron's activity is highest for a movement in its own preferred direction and decreases progressively as a function of the cosine between the neuron's preferred direction and the direction of the planned movement. The sum of activity of all the neurons in the set produces a resultant vector that indicates the direction and magnitude of the muscular forces required to produce the planned movement. This is illustrated in Figure 2.9. The fact that these neurons represent planned action is evident because their firing rates precede movement by approximately 150 milliseconds (ms). That these plans are symbolic is evident from the fact that this representation is very different from what is required by motor neurons sending signals to individual muscles [Georgopoulos95].

One of the most thoroughly studied motor representations is the motor map in the superior colliculus that controls saccadic eye movements. The sets of neurons that control eye movements vary systematically across the surface of the superior colliculus. The neurons that generate motor commands to direct the gaze at a particular part of the visual field are registered with the visual and auditory sensory map that also exists on the surface of the superior colliculus. The motor commands computed at each point by these sets of neurons define the motion direction and amplitude, or velocity, required to steer the optical axis to that point in the retinal field. These commands are clearly abstract and symbolic. Before they can effect movement, they must first be decomposed into simultaneous coordinated movements of head and eyes, and then be further transformed into commanded forces and velocities for various muscle groups [Sparks and Groh95].

The ability of the mind to form associations between written or spoken words and pictures or images of objects in the world has been studied extensively. Figure 2.10 shows where the regions of the brain involved in language are located. Visual information is integrated with auditory information, and language understanding takes place in Wernicke's area, which lies at the intersection of the parietal, temporal, and occipital areas of the cortex. Wernicke's area is tightly coupled with Broca's area, which contains the circuitry involved in generating grammar and syntax. Broca's area is located between the frontal cortex where planning takes place and the motor regions that control movement of the lips, tongue, and larynx.

Recent work by Martin and others using positron emission tomograpy (PET) images has revealed some clues as to how meaning or semantic information about a particular object is represented in the human brain [Martin98, Martin et al.95,96]. It had previously been observed that damage to the human brain may cause selective loss of knowledge about specific categories of objects. One patient may have difficulty in identifying or naming animals, whereas another may have selective difficulty identifying manufactured objects such as tools. Martin and his co-workers used PET technology to map the regions of the normal human brain that are associated with naming animals and tools. They found that naming animals selectively activated the left medial occipital lobe, a region associated with the earliest stages of vision. In contrast, naming tools selectively activated a left premotor area also activated by imagined hand movements, and an area in the left middle temporal gyrus also activated by the generation of action words.

Thus the brain regions active during object identification are dependent, in part, on the intrinsic properties of the object presented. Different attributes are stored in different locations, so that different object classes excite the network

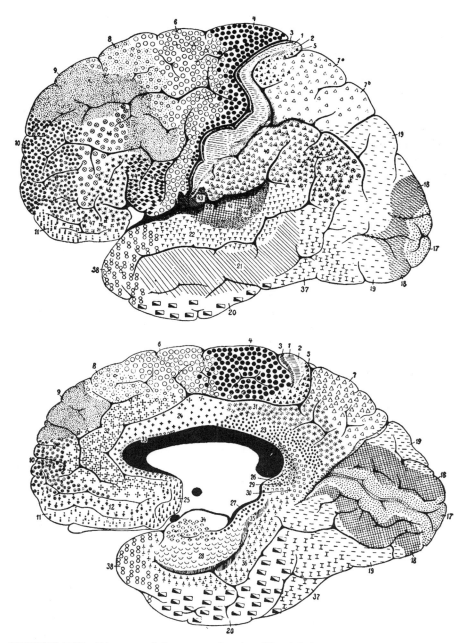

FIGURE 2.10 Diagram of the cortex showing Wernicke's area at the intersection of areas 39, 40, 41, and 42 and Broca's areas in parts of areas 4 and 6 that are dedicated to speech (i.e., the lips, jaws, tongue, and larynx), as shown in Figure 2.5. [From Brodmann14.]

in different ways that are distinguishable by PET imagery. For example, attributes that describe animals tend to be stored in locations associated with the computation of shape, color, and motion, whereas attributes that describe tools tend to be stored in locations associated with action words describing how they are used. Martin's work demonstrates that object names and attributes are stored in a distributed network of discrete cortical regions. These regions can be observed to be active when the object is perceived visually, or its name is heard or produced either vocally or silently.

Martin's hypothesis is that semantic information about an object is stored as a distributed system in which the attributes that define the object are represented in, or near, the same tissue that is active during perception. For example, information about the color of an object (e.g., that kangaroos are tan rather than blue) is stored near regions active during color perception, while information about the pattern of motion associated with an object (e.g., that kangaroos hop rather than gallop) is stored near regions active during the perception of motion. It is known that the perception of color and motion are mediated by separate systems that include distinct regions of the occipital cortex. The anatomical location of these regions in humans has been identified in PET studies [Corbetta et al.90; Zeki et al.91]. It is also known that patients with focal cortical lesions indicated that retrieval of information about each of these attributes could be disrupted selectively, suggesting that semantic information about color and motion may be stored in separate brain regions.

Subjects were shown black-and-white line drawings of common objects. During one PET scan they were asked to name the objects. During a second scan they were asked to retrieve the name of a color commonly associated with the objects. During a third scan they were asked to name an action commonly associated with the objects. For example, in response to an achromatic line drawing of a child's wagon, subjects responded "wagon," "red," and "pull" during the object naming, color word generation, and action word generation conditions, respectively. A second experiment was also run that was identical to the first except that the subjects were presented with the written names of objects, rather than pictures. The two experiments yielded similar results.

This suggested that the organization of information storage in the brain follows a design similar to that of the organization of sensory and perhaps motor systems as well. Thus information about a single object (e.g., a wagon) is not stored as a single entity in a single place. Rather, the attributes and features that characterize the object are stored separately in sites that are active during perception of the object. More recent work has suggested that regions specifically associated with perception are themselves not reactivated when perceptually based information is retrieved. Thus it appears that retrieving information about a specific object attribute such as color required activation of a region of the brain situated close to, but not including, the neural circuitry involved in the perception of color.

Of course, PET images provide only gross measurements of neuronal activity. PET cannot resolve the precise synaptic sites that associate semantic meaning with visual images or that generate verbal or written responses to questions about abstract concepts. However, the PET images do suggest that object attributes are represented close to, but not in, the sensory and motor regions of the brain that were active when information about that object was perceived. Thus the storage and retrieval of attributes is parallel to the regions in which the attributes are computed during object recognition and used during the planning of behavior.

Thus, even though there is much that remains a mystery, we know a lot about how symbolic information is processed and stored in the brain. And we are learning more at a rapid pace as the technology of brain imaging is exploited and improved. The examples given above are only a small sample of a massive and rapidly growing literature. Experiments are being conducted in hundreds of laboratories throughout the world. Neuroscience journals are filled with detailed reports of studies of the computational mechanisms and representations of knowledge in the brain. It is steadily becoming more clear how representations of knowledge are derived from sensory data and used to generate motor commands in the brain.

REPRESENTATION OF THE REAL WORLD

The world in which we live is rich and complex far beyond our ability to describe in words or symbols and even beyond our ability to capture in an image. When we look at a large tree, we see hundreds of thousands of leaves, twigs, and branches, all of which converge to a trunk that disappears into the ground. The bark is furrowed and textured in patterns that can, at best, only be summarized by language. The upper branches weave a complex web of branches, twigs, leaves, and seeds—some alive and strong, others dead and weak. A squirrel running through the treetops has little need for a sophisticated language, but definitely must be able to represent and compute the position and motion of objects (such as branches and rival squirrels) and assign attributes to them (such as "safe for holding weight" or "to be pursued" or "to flee from").

It is often said that "a picture is worth a thousand words," but a typical out-door scene is richer and contains more information than a thousand pictures. For example, a typical monkey's visual system processes thousands of images in generating a representation of the jungle canopy as the monkey forages for food, fights with competitors, or flees from predators through the treetops. Closer to home, consider the complexity and variety of sensory input that a typical human experiences during the course of a simple backyard barbecue in an average suburban neighborhood. A person might see flowers, bushes, and trees blowing in the breeze. Buildings, windows, doors, grass, and rocks would be illuminated by sunlight and dancing shadows. Birds, insects, and falling leaves might fly through the air. There might be patio furniture, pets, and people eating and drinking and moving about. One might hear birds chirping, the wind blowing, traffic on the street, children at play, and people talking and laughing. One might smell hamburgers cooking, see and smell charcoal smoke, taste food and drink, feel the warmth of the sun, and experience the chill of a dip in the swimming pool.

The central nervous system assimilates all of this sensory information simultaneously and apparently without effort or thought. Immediate sensory experience is combined automatically with prior knowledge of everything one knows about how the world works and how plants, animals, people, and insects normally behave. Prior knowledge includes all that the brain has learned from an entire lifetime of experiences at home, in school, in church, at work, and play since the day one was born.

Somehow all this information is represented internally in a form that enables us to behave successfully in the complex, dynamic, and uncertain environment of the natural world. We easily move about a world filled with stationary and moving objects without collisions. We manipulate objects and tools, gossip with friends, perform complex tasks, make plans, and achieve goals. We can read and write and understand abstract principles of mathematics and science. We can learn from history and plan activities for the future. We can make plans for the rest of today, for tomorrow, next week, next year, and after retirement. We can believe in God and imagine our status in the afterlife. How is this done? What kind of representations can support these cognitive and behavioral capacities? What kind of mechanisms can assimilate all this information and make it available for planning and decision making in real time?

Surely, our representation cannot be a simple one, or it could not handle so much complexity and subtlety, or capture such exquisite detail. Clearly, the richness that we observe in nature must be represented internal to our mind (at least momentarily) or it could not be perceived at all and could not be used to generate behavior. Certainly, there is some mechanism by which what is directly observed is combined with, and augmented by, what is known a priori. Otherwise, we would not be able to recognize our friends, make conversation about previous experiences, or even to know who we are or how we got to where we are. Without the ability to meld present experience with prior knowledge we would all suffer the equivalent of complete and perpetual Alzheimer's disease.

It is clear that either symbolic or iconic representation alone is inadequate to capture all that we can know about the world. More than symbols are required for the kind of reasoning, planning, and behavior that humans do in everyday life. More than images were required to survive the fierce competition for survival in the jungle and savanna that resulted in the development of human intelligence. Human knowledge representation is capable of supporting a wide range of intelligent behavior, from hunting, home building, farming, and tribal warfare, to all manner of cooperative endeavors and competitive conflicts that have occurred during the rise of civilization and the building of empires. The representation of knowledge required to support human levels of performance in commerce, industry, science, politics, sports, warfare, and everyday living obviously go far beyond those possessed by insects, birds, and lower mammals. Human behavior requires the representation of pictorial, iconic, symbolic, and procedural forms of knowledge; and the combination of these forms in a computational architecture that is capable of representing the spatial and temporal richness and complexity of the natural world. Successful behavior requires storage and retrieval mechanisms that are fast enough to support real-time perception, learning, situation analysis, decision making, planning, and reactive control. It requires optimal use of finite resources of memory and computational power in order to survive in competition with nature, other species, and other members of our own species.

To understand the full range of human intelligence, we must begin with an appreciation of the fact that the processing of images, arrays, and maps is among the most fundamental of the brain's computational functions. Iconic representation and parallel processing are the underlying mechanisms upon which the higher-level functions of perception, cognition, and behavior (including language behavior) are based. However, images and maps alone are not sufficient. Symbolic representation and serial processing is also necessary. Rules, relationships, and abstract concepts such as mathematics, logic, and language require symbolic representations. The manipulation of symbols clearly plays a major role in in human behavior. Thus both iconic and symbolic representations and parallel and serial processing are integral components of the mind and brain. Both of these two forms of representation and computation must be tightly coupled and fully integrated to produce the full range of behavior that is characteristic of human beings.

REPRESENTATION IN ARTIFICIAL INTELLIGENCE AND ROBOTICS

The issue of how to represent knowledge about the world is a subject of active research in the field of computer science and a source of heated debate in artificial intelligence (AI) and robotics. Among researchers working on language, there is a strong tendency to represent everything symbolically in a form that can be expressed by logical theorems, expert system rules, or linguistic grammars [Newell and Simon63,72, Newell et al.58, Simon57,62,91, Chomsky88, Laird et al.87]. Many AI researchers largely or completely ignore images as a mechanism for knowledge representation. Many entire books on knowledge representation written from the AI perspective do not even mention images or maps [Galambos et al.86].

AI representations designed for laboratory experiments are often confined to carefully constrained environments such as the "blocks world" [Winograd72, Winston84, Winograd and Flores86]. However, one need only look around the room, walk down the street, or venture into the woods to realize how sterile blocks world representations are and how distant from reality. Laboratory experiments with six-degree-of-freedom robot kinematics and dynamics are a far cry from the real-world problems of muscles controlling the skeleton of a cat or a deer running through the forest or of a bird flying between tree branches. Understanding relationships in the typical semantic network is a long way from coping with the dynamic problems an ape encounters in raising a family in the rain forest, or even the problems a crab experiences in scavenging for food on a coral reef.

The AI community has been soundly criticized by Dreyfus[72,92], Searle[80], Kickhard and Terveen[96], and others, who point out that symbol manipulation alone is inadequate even for understanding language, much less for supporting the full range of human perception, cognition, and behavior. Much of Dreyfus's criticism is focused on AI researchers such as Simon, Newell, Minsky, Schank, Winograd, and others, who extrapolated from initial successes with symbol manipulation, logic, and formal languages to more complex situations in the real world. Dreyfus argues that the methods that produced early progress in problem solving, game playing, and formal languages do not scale up to solve to deeper problems of understanding real-world situations. Searle argues that syntatic competence alone can never produce systems that understand natural language. He uses his famous "Chinese Room" thought experiment to demonstrate that a pure symbol manipulator can never really understand what a statement in a language means, and hence can never produce satisfactory language translation. Both Dreyfus and Searle insist that language understanding requires far more then rules for symbol manipulation.

In recent years, partly as a result of the difficulties encountered in automated language translation, and partly in response to the criticisms of Dreyfus and Searle, many AI researchers have retreated from trying to duplicate human levels of performance in knowledge representation and understanding. In many labs, current efforts are more directed toward building primitive behaviors such as are exhibited by insects than in understanding human intelligence. In some respects, this is a healthy turn of events. Certainly, there is much to be learned from attempting to understand the computational mechanisms that enable low-level behaviors such as are exhibited by houseflies and cockroaches. The insect world is a fascinating place filled with incredible examples of physical dexterity and engineering prowess. Interesting behaviors can be observed in the ant, the bee, the spider, the dragonfly, and the butterfly. There is undoubtedly much to be learned from attempting to build machines that can duplicate the dexterity and behavioral skills of insects. We should at least be able to offer a plausible explanation for insect behavior before we make too many claims about the processes that comprise the human mind.

Unfortunately, the retreat from symbolic representations alone has, in some quarters, turned into a general retreat from all forms of knowledge representation. Brooks[90] has gone so far as to reject internal representation altogether. Following the lead of Gibson[50,79], Brooks argues against explicit internal representations based on the reasoning that the world is its own representation and can be sensed when necessary for making behavioral choices. This line of research focuses on reactive behaviors as the foundation of intelligence and emphasizes the role of stimulus–response learning in the tradition of behaviorism.

However, to most people it is obvious that there are many situations where behavioral goals and plans require information not directly or immediately available from sensory input. For example, sensory information alone is insufficient to decide whether an occluded object still exists. Without a map it is impossible to tell whether a winding road leads to a desired destination. Without stored knowledge, an observed object cannot be recognized as having been seen before. The minimalist approach can support only the most rudimentary forms of stimulus–response behavior. For higher levels of intelligent behavior, complex dynamic internal representations are required.

Unfortunately, the Dreyfus–Searle critique is primarily negative. No suggestion is made of where to search for additional mechanisms beyond symbols that might suffice for representing the complexity of the natural world. Rather, Dreyfus and Searle conclude pessimistically that true machine intelligence is beyond the ability of current scientific knowledge to achieve. They invoke a form of Cartesian dualism by arguing that if mechanisms of symbol manipulation cannot produce human levels of perception and cognition, there must be something in natural intelligence that lies beyond the realm of computational processes. Similar reasoning has led Penrose[89] to speculate about quantum effects in microtubules to explain the properties of mind that exceed our current understanding.

What Dreyfus, Searle, and Penrose (along with the mainstream AI researchers that they criticize) fail to appreciate is the richness and complexity of knowledge representation and cognitive reasoning powers that can be realized in the domain of images and maps. They also neglect the time and frequency domains and ignore the computational power of modern approaches to dynamic systems analysis, signal processing, and control theory. In neglecting the representational and computational powers represented by images, they miss what can be achieved when the massive parallelism of the image domain is augmented by the precision of mathematics and logic in the symbolic domain. In neglecting modern approaches to system dynamics, they miss what can be achieved when multiple control loops are closed at multiple levels of resolution in time and space simultaneously. We believe that these omissions constitute a fatal flaw in the critic's arguments and lead to the overly pessimistic conclusion that properties of the mind are beyond the reach of electronic computation.

We argue here that the failures of artificial intelligence can be readily explained without retreating to Cartesian dualism or proposing exotic theories of quantum effects in neuron microtubules. We believe that the difficulties encountered by pure symbol manipulation systems in dealing with real-world planning, reasoning, and situation analysis can be remedied. We believe that the solution lies in combining the power of parallel representation and computation in the image and map domains with analysis in the time and frequency domains in a multilevel architecture of dynamic recursive loops.

Images are by far the oldest, most highly developed, and most ubiquitous forms of representation in nature. Dynamics are crucial to all forms of behavior. Multiresolutional closed-loop control dynamics are central to all complex organisms that pursue goals in the real world. To ignore images and dynamics in the quest for understanding knowledge is like ignoring the concepts of space and time while attempting to understand physics. Even in the world of insects and crustaceans, images are fundamentally important. The compound eye of snails, insects, and spiders produces an array of signals that comprise an image. Early studies of lateral inhibition were done on the compound eye of the horseshoe crab [Ratliff and Hartline59]. Insect vision has the ability to detect edges and compute image flow. Houseflies can compute time to collision with a landing site [Borst90]. Maps are also fundamental. Bees clearly have some internal representation of space and can communicate navigational concepts such as heading with respect to the sun and range to a source of food. There is evidence that many creatures, including some insects, use dead reckoning to construct an internal map of their environment that they use for navigation. Birds, bees, and a number of insects, such as the African desert ant, all have the ability to return directly to their nests from the point where they find food, even after wandering through lengthy and circuitous paths while hunting. Wasps locate their nests by landmark navigation. Rats swimming in a pool of opaque water can find submerged resting places by reference to landmarks [Morris81]. Migratory birds can navigate at night by recognizing the pattern of stars in the sky. Insects, fish, reptiles, amphibians, birds, and mammals all use visual images for generating behavior.

Images are important not only for vision, but for touch and hearing. Tactile sensors are arranged in arrays on the surface of the skin, and tactile images are mapped onto the surface of the brain. Many species use tactile feelers or whiskers to augment their visual perception. Acoustic signals are also represented on the surface of the brain as an array that is related to the spatial position of the source of the sound on the egosphere [Jay and Sparks84,87, Sparks and Groh87].

Many researchers in the field of robotics have explored the use of images and maps for representing knowledge about geometry and spatial relationships needed for path planning for manipulation and locomotion. Many robotic systems represent spatial relationships within and between objects as two-dimensional maps, potential fields, or three-dimensional grids, or octrees [Moravec88, Bornstein and Koren91, Khatib86, Matthies and Elfes87, Lozano-Perez and Wesley79]. Some researchers in machine vision have even explored the relationship between images and symbols by representing symbolic entities iconically in images or maps as well as in lists or frames containing states, attributes, and pointers [Riseman and Hanson90]. Others have developed algorithms for extracting geometric information such as generalized cylinders [Binford82], geons [Biederman90], and shape of image contours [Brady and Yuille87, Koenderink and Van Doorn79].

Representation of knowledge about the shape and size of objects and structures using diagrams and drawings is well developed in the fields of manufacturing and construction. Object geometry is typically represented in computer-aided design (CAD) databases by symbolic data structures such as points, lines, arcs, vertices, vectors, polygon surfaces, and volumes. Industrial machines and robots use symbolic knowledge represented in computer programs to manipulate parts and tools, perform inspections, and assemble products. Machine tool behavior can be specified by RS274 programs, cutter location (CL) data files, or computer-aided manufacturing (CAM) programs. Inspection programs can be expressed in the dimensional measuring interface inspection system (DMIS) language. Karel and VAL are popular robot programming languages. The standard for the exchange of product model data (STEP) defines a number of standard formats for representing knowledge of how to manufacture or construct a wide variety of products and structures [STEP90].

We agree with critics of artificial intelligence such as Dreyfus, Searle, and Penrose in arguing that symbolic computation alone can never produce human levels of intelligent behavior. However, we believe the critics are wrong in predicting that digital computation is fundamentally incapable of ever producing human levels of intelligence. We believe that parallel multilevel computation operating on a representation that fully integrates iconic, symbolic, and dynamic knowledge can model space and time, entities and events, tasks and skills, plans and behavior, logic and language, physics and mathematics, morals and values, with sufficient richness and complexity to support human levels of performance in sensation, perception, cognition, and commonsense behavior.

Of course, many fundamental questions remain. How is knowledge perceived? Where and how is it stored? What kinds of data structures are used to represent it? How is it retrieved and used by mental processes to generate expectations, focus attention, recognize and classify objects and events, analyze situations, and plan future actions? There is much that we still do not know, and much that will remain a mystery for many years. Nevertheless, a computational theory of mind based on neurological mechanisms in the brain appears to offer the best avenue for future exploration. At present, there appears to be no aspect of the mind that lies beyond the reach of this hypothesis. In particular, there appears to be no need to assume that phenomena such as subjective feelings, religious experience, and free will cannot be accounted for by computational processes within the neuronal substrate. In principle, there seems to be no need to postulate a Cartesian dualism that involves spiritual elements, or to invoke to exotic mechanisms such as quantum effects in microtubules to explain the phenomena of mind. We believe that the mechanisms of mind can be adequately explained by computational theory, albeit computation on a scale and involving a degree of complexity beyond anything that has been implemented to date.

3 Perception

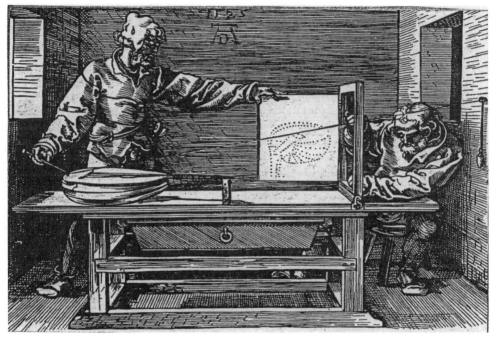

A Man Drawing a Lute by Albrecht Dürer from *The Complete Woodcuts of Albrecht Dürer*. [By permission from Dover Publications.]

Perception is a window onto the world. Perception is finding out, or coming to know, what the world is like through sensing. Perception enables us to acquire and confirm knowledge and to build and maintain an internal representation of external reality. Perception extracts from the sensory input the information necessary for an intelligent system to understand its situation in the environment and adapt its behavior so as to act appropriately and respond effectively to perturbations and unexpected events in the world.

It is widely believed that perception is the result of a set of computational processes that transform sensor signals into knowledge about entities,

57

events, and situations in the world. These processes include attention, grouping, segmentation, filtering, hypothesis generation and testing, classification, and recognition. Perception establishes a correspondence between things in the external world and representations of those things in the internal world model of the intelligent system. Establishing this correspondence is sometimes called *symbol grounding*. It provides the foundation for semantic meaning. It enables us to name things, track them over time, and act on them purposefully. It enables us to understand speech and written language and to have a meaningful conversation with others about what interests us. It enables us to understand the state of the physical world and our relationships within society.

The idea that perception results from computational processes in the brain goes back at least to von Helmholtz [Southall24]. It remains the leading hypothesis among vision researchers today [Hubel88, Marr82, Barlow72, Rosenfeld and Kak76, Biederman90, Ullman96]. Yet there remain many questions regarding the exact nature of these processes. How are images we see transformed into what we perceive about the world? How are sounds we hear transformed into concepts and ideas? How is what we sense transformed into what we feel? How do we identify objects and situations? How do we classify people, places, and things? How do we recognize what we have seen before, even though we have never seen it from the same perspective, in the same light, or in the same situation? How do we identify something we have never seen as belonging to a class? How is the transformation from external reality to internal model accomplished?

Consider, for example, a simple act such as perceiving a cat on a sofa. How do the signals from millions of photodectectors in our eyes get transformed into knowledge that there is a cat is on a sofa? What happens when we perceive a cat? Do we observe a furry shape in the form of a cat? Do we recognize a volume in space as being occupied by a cat? Do we establish the truth of a proposition that a cat exists at a particular place at a particular time? Is there an image in our brain consisting of a set of pixels in the shape of a cat? If so, what interprets this image as a cat on a sofa? Could each pixel in the retinal image of the cat have a pointer to a symbolic data structure labeled "cat" and each pixel imaging the sofa have a pointer to another data structure labeled "sofa"? Could there exist a *situation data structure* (or semantic net) in which pointers establish the "on" relationship between the cat and sofa? If so, how would these pointers get set? How would they be maintained when the gaze shifts from place to place in the scene as the eye moves relative to the world? How could the world appear to remain stationary when the image of the world is moving about on the retina?

What happens when perception is wrong? A furry shape on the sofa might not be a cat. It might be a toy replica of a cat, or a bundled-up sweater of the same color, size, and shape as a cat. In poor light, a hat or a pillow might be mistaken for a cat. Perception is something that occurs internal to the mind of the observer. It may be mistaken. On a dark and windy night, a person alone in a strange house may mistake a creaking sound for a burglar or a ghost.

Perception does not necessarily represent truth about the world. A dictionary defininition of *truth* is "being in accordance with the actual state of affairs." In perception, truth is correspondence between an internal representation and the actual state of the external world. Truth is an internal model of external reality that can be relied on to make accurate predictions about the world. This raises a number of questions regarding accuracy and reliability of models and predictions. How is accuracy defined? At what resolution is accuracy measured? How accurate is good enough? What is reliable? What is the measure of reliability? How reliable is good enough? How does one verify the accuracy and reliability of an internal mental model of the external world? These questions raise both philosophical and engineering issues. The philosophical issues have provoked debate among philosophers for centuries. The engineering issues can be addressed by mathematical procedures. Signal processing theory provides methods for comparing observations based on sensors with predictions based on an internal model [Weiner49, Kalman60]. Correlation and variance can be computed by well-defined computational algorithms. Levels of confidence in the model and in the sensory data can be computed by recursive estimation. Decision theory provides methods for choosing among uncertain alternatives [von Neumann and Morgenstern44]. As confidence in the internal model rises, the model may be considered more reliable than sensory observations. As confidence in the model falls, sensory observations may be considered more reliable than the model.

Perception is not a single process. It is a multiplicity of processes (e.g., vision, hearing, touch, proprioception, internal, smell, taste) that operate simultaneously and in parallel at multiple levels in a hierarchy of processing modules (see, e.g., Figure 2.4). At each level, perception involves parallel array processing as well as sequential symbolic processing. Perception does not take place all at once. It occurs in stages. At an early stage, we are aware of a rich tapestry of sensations consisting of sights, sounds, odors, tastes, touch, force, acceleration, pressure, and vibration. In later stages, we focus our attention and filter our sensory input. We select from the environment what we feel is important. We operate on the sensory input to detect features, segment regions, cluster entities, and abstract information about objects, events, and situations [Ullman96].

Perception results from multiple processes that function both individually and collectively to generate a coherent model of the environment. It is clear that different aspects of perception are accomplished by different computational processes in different parts of the brain. For example, the vision system appears to contain at least two distinct processing channels: one for computing position and motion, the *where* channel, and another for shape and class membership, the *what* channel [Van Essen and Deyoe95]. Figure 3.1 diagrams the convergence and divergence of information in different modules within the image processing hierarchy, starting with the small parvocellular (P) and large magnocellular (M) cells in the retina and lateral geniculate and ending with spatial relationships (where) in the posterior parietal cortex and object recog-

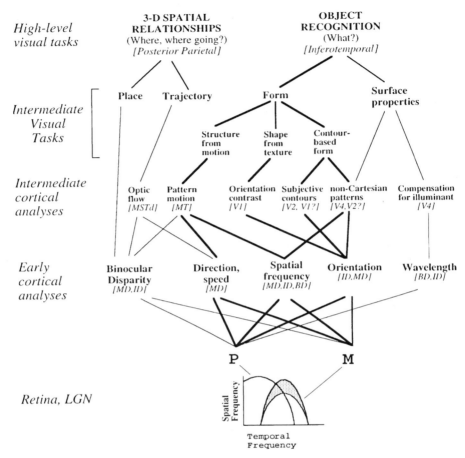

FIGURE 3.1 Functional diagram of the visual processing system. The types of processing algorithms that reside in various regions in the visual cortex are shown along with the communications pathways that convey output and provide input to the computational modules. The P cells provide good spatial contrast; the M cells are more sharply tuned to temporal differences. [From Van Essen and Deyoe95.]

nition (what) in the inferotemporal cortex. Figure 3.1 is a simplified version of Figure 2.4.

Each perceptual process makes its own individual contribution to the whole. For example, one can see the outline of an object without recognizing what it is. One can hear a string of words without understanding what they mean. One can feel or smell or taste something without knowing what it is. One can visually detect a furry shape without recognizing it as a cat. One can classify an animal as a member of the cat family without recognizing it as a particular cat or knowing its name or where it lives. A full understanding of an image may require that segmented surfaces and objects be overlaid with labels and class names and that these labels be interconnected in a semantic network of relationships and situations. A full understanding of a segment of human speech may require that words and phrases be associated with labels and class names, and that classes be interconnected by contextual relationships that denote meaning.

Our understanding of what we see begins with signals from rods and cones in the retina. It grows as layers of neurons detect surface patches, edges, and discontinuities in brightness, color, range, and motion. It culminates with the classification and recognition of surfaces, objects, events, and situations. At many different levels, the visual input is manipulated by gestalt hypotheses that segment the image into surfaces with edges and contours, and finally, into objects that have attributes such as shape, size, color, texture, temperature, position, orientation, and motion [Marr82].

Understanding what we hear begins with signals caused by vibrations in the cochlea. It grows as layers of neurons detect amplitude, frequencies, impulses, and phase differences between the ears. It culminates with the classification and recognition of sounds that convey meaning. At various levels, the auditory input is manipulated by gestalt hypotheses that cluster the acoustic signal into syllables, words, phrases, concepts, and ideas. The visual recognition of *what* information regarding written symbols and words merges with the auditory recognition of *what* information regarding spoken words and phrases in Wernicke's area, where key elements of language understanding are located. The visual measurement of *where* merges with the tactile measurement of *where* information in the parietal somatosensory cortex, where spatial and temporal relationships between entities and events in the world are computed. The details of exactly where these interactions occur are being revealed by the new technologies of PET and functional MRI [Carter98].

Perception is by no means a one-way flow of information. There is solid anatomical and physiological evidence that information flows downward and horizontally as well as upward during the perceptual process [von Holst and Mittelstaedt50]. The information pathways shown in Figure 2.4 convey information both upward and downward as well as laterally between computational modules at the same level [Van Essen and Deyoe95]. Perception depends on what we experience, what we know, and what we intend. What we experience is interpreted in terms of what we know about the world. Where we look, what we touch, and what we listen to depends on what our intentions are and what we care about. From a computational viewpoint, the downward flow of information generates perceptual expectations, predictions, and priorities based on task goals. Downward information flow controls the focus of attention and influences the interpretation of sensory experience. The downward flow of information enable us to project our expectations, hopes, and fears onto the world we perceive. The lateral flow of information within the sensory processing and world modeling processes enables filtering, correlation, and recursive estimation. The lateral flow of information from perception to behavior provides feedback for error compensation and learning.

SENSORS

Perception begins with signals from sensors. Intelligent systems typically have large numbers of sensors that enable them to sense the environment. In biological systems, sensors are specially adapted neurons that transform input energy into signals represented by voltage levels across cell membranes. Each sensory neuron produces a signal that is a function of the intensity (or time rate of change of intensity) of the physical phenomenon that it measures. For example, in the visual system, rods and cones in the retina transform radiation in the visible portion of the electromagnetic spectrum into electrical signals that represent intensity of energy at different parts of the visual scene that is imaged on the retina by the optical system. This is illustrated in Figure 3.2. Each rod or cone measures the light energy focused on the area it occupies on the retina in the spectral band to which it is sensitive radiated in its direction from a corresponding surface patch in the world.

Commands to the lateral and medial rectus muscles determine the azimuth of the eyes relative to the head. Commands to the superior and inferior rectus muscles determine the elevation of the eyes relative to the head. Linear and rotational acceleration sensors in the vestibular system generate signals that can be integrated to infer position, velocity, and orientation of the head relative to an inertial coordinate frame. Stretch sensors in the neck measure the position of the head relative to the body. This is illustrated in Figure 3.3.

Auditory sensors in the ears transform pressure vibrations in the air into signals that represent sound energy in the audio-frequency band to which they are sensitive. Tactile sensors in the skin transform pressure, vibration, and temperature into signals that represent touch, vibration, temperature, pain, and itching. Taste sensors in the tongue transform chemical properties of food into signals that represent taste sensations. Olfactory sensors in the nose transform odors into signals that represent smell. Internal sensors transform temperature, blood pressure, muscle stretch, tendon tension, and blood chemistry into signals that represent heart rate, flow of blood, breathing, intake of food and water, state of digestion, and need to discharge waste products. Proprioception sensors transform mechanical properties of stretch, tension, position, and velocity in the muscles, tendons, and joints into signals that represent the state (i.e., position, orientation, motion, and force) in the torso, limbs, eyes, hands, and fingers.

In artificial systems, sensors consist of transducers that transform input energy into signals represented by voltage levels on wires. For example, a typical charge-coupled device (CCD) camera consists of a two-dimensional array of photodetectors that sense the intensity of incoming radiation in the visual or infrared spectrum. Each element in the array is a picture element, or

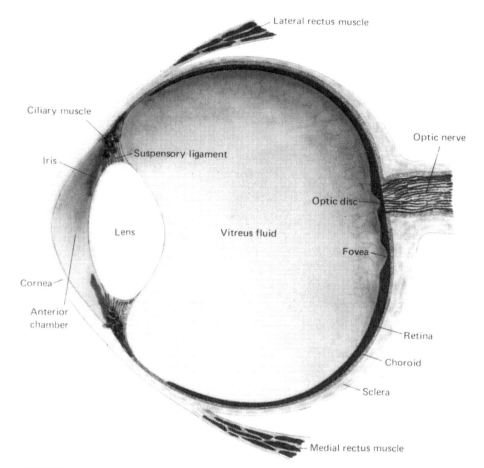

Lateral rectus muscle

Ciliary muscle

Optic nerve

Suspensory ligament

Iris

Optic disc

Optic disc

Lens   Vitreus fluid

Fovea

Cornea

Retina

Anterior
chamber

Choroid

Sclera

Medial rectus muscle

FIGURE 3.2 Cross section of the human eye. Light from the external world is focused by the lense on the retina. The focus of the lens is controlled by the ciliary muscle. The aperature is controlled by the iris. [Kaufman79, from *Perception*, edited by Kathleen Akins. ©1996, used by permission of Oxford University Press, Inc.]

pixel, that carries a voltage or electric charge. Black-and-white cameras consist of pixels that integrate the intensity across the entire visual spectrum. Color cameras typically have three arrays of pixels that are differentially sensitive in the red, green, and blue regions of the spectrum. Forward-looking infrared (FLIR) cameras have a thermal sensor at each pixel. Laser range imagers (LADARs) measure the time of flight for a pulse of light emitted from each pixel to bounce off a surface in the world and be detected in the LADAR. LADAR and imaging radar produce range values at each pixel.

   Sensors may be grouped into a variety of sensory subsystems. For example, a vision subsystem may consist of one or more cameras, including

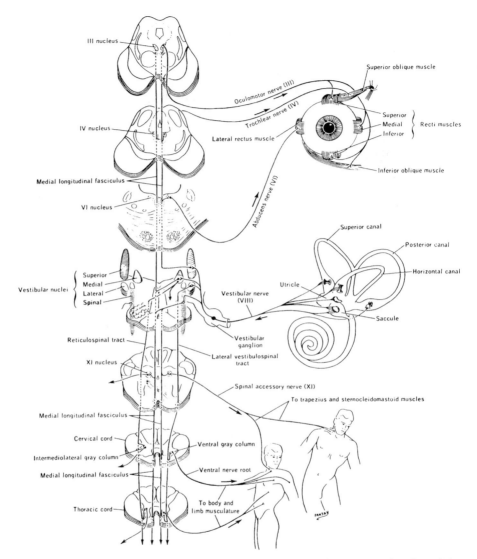

FIGURE 3.3 Diagram of the vestibular system showing the computational modules and interconnecting pathways that constitute the vestibular-occular reflex. The pointing direction of the eye in the horizontal plane of the head is controlled by the lateral and medial rectus muscles. Pointing of the eye in the vertical plane of the head is controlled by the superior and inferior rectus muscles. The roll of the eye about its optical axis is controlled by the superior and inferior oblique muscles. [House and Pansky60, reprinted by permission of McGraw-Hill.]

black-and-white or color CCDs, FLIRs, and LADARs organized as monocular, stereo pairs, or sets of cameras with different magnifications and nested fields of view. An intelligent machine may also have acoustic sensors that detect sounds in the environment. Acoustic sensors may include microphones, hydrophones, seismometers, and sonar. These produce signals that are proportional to the amplitude of the acoustic energy in the world. Acoustic sensors can be configured in arrays to estimate the direction of arrival of the incoming energy. Sonar transducers transmit and detect reflected sound energy. Tactile sensors may measure touch, vibration, force, temperature, or pressure.

An intelligent vehicle system may use an odometer, a speedometer, accelerometers, tilt meters, gyros, and global positioning satellite (GPS) receivers to estimate position, velocity, heading, and orientation relative to Earth. Encoders and resolvers may measure position, velocity, acceleration, force, torque of actuators and motors, or indicate the pointing direction of pan/tilt units for cameras and range finders, or represent other parameters, such as vergence, focus, iris, and zoom. Other types of sensors may measure steering angle, gear setting, wheel slippage, engine speed, coolant temperature, oil pressure, fuel level, road vibration, magnetic fields, electromagnetic waves, nuclear radiation, and presence of chemicals in the air or on the ground.

An intelligent manufacturing system may have sensors that measure position and orientation of parts; velocity, force, and vibration of tools or fixtures; and temperature, hydraulic pressure, and operating state of machines. Touch probes may measure position, orientation, and dimensions of parts and part features. Vision sensors may measure properties of surfaces, the transparency of liquids, and the condition of materials. Vision may also be used to count objects, detect patterns, or measure the distribution of materials. Sensors may read bar codes and identification tags. Chemical plants, refineries, and power plants may have sensors that measure temperature, pressure, flow rate, volume, and state of chemical reactions.

Resolution

Outputs from sensors are typically in the form of analog voltages or currents. Analog signals can be processed in analog computers or used directly in analog feedback loops. However, analog signals are subject to noise interference and suffer from dissipation and distortion during transmission over long distances, due to resistance, capacitance, and inductance. In biological systems, analog sensory signals represented by cell membrane voltages are sometimes used locally for analog computations, but are more often are transformed into impulse rates (or interpulse intervals) of action potentials for transmission long distances over axons. In artificial systems, voltages from sensors are typically sampled and digitized by analog-to-digital converters into binary numbers before being transmitted long distances or processed in a digital computer.

The rate at which sensory signals are sampled establishes the resolution along the time line and hence constrains the bandwidth of signals that can be represented. The *Nyquist criterion* states that the highest-frequency signal that can be represented by sampling must be less than one-half the sampling rate. In practice, signals are typically filtered to eliminate frequencies higher than one-tenth of the sampling frequency to prevent aliasing. The size of quantization steps in analog-to-digital converters establishes the resolution in amplitude. This limits the precision with which measurements can be represented. Signals from photodetectors typically are quantized with 8 bits (256 levels) of resolution. Pressure, force, or tension sensors may be quantized with 10 bits (1024 levels) of resolution. Position sensors may be quantized with 12 to 16 bits (4096 to 65,536 levels) or more of resolution.

The spacing of sensors in an array such as in a camera, in a retina, or on the skin establishes the spatial resolution of the sensor array, and hence the level of detail in, or magnification of, the image that can be represented. The size of the array establishes the field of view of the image, or the range of the map represented by the array. The size divided by the resolution establishes the density of the image. Low-density image arrays may contain 128×128 ($\approx 16,000$) pixels or less. High-density images may contain 1024×1024 ($\approx 1,000,000$) pixels or more. Variable resolution may be used to produce images with high spatial resolution in dense regions of the sensor array and low resolution in sparse regions of the sensory array. For example, the eyes of many biological species have closely spaced photoreceptors in the fovea and widely spaced sensors throughout the remainder of the retina. In the human eye, the density of vision signals leaving the retina decreases roughly as the logarithm of the distance from the center of the fovea. This produces a high-resolution highly magnified image in the fovea, with lower resolution and lower magnification in the periphery. This is called *foveal/peripheral vision*. A similar effect can be achieved in artificial systems through a variety of mechanisms, such as log-polar CCD arrays, optical lenses with nonuniform magnification, zoom lenses, or multiple cameras with lenses of different focal lengths.

Sensors may be tuned (i.e., maximally sensitive) to portions of the frequency spectrum of the incoming energy. For example, red, green, and blue cones in the retina are sensitive to energy in three different bands of the visible spectrum. Rods are much more sensitive than cones to low light levels but generate only gray-scale images. In the ears, the cochlea is a tuned cavity designed so that vibration sensors at different places in the cochlea are tuned to specific frequencies of sound. In the skin, different kinds of sensors have evolved to detect different types of tactile sensation, such as pressure, vibration, temperature, and pain. In the smell and taste systems, sensors are tuned to detect chemical properties of molecules with which they come in contact.

Of course, sensing is only the input to perception. Sensory signals must be processed and transformed before they can be used for generating and controlling behavior. Critical to these computations are transformations between coordinate systems. Measurements made by each sensor are in the coordinate system of that sensor. Actions taken by the behavior generation system are in the coordinate systems of each muscle and actuator. In low-level servo loops, there may be a one-to-one relationship between the coordinate systems of sensors and actuators. Examples are stretch receptors in the muscles and encoders attached to output shafts of servomotors. However, for higher-level control loops, sensors and actuators are typically in different coordinate systems. For example, tactile sensors in the skin that detect an itch are in a different coordinate system from the muscles that scratch the itch. The eyes are in a different coordinate system from the hands and feet that are controlled by visual information. Information from sensors must often be transformed several times into different coordinate systems as it flows from sensing to behavior [Dickmanns99]. Multiple coordinate transformations may also be required as information flows downward from high-level goals to low-level actions of manipulation, locomotion, and gaze control.

Ambiguity and Noise

Sensory input is almost always ambiguous. It is rare (except in the case of simple servo loops) that knowledge about the world necessary for controlling behavior can be directly measured. For example, a two-dimensional image on the retina is completely ambiguous with regard to the distance (or range) to objects in the image. Photodetectors in a camera or the retina of an eye measure only the intensity of light falling on them. A signal from a single photodetector is completely ambiguous with respect to the shape, size, distance, or class of the object emitting the radiant energy. The visual signal from an entire camera array is ambiguous with respect to the distance, size, and nonvisible characteristics of an object. Perceptual processes must infer this kind of information by interpreting sensory data in the context of signals from other sensors (such as acoustic, tactile, smell, taste, vestibular, and proprioceptive sensors) and information from the world model (such as knowledge of physics, math, logic, and characteristics of the class or classes of which the observed object is recognized as being a member). Range can be inferred from stereo disparity between images in two eyes and vergence between the eyes, from motion parallax, or from other characteristics of the scene, such as shading, texture, and occlusion boundaries. The size of objects can be inferred from angular size in the image multiplied by the estimated range. Computation of the velocity of perceived objects can be inferred from image flow measurements if the linear and angular motion of the eye or camera is known. Even the most elementary visual feature detection requires inference. When we visually perceive edges between surfaces in the world, these are inferred from gradients in intensity or color images in our visual cortex. We can infer that sounds are coming from a particular direction, by computing phase and intensity differences between sound waves in our two ears. In tactile sensing, the shape and surface properties of objects can be inferred from pressure, temperature, vibration, and pain measurements in the skin.

Acoustic signals are also ambiguous. The sound of a footstep may indicate the presence of a loved one or an assassin. In ordinary speech, there are many words that sound the same but mean different things. For example, "attack" and "a tack" are pronounced exactly the same but have completely different meanings. Yet the two are rarely confused, because of the context in which they are used. Apparently, the brain solves the problem of ambiguous speech by hypothesizing several possible meanings and choosing the one that best explains the totality of sensory input integrated over a window of time and space.

Often, the chain of perceptual inference is much longer. For example, I infer when my car needs gas by visually observing my gas gauge pointer near empty, not by direct observation of the liquid level in my tank. I have never been to Africa, but I infer from reading the newspaper that members of the Dinka tribe in the Sudan have lost their traditional herds of cattle because of the a war that has raged for decades between the government in the mostly Islamic north and armed groups in the Christian and animist south. This perception arrives in my world model through a long chain of inference that begins with the eyes and ears of a reporter. It meanders through a series of entries in notebooks, historical records, keystrokes on a word processor, markup by editors, and mechanical processes involving printing presses and delivery trucks, until a person brings a newspaper to my house. Finally, the information appears as a string of characters on the printed page that is imaged on my retina. In the modern world, much of what people perceive is not observed directly, but is inferred from reports via print or electronic media.

Sensory data may be corrupted by noise. For example, signals from photodetectors in cameras typically are noisy, especially under low light levels. Acoustic energy arriving at the ear typically is embedded in a cacophony of sounds originating from a wide variety of sources, most of which are irrelevant to what the listener is trying to hear. Sensory input may even be actively deceptive. Prey creatures often use camouflage and deceptive behavior to confuse perceptual processes in predators.

Hypothesize and Test

To cope with ambiguity and noise in the sensory input, both biological and artificial intelligent systems have developed sensory processing methods that extract the information needed to construct reliable world models despite ambiguity and noise in the sensory input. The most successful of these methods is hypothesize and test. Hypothesize and test implies the ability to formulate likely hypotheses and test how well each hypothesis predicts experimental observations. This is the central concept in recursive estimation [Weiner49] or Kalman filtering [Kalman60]. The technique of hypothesis and test can be found in all but the most primitive intelligent systems. In human society, hypothesize and test has been refined into what is known among scientists as the scientific method, and among lawyers as rules of evidence.

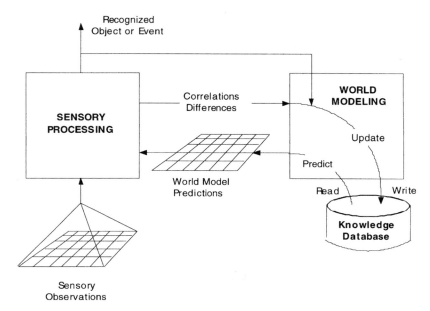

FIGURE 3.4 Hypothesize and test in an intelligent system. The world modeling process uses the current best estimate of the state of the world stored in the knowledge database to predict the current sensory input. This prediction is compared with sensory observations in sensory processing. Correlation and variance are returned to the world modeling to update the knowledge database. This is a recursive estimation process. [From Albus et al.87.]

A simple illustration of hypothesize and test is shown in Figure 3.4. The world model process hypothesizes a prediction based on the set of estimated state variables that reside in the knowledge database. The world model predictions are compared against sensory observations that come directly from sensors or from lower-level perceptual processes. Correlation between sensory observations and world model predictions indicate the degree to which a particular hypothesis is correct. If correlation exceeds threshold, the hypothesis is labeled as verified, or confirmed. When the hypothesis is verified, the focus of attention can be narrowed and residual difference values can be used to update estimated state variables to improve the hypothesis. Differences or correlation offsets indicate error between the world model prediction and perception observation. If correlation falls below threshold, the hypothesis is rejected, focus of attention is broadened, and a new hypothesis is generated.

Recursive estimation is a hypothesize and test process that tightly couples observations with predictions to extract a reliable estimate of the state of the world. Using recursive estimation, perception and world modeling work together to establish and maintain a correspondence between the real world and knowledge about the world in the intelligent system representation. Knowledge at time t is a combination of prior knowledge at time $t - \Delta t$, plus current observations at time t from sensors. This can be expressed as

$$KD(t) = KD(t - \Delta t) + \text{observation}(t) \tag{3.1}$$

where $KD(t)$ is the current best estimate of the state of the world at time t and observation(t) = sensory input(t)

THE GOAL OF PERCEPTION

In our view, the goal of perception is to provide the intelligent system with a dynamic model of the world that is accurate, reliable, current, and filled with the knowledge required to accomplish behavioral goals. To be useful, knowledge in the world model should be in a convenient coordinate system and in a format that is easy for the behavior generating system to use. To be efficient, knowledge should be represented at a resolution and over a range in space and time that is optimal for planning and controlling behavior.

Our view of the goal of perception leads to a somewhat different approach from that of many researchers in the field of image understanding. For most image processing research, the goal is to describe the image in a way that is somehow simpler than the original input. For example, Duda and Hart[73] state: "Our primary interest is in methods that lead to simpler characterizations of figures than enumeration of points." Riseman and Hanson[90] state that "transforming [image] data into a much smaller set of image events [or tokens] is the objective of segmentation processes." Ballard[97] suggests that "reducing the size of the [sensory] state space is central to natural computation." In the field of robot vision, it is common practice to attempt to reduce the image to a set of geometric entities such as edges, lines, curves, and surfaces with a few descriptive parameters, and to describe objects in terms of generalized cylinders, geons, superquadrics, or networks of polyhedral surfaces [Biederman90]. Typically, at each stage of processing, information is thrown away, and the image is simplified or generalized into fewer and fewer higher-level symbolic entities. At the highest level, the image is discarded altogether and further processing takes place in the symbolic domain of logic, mathematics, and language.

The model of perception that we present in this book takes a fundamentally different approach. Our model does not attempt to reduce the original image, but to enhance it with a priori knowledge and information gathered from many different sensors integrated over time and space. In our model the goal of perception is not a simpler representation but a richer one that increases, not reduces, the dimensionality of the perceptual space. Our model proposes adding to the original image a multiplicity of additional images consisting of computed attributes, cost/benefit values, and pointers to entities and classes.

All of these are registered and overlaid on the original image. With each additional overlay, each pixel acquires an additional attribute or pointer. The classical model of the vision processing system is a pyramid, with smaller numbers of lower-resolution elements at the upper levels. Our model proposes instead that the visual pathway more closely resembles a stack of images, with many layers consisting of attribute images, entity images, class images, and worth images, such as shown in Figure 7.8 [Albus98].

There is considerable evidence to suggest that this is a plausible model of how the visual processing system is organized in the brain. It is well known that the brain uses a multiplicity of neurons to compute a multiplicity of attributes for each pixel [Ullman96, p. 152]. Studies of the neuroanatomy and neurophysiology of the visual cortex reveal that the visual field is replicated many times in many different places in the brain and that these various representations are maintained in registration with the retinal image. Certainly, there are enough neurons in the brain to support this model. The visual cortex contains 600 times more neurons than there are inputs from the lateral geniculate nucleus [Van Essen and Deyoe95]. If each neuron computed an attribute, or contributed to a pointer in a relational network, there could be hundreds of attribute, entity, class, and worth images in the visual system. It is well known that in the primary visual cortex, attributes such as edge orientation, image flow, and stereo disparity are each computed separately by different neurons that are topographically registered with the pixels in the original image. Hubel and Wiesel[74] showed that the neurons in the primary visual cortex are organized such that each pixel is represented by many different neurons, each of which is sensitive to lines and edges at a preferred orientation. This effectively produces edge and line orientation attribute images. Barlow[72], Van Essen and Deyoe[95], and many others have shown that this principle extends to other kinds of attributes, such as stereo disparity, range, color, motion, and texture. For each attribute, each pixel has many neurons that respond more or less preferentially to that particular attribute. The set of neurons representing each attribute forms an attribute image that is overlaid in registration with the original image.

The visual field is also represented at higher levels in the visual cortex, where pixels are grouped into entities such as boundaries, surfaces, objects, and groups with computed attributes such as shape and symmetry [Francis et al.94]. Single cell recordings in the inferotemporal cortex reveal populations of cells that respond to complex shapes such as faces and hands. A particular population of cells in the anterior inferotemporal cortex responds selectively to face images. Other cells in a nearby region respond to hand images. Some of the face-sensitive cells respond to complete faces, others to face parts [Perret et al.82,85, Young and Yamane92]. Many show considerable invariance to size, position, and orientation in the image plane. In general, object shapes are represented by the combined activity of cells either tuned broadly to different shapes or tuned to different parts of the entire shape [Ullman96, p. 151].

ENTITIES

When we look at the world, we don't see arrays of pixels. We perceive entities: surfaces, objects, groups, places, situations, and relationships. When we listen, we don't hear strings of neural impulses from acoustic sensors in the ears. We perceive sounds: chirps, growls, words, notes, chords, songs, and patterns of speech. We feel surfaces, smell aromas, taste flavors, experience pain and pleasure. We perceive the world in terms of entities, events, and situations, not patterns of neural impulses. We perceive points of light in the sky as stars. We perceive objects such as buildings and trees, not clusters of pixel attributes. We see specks of dirt, edges of roads, corners of boxes, and forks of tree branches. We perceive clumps of grass, bushes, animals, and clouds as objects. We perceive posts, fireplugs, automobiles, people, hats, chairs, buildings, windows, stairways, and steps. These are represented in the brain by complex data structures in which we perceive patterns and shapes having symmetry and unity of form. We perceive things that can be given names. We perceive *entities*, events, and situations.

Entity: something with continuity of geometric structure that is perceived in space

It is important to distinguish between entities that exist physically in the world and entities that exist mentally in the mind. We designate entities in the world as *external entities* and entities in the mind as *internal entities*.

External entity: an entity that exists in the real world external to the mind

External entities typically are clumps of matter that occupy space and are physically or energetically connected in some way. External entities typically have mechanical structure or are subject to forces that cause them to behave as a unit. External entities often correspond to things that can be measured or counted, such as pebbles, grains of sand, or drops of water. External entities typically are physical things that have states and obey the laws of physics. They often have mass and energy.

External entities may not be uniquely defined or precisely measurable. There are many patterns and combinations by which matter that exists in the world can be organized into entities. The choice of how the world is perceived to be organized occurs in the mind of the observer. For example, one can perceive a sand dune or a mountain as an entity, even though there is no precise boundary where it begins or ends. One may perceive a bush or tree to be an entity, even though it consists of hundreds or thousands of leaves, branches, twigs, and roots. One may perceive a vine as an entity, even though it climbs on and intertwines with the branches of a tree or bush. One may perceive a river as an entity, even though it has many tributaries and it sometimes floods and sometimes dries up. One may perceive a lake as an entity, even though it has miles of coastline consisting of rocks and inlets with waves that surge and ebb. One may perceive groups of entities that are not physically connected but possess common attributes such as spatial proximity or similar motion. For example, a cloud of smoke, a swirl of dust, a school of fish, a flock of birds, a wave of water, a crowd of people, a stream of traffic, a forest, or a storm may be perceived as an entity even though each consists of many parts that are only loosely connected. Thus, even though external entities are physical things that exist independently of the mind, their perception as entities takes place within the mind of the observer.

On the other hand, internal entities are entirely mental constructs that exist only in the mind. Internal entities are categories that our minds impose on the sensory input to make it explainable and predictable.

Internal entity: a data structure in the mind that may represent an external entity

Internal entities are things we perceive about the world. They exist only in our minds. Internal entities are data structures that the mind constructs to represent knowledge about the world. An internal entity may consist of a labeled group of pixels in an image or a symbolic frame that contains attributes, state, and relationship pointers. An internal entity may represent an external entity such as a rock or a tree, or the surface of a road or a building. However, the internal entity is only a data structure that contains information about the external entity. It is simply the mind's best estimate of what exists in the world.

Of course, perceptions are not always correct. Perceived entities may not accurately represent the structure of the world. We are often mistaken in what we perceive. There is no guarantee that representations in the mind correspond to anything real in the external world. For example, we may perceive an edge entity consisting of a group of edge pixels in an image of a sphere. But a sphere has no edges. The edge entity we perceive is the set of points on the retina where the projected image of the sphere occludes the background. We may perceive patterns in the stars, in the clouds, and in tiles on the bathroom floor. What we perceive depends on our perspective and state of mind. Our perceptions are a function of our viewpoint, our prior knowledge, our interest, and our intent.

Entities in the mind can represent abstract concepts such as truth or justice. Internal entities can represent things that cannot be observed directly, such as atoms, magnetic fields, quarks, and black holes. Entities in the mind can even represent things that may not exist at all, such as spirits, gods, angels, ghosts, or demons. Children often create imaginary friends. Internal representations are hypothetical constructs that the intelligent system generates to interpret sensory input. Whether or not there exists any corresponding external reality is only a hypothesis. Optical illusions are a common result of incorrect hypotheses. Camouflage and deception are used widely to create representations in the mind of observers that differ from reality in the world. Camouflage causes unsuccessful behavior by the deceived and increases the probability of survival of the deceiver.

So how do we know whether our representations in the mind have any corresponding external reality? The short answer is, "We don't." This is why Hume concluded that it is impossible to prove the existence of a real world. As Kant explained, at best we can only estimate the probability that our internal entities correspond to external realities. Ultimately, it is less important whether an entity hypothesis is true than whether it has functional utility, that is, whether it produces a representation of the world that is useful in generating successful behavior. The functional utility of a hypothesized entity can be measured by its ability to explain the past and predict the future. The criterion is whether an entity hypothesis produces significant performance benefits. An intelligent system that can accurately predict even a few seconds into the future is more likely to survive and achieve behavioral goals than one that can react to events only after they occur. A system that is able to predict hours or days into the future has an enormous behavioral advantage. Reliable prediction enables the intelligent system to make plans and take preemptive action to avoid danger, maximize benefit and payoff, minimize cost and risk, and win in competition with other systems that have inferior predictive powers. Whether internal entities actually correspond to external reality is largely irrelevant and in many cases may be undecidable. What really matters is whether hypothesized entities increase the reliability of predicting the future and hence the probability of successful behavior.

From an engineering viewpoint, we know empirically that the external world has structure that can be observed by sensors. We observe from experience that the ability to perceive and reason about entities, events, and situations in the world is beneficial to survival. It has proven useful to hypothesize entities, events, and situations as a means of explaining the world and predicting what will happen next. When predictions based on hypothesized knowledge in the mind correlate with observations, the probability that the hypothesized entities, event, and situations correspond to external reality is increased. When predictions based on hypotheses in the mind fail to correlate with observations, the probability that the knowledge in the mind is correct is reduced. If the probability of correctness rises above an upper threshold, the hypothesis may be accepted as true and the system may act accordingly. If the probability of correctness falls below a lower threshold, the hypothesis may be rejected as false and an alternative hypothesis may be generated.

ENTITIES AND ATTRIBUTES

Entities have *attributes* that describe their properties.

Entity attribute: a property of an entity

Attributes are typically things that can be measured about the entity. A *pixel entity* may have attributes such as intensity, color, spatial or temporal gradients, range, position in the image, and image flow rate and direction. An *edge entity* may have attributes such as length, curvature, position, velocity, orientation, magnitude, and type (e.g., range-edge, brightness-edge, color-edge, texture-edge, intersection between surfaces). A *surface entity* may have attributes such as area, curvature, position, velocity, orientation, temperature, color, and texture. An *object entity* may have attributes such as volume, shape, position, velocity, orientation, temperature, mass, and behavior. A *group entity* may have attributes such as volume, density, shape, center of gravity, mean velocity, variance about the mean, and relationships between group members.

All measurements of attributes depend on the resolution of the measuring instrument. At one resolution, a surface may seem perfectly smooth, but at a higher resolution the same surface may appear extremely rough. For fractal edges, measure of length may depend on the resolution of the measurement. For example, the length of the coastline of the British Isles depends on resolution of the instruments used to make the measurement. The attributes of position, velocity, energy, momentum, and temperature are often called *state*.

State: the condition of an entity or process at a point in time (usually, the present)

The state is a set of attributes that represents the dynamic properties of an entity, such as its position, orientation, velocity, temperature, energy, or momentum at a point in time. The meaning of the word *state* depends on the type of entity to which it refers. For example, the state of a physical mass typically refers to its position, velocity, and acceleration. The state of an electric circuit may refer to voltage or current levels or rates of change at specific points in the circuit. The state of a finite-state machine refers to the node in a state graph, or the line in a state table, that is active. The state of a software process may represent the point of execution in a flowchart, the contents of an instruction counter, or the currently executing statement in a program at a point in time. The state of a chemical reaction may be the temperature, pressure, density, volume, chemical composition, or flow rate of the participants in the reaction at a point in time. The state of a control system is the current condition of the system. In the ideal case where there exists a perfect system model and complete knowledge of future inputs, knowing the current state enables the prediction of all future states. The state of a situation may include the relationships that exist between entities in an environment at a point in time.

In addition to attributes and state, entities can have behaviors. Inanimate entities have simple behaviors that can be described by Newton's laws of motion. Animate entities can exhibit many different types of behavior, depending on circumstances and the type of entity. All animals have innate reflexive behaviors that are reasonably predictable. Most animals have conditioned reflex behaviors that vary with experience. Higher mammals and humans have behaviors that are extremely complex but often characteristic. Entities may also belong to classes and participate in relationships. Class membership may be determined based on attributes, state, or behavior. Relationships may be determined by spatial and/or temporal configuration or by logical or mathematical correspondence.

States and attributes can be represented in the brain by spike rates on axons, electrical potentials in neuronal membranes, and concentrations of hormones and transmitter chemicals at a point in space and time. States and attributes may be represented in a computer by the contents of registers or by the values of state variables. Behaviors can be represented by procedures or knowledge of how entities react to events and situations. Class membership can be represented by labels, class names, or pointers. Relationships can be represented by networks of entities linked by pointers.

Thus perception does not reduce the complexity of the sensory input. Rather, it enhances and enriches it, interprets it, and classifies it so that it contains much more information than is present in the sensory input. Perception does not reduce the sensory input to a small set of tokens. Instead, it overlays it with a multidimensional world model that captures the richness and complexity of external reality. Entities, attributes, names, and values are inferred and overlaid on the sensory input so that behavioral decisions can be computed easily and quickly. Detail is not lost but enhanced with significance, worth, and meaning.

We believe that this model is supported by evidence from neuroanatomy and neurophysiology as well as by introspection. When we look at the world subjectively, we perceive much more than an array of signals from the retinal image. Our perceptual experience is far richer and more complex than the input image. When we look at the world, we see entities with attributes of shape, texture, and motion. We see objects, animals, and people that we can recognize and characterize. We see objects and situations that we can classify and predict.

We also perceive objects, events, and situations in great detail, not as low-resolution abstractions. When we recognize an object such as a cat, we do not see it as a vague blob or blurred token. We see it in sharp definition. We see its whiskers and the fine hairs on its back and ears. We see the details in the iris of its eyes. We perceive its posture, the twitching of its tail, the scanning motion of its ears. We can generalize all this complexity under a single symbolic label "cat," but the detail is perceived as well. This label applies to all of the sharply defined and finely detailed regions in the retinal image of the cat, including the fine details of hair, eyes, ears, nose, mouth, and tail. High levels of detail in perceived objects is not what would be expected from a perceptual pyramid that throws away information at each successive level of processing. Perceived detail is an effect produced by an internal representation that builds up information that becomes richer and more complex at each higher level of processing.

Once this complex multifaceted world model is constructed, it is relatively simple to recognize objects, events, and relationships and generate behavior that has a high probability of success. Once we have established linkages between images and entities, computed attributes and states, established class memberships, and defined relationships between entities, classes, and situations, it is straightforward to associate emotional values such as love, fear, hate, or disgust with what we perceive. We can assign worth to objects in the visual image and, by reference, project worth onto the corresponding objects in the world.

Once a complex multifaceted world model is constructed, it is relatively simple to recognize objects, events, and relationships. Once we have established linkages between images and entities, we can compute attributes and states, establish class membership, and define relationships, classes, and situations. It becomes possible to project emotional values such as love, fear, hate, or disgust onto what we perceive. We can assign worth to objects in the visual image and, by reference, project worth onto the corresponding objects in the world. We can also make decisions and generate behavior that has high probability of success.

REASONING WITHIN A RICH WORLD MODEL

Our proposed model of perception makes it possible to reason in both the symbolic and image domains. Regions in images can be given symbolic labels and assigned to classes. Symbolic representations of entities and events can be referenced to locations in the image domain. For example, we can compute attributes of individual cats in the image domain and compare them with attributes of cat classes in long-term memory. We can perform logical inferences about cats and their attributes and overlay our results on the image. Many questions, such as "Is that thorn sharp?", "Is that dog likely to bite?", "Is there a bee on that flower?", "Is that tiger looking at me?" can be answered with a simple yes or no. But the problem of extracting the truth or falsehood of such a proposition is not made easier by a parsimonious or fuzzy representation of the world. The fact that a decision to act can be encoded as a few bits in a task command does not mean that the world model need contain only those few bits. Making wise behavioral decisions requires a rich, high-dimensional, accurate, current, and reliable representation of the world in both symbolic and image domains from which detailed analyses of complex situations can be made.

The type of information that is important for behavior varies, depending on the behavior and the situation. For example, in military planning it is important to know such information as: Where are the friendly and enemy forces? Where is the high ground? Where are the roads, rivers, woods, swamps, and mine fields? What areas are visible to enemy observers? What areas are within range of enemy guns? Where are the barriers to movement? Where is it dangerous to go? Where is it safe to hide? Military planners typically make extensive use of detailed maps, charts, and physical models to plan operations, to distribute resources, and to design battle scenarios. Battle plans typically include icons representing resources overlaid on terrain maps and charts that provide elevation, ground cover, and features such as roads, bridges, streams, buildings, towns, and mountains depicted as two-dimensional arrays of pixels. Military plans may be supplemented with symbolic text and implemented using verbal or written commands, but they are generally conceived and analyzed in the image domain and displayed as arrays of pixels representing graphic images and icons.

Path planning for vehicle navigation typically requires map representations that contain thousands of details about terrain features and landmarks. Without such detail, it is not possible reliably to compute cost, risk, and benefit for alternative paths. Path optimization in a complex environment often requires search algorithms that traverse dense pixel arrays and compute complex cost functions over convoluted trajectories by integrating the cost and benefit of all the pixels traversed.

In the fields of manufacturing and construction, plans are often expressed as drawings, with many precise details overlaid with text defining materials, dimensions, tolerances, and manufacturing procedures. Path planning for machining, assembly, or inspection tasks typically requires both iconic and symbolic data structures. Drawings of parts to be manufactured or structures to be constructed are typically more useful for reasoning and planning behavior than purely textual or verbal descriptions. Instructions for assembly of products typically include both drawings and text. Often, the amount of information required is enormous. For example, a set of STEP data files that capture all the information necessary to manufacture a complex product such as an automobile or aircraft engine may require files containing gigabytes of data. A complete descriptions of all the details embodied an entire airplane may contain trillions of bytes of data and many thousands of drawings.

Even in the most abstract domains of mathematics and logic, intuitive insights are often based both on graphs, charts, and pictures, as well as on equations expressed as strings of symbols. For many people (the author included), understanding of mathematical equations and physical principles is facilitated by pictures and diagrams. For example, the Bohr model of the atom is best represented by a picture or diagram. Electromagnetic fields are best understood by three-dimensional arrays of tiny arrows or lines. Even purely abstract mathematics is typically represented by formulas written or drawn on a sheet of paper or a white board. Abstract symbol manipulation is typically done by writing symbols in a form such that one can imagine physically moving symbolic characters from place to place to perform operations such as factorization, grouping, substitution, addition, multiplication, and division. Abstract concepts such as four-dimensional (and higher) space are often difficult to grasp mentally precisely because they cannot be represented pictorially.

Of course, symbolic representations are necessary for many higher-level reasoning processes. Symbolic information is crucial for cognitive processes involving logic, mathematics, or language. Symbolic formulas and equations are powerful methods for encoding scientific and engineering knowledge. Symbolic language is the primary means of human communication. Language is capable of expressing many concepts that are difficult or impossible to render pictorially. Language symbols and mathematical equations can express ideas with precision and subtlety that cannot be achieved graphically. Text is often needed for labeling regions and points on a map or for explaining objects, features, and relationships in a drawing. Images are a much more primitive and basic form of representation. Only humans make significant use of language, while virtually all creatures use images for making their way in the world.

Commonsense reasoning requires both iconic and symbolic representations of the world. Generating appropriate behavior in everyday tasks requires both pictorial and symbolic data structures that are linked together. Regions in the image must be labeled with pointers to symbolic representations, and symbolic representations must have pointers back to regions in the image. This increases the dimensionality of the representational space and multiplies the amount of memory required for representing the world, but it vastly simplifies the problems of object recognition and scene understanding. It makes image segmentation and classification much easier, and it makes the problem of generating intelligent behavior much more tractable.

ATTENTION

A rich and comprehensive representation of the world raises an issue of complexity. The world is infinitely rich with detail. The natural environment contains a practically infinite variety of real objects, such as the ground, rocks, grass, sand, mud, trees, bushes, buildings, posts, ravines, rivers, roads, people, animals, vehicles, and machines. The environment also contains elements of nature, such as wind, rain, snow, sunlight, and darkness. All of these objects and elements have states and may be part of, or cause, events and situations. The environment contains a practically infinite regression of detail, and the world itself extends indefinitely far in every direction.

Yet the computational resources available to any intelligent system are finite. No matter how fast and powerful computers become, the amount of computational resources that can be allocated to any practical sensory processing system will be limited. Even the human brain is computationally limited. The number of neurons in the brain is very large but finite, and the amount of energy that can be allocated to sensing, reasoning, planning, and controlling behavior must be balanced against other demands, such as body strength, endurance, and nutritional needs. Natural selection tends to eliminate creatures that are less than optimally adapted to their environment. Therefore, the brain cannot afford to allocate more neurons than necessary to building a model of the world.

Fortunately, the world is not uniformly important. Some things are more important than others. At any point in time and space, most of the detail in the environment is irrelevant to the immediate behavioral task of the intelligent system. For example, in a task of manipulation, details about the object to be manipulated are important, as are the tools and materials needed to perform the task. But most of what else can be observed about the rest of the world is relatively unimportant. During a hunt, attention is focused on the prey, and most everything else is irrelevant [Watanabe98]. During an escape, the state and attributes of the predator and the location and type of a place of refuge are important. Everything else matters little.

Therefore, the key to building practical intelligent systems lies in understanding how to focus attention on what is important and ignore what is irrelevant. Attention is a mechanism for allocating sensors and focusing computational resources on particular regions of time and space. Attention allows computational resources to be dedicated to processing input that is most relevant to behavioral goals. Focusing of attention can be accomplished by directing sensors toward regions in space classified as important and by masking, windowing, and filtering out input from sensors or regions in space classified as unimportant.

In most biological creatures, visual photoreceptors are concentrated to produce high resolution and high magnification in a foveal region, and distributed less densely in the periphery. In humans, the fovea is a tiny region only about $0.5°$ in extent (about the size of one's thumbnail when held a arm's length). Yet about one-third of all the processing neurons in the visual system are dedicated to foveal vision. One-half of the primary visual cortex is dedicated to processing only the central $10°$ of the visual field [Wandell95]. Similarly, tactile sensors tend to be closely spaced to produce high-resolution tactile sensing in the fingertips, lips, and tongue, and distributed less densely in other regions of the skin.

Attention steers the high-resolution sensory regions of the visual and tactile systems so as to apply the maximum number of sensors to objects and regions of importance. We automatically look (i.e., point our fovea) directly at the current point of attention. This positions the highest-resolution region of the visual field on the task object. When we attend visually to an object, our eyes focus and our pupils widen so as to reduce the depth of field and blur the background. When we focus tactile attention, we touch objects with our fingertips or put them in our mouth so as to apply our tactile computing resources to what we are most interested in. Thus, attention causes the perceptual system to generate a highly detailed representation of those parts of the world that are important to behavioral tasks and to limit the detail elsewhere. As a result, the perception system makes the largest possible number of measurements of the most important entities and events in the environment, and ignores or relegates to the background (i.e., process at lower resolution) those entities and events that are less important.

A hierarchical control architecture can facilitate the focusing of attention. Higher levels with broader perspective and longer planning horizon can determine what is important to achieving high-level goals, and this information can then be passed down to lower levels in the form of priorities, modes of behavior, and objects of interest. Lower levels can use this information to focus attention on objects and tasks that are important to higher-level goals. At each level, planning resources can be focused on selecting subgoals that contribute to the high-level goals. At each level, attention can be used to mask, filter, and window sensory data and to focus sensory processing and world modeling resources on objects and events that are important to high-level goals. This enables the entire intelligent system to think and act toward a single unified set of goals, with behavior throughout the entire hierarchy down to the lowest level optimized and focused toward achieving those goals.

WHAT IS IMPORTANT?

The job of deciding what is or is not important is performed in the brain by the limbic system. Evaluations performed by the limbic system define what objects and events are good or bad, right or wrong, attractive or repulsive, valuable or worthless. In primitive brains, the limbic system consists primarily of neurons that analyze smell, taste, and pain. These basic evaluations enable the most basic of decisions, such as what to eat and what to avoid (i.e., eat what smells and tastes good, don't eat what smells or tastes bad, and avoid whatever is associated with pain or nausea). In higher-level creatures, the limbic system has evolved to perform much more sophisticated analyses that involve emotions and motivating factors such as anger, fear, love, hate, and appreciation of beauty and harmony. Evaluations generated by the limbic system are used by behavior mechanisms to decide what goals to pursue, and by attention mechanisms to allocate computational resources. Goals evaluated as good are pursued and attention is focused on entities and events that are judged important to achieving those goals.

The problem of deciding what is important can be addressed from two perspectives: top down and bottom up. Top down, what is important is determined by behavioral goals selected on the basis of evaluations by value judgment processes in the limbic system. The intelligent system is thus driven by high-level goals and priorities to focus attention on objects specified by task goals or identified by task knowledge as necessary to accomplish task goals successfully. High-level goals and priorities can also generate expectations of what objects and events are likely to occur during the task and which of these are most important for achieving the goal.

Bottom up, what is important is the unexpected, unexplained, unusual, or out of limits. For example, objects that move are more likely to be important than those that do not. An object that suddenly grows in apparent size in the visual field (suggesting a rapid approach) is more likely to be important than is an object that remains constant in size. Loud, sharp, unexpected sounds are more likely to be important than dull, monotonous sounds. Events that cause fear or rage are more likely to be important than those that produce boredom. Pain is important because it indicates bodily damage. Anything that is unexpected is important because it indicates that the world model is incomplete or incorrect and needs to be updated.

In most cases the region in space and time that is most relevant to the behavioral choices of an intelligent system centers around the "here and now." Each intelligent system always resides at the center of its own egosphere, in both space and time. The relevance of objects in the world is typically inversely proportional to their spatial distance from the egosphere origin "here (i.e., range = 0). The relevance of events is also typically inversely proportional to their temporal distance from the egosphere origin "now" (i.e., $t = 0$). As objects approach the spatial center of our egosphere or events approach $t = 0$ in time, they become more important and attention must be paid to them. Objects and events that lie far away in space and time can usually be safely ignored.

But not always. Sometimes distant objects and remote events can be very important. There are occasions where the ability to perceive distant objects and predict events far in the future has significant survival benefit. However, this capability incurs computational costs. There are many more objects in the distance than nearby and astronomically more possible scenarios in the long-term future than in the immediate future. The number of objects in the egosphere goes up with the square of the egosphere radius for creatures that live on the ground. It goes up as the cube of the egosphere radius for creatures such as fish and birds that live in a three-dimensional world. The number of possible future scenarios increases as the number of alternative actions in each resolution element in time raised to the power of the number of resolution elements on the time line. Clearly, it is not feasible to analyze all possible long-range futures. Thus the ability to see far away magnifies the need to focus attention and discriminate between what is important and what is irrelevant. Only likely future scenarios should be considered seriously, and only the most important ones should be analyzed in any detail.

In all cases, knowledge of what is important and the ability to focus attention on what is important are critical to the design of intelligent systems. They are fundamental to a theory of mind. To build practical cost-effective intelligent systems, we must understand how to design sensory processing systems that mask, window, and filter the sensory data to extract information that is relevant to what is important in the world. We must understand how to build value judgment systems that can distinguish between what is important and what is not. And we must understand how to design behavior generation systems that can allocate behavioral resources to accomplish what is important to achieve.

We turn next to the issue of behavior.

4   Goal Seeking and Planning

The construction of the obelisk of Caligula in St. Peter's Square in Rome on September 10, 1586. Eight hundred men and 140 horses worked on winches under the supervision of architect and engineer Domenico Fontana. [By permission from Bibliotheque Nationale, Paris.]

The brain is first and foremost a control system. The brain is a well-organized society of computational processes designed by natural selection to generate and control behavior. All brains that ever existed, even those of the tiniest insects, generate and control behavior. *Behavior* is the sequence of actions that a control system produces. Action results from actuators that exert influence on the world. Output from computational processes in the brain causes actuators to generate forces that move arms, legs, hands, and eyes. Actuators point sensors, excite transducers, manipulate objects, use tools, and steer and propel locomotion. An intelligent system may have tens, hundreds, thousands, even millions of actuators, all of which must be coordinated to perform tasks and accomplish goals. Natural actuators are muscles and glands. Machine actuators are motors, pistons, valves, solenoids, displays, and transducers for acoustic and electromagnetic radiation.

A goal is a desired result, or desired state of the world, that behavior is designed to achieve or maintain. Goal seeking is the raison d'être of intelligent behavior. The ultimate goal is the intelligent system's reason for being. A goal may come from outside the intelligent system in the form of a command or be generated internal to the system in response to a need, drive, or urge. An example of an external goal might be an order from a superior in a military chain of command to be at a specified place at a specific time. An example of an internal goal might to find food or shelter in response to hunger or feeling cold. A measure of intelligence is the degree to which the system is successful in achieving or maintaining its goals. Natural selection favors behaviors that are most likely to result in the survival and reproduction of the individual's genes. The brain is thus a mechanism for realizing the gene's goal of maximizing the likelihood of its own reproduction.

Of course, the ultimate goal of survival and propagation cannot be achieved by any single action or any fixed series of actions. To survive and propagate in the natural world, a creature must decompose its ultimate goal into subgoals such as find food, acquire territory, build shelter, escape predators, attract a mate, rear a family, and adapt to the changing seasons. For example, survival may require hibernation or migration toward the equator during the winter. Each of these subgoals must be further decomposed into lower-level shorter-term sub-subgoals, involving more localized and specific behavior. At each level of decomposition, a goal from a higher level causes behavior consisting of actions designed to achieve strings of lower-level goals that work together to accomplish the higher-level goal. At each level, feedback from the environment can be used to compensate for errors and perturbations so that goals are accomplished despite mistakes and unexpected events. Eventually, at the lowest level, strings of neural impulses are delivered to muscles and glands that act on the world.

The structure of the brain has evolved to facilitate this principle of task and goal decomposition. In humans, plans for the long-term future are developed in the forebrain, the most recently evolved brain structure. The frontal cortex uses reason and logic to manipulate abstract models of the world, including maps, symbols, rules of mathematics, and laws of physics for synthesizing high-level concepts and plans. The frontal limbic system also uses high-level value judgment to make decisions based on social customs, religious beliefs, and legal obligations. In the frontal lobes, plans may be developed for organizing a hunt, building a house, writing a book, planning a military campaign, pursuing a career, or performing scientific research [Farah95, Robin and Holyoak95].

Plans developed by the forebrain are expressed in terms of sequences of goals and actions for which there exist behavioral skills residing in the evolutionarily older premotor cortex and associated limbic regions. Behavioral skills may enable activities such as capturing prey, escaping a predator, winning a fight, consummating a sexual encounter, going to the store, getting a haircut, buying a car, or going out for dinner. Plans developed by the premotor regions may be expressed as sequences of simple tasks that can be performed by the still older primary motor cortex and the basal ganglia. Simple tasks may consist of activities such as climbing over a log, following a path, putting on a coat, or opening a door.

Plans developed by the primary motor cortex and basal ganglia are expressed in terms of elementary movements for various parts of the motor system, such as the arms, legs, and torso. Elementary movements may consist of activities as such as reaching, grasping, lifting, throwing, stepping, and jumping. Elementary movements are further decomposed and coordinated with balance, posture, and orienting information in the midbrain and cerebellum [Georgopoulos95]. At the lowest level in the spinal cord, there are computational modules containing the final motor neurons that control stretch in the muscles, tension in the tendons, and position of the joints. These produce the output drive signals to muscles that cause limbs and digits to move in coordinated ways so as to accomplish the behavioral goals generated at higher levels.

There is a sensory processing hierarchy that runs parallel to this behavioral hierarchy. Modules in the sensory processing hierarchy operate on signals from sensors to extract features, detect patterns, recognize events, classify entities, analyze situations, and recognize concepts. At each level, feedback from sensory processing is integrated with commands from higher levels to modify behavior so as to accomplish high-level goals despite unforeseen events in the world. The sensory-motor system contains many parallel and cross-coupled hierarchies of computational modules. Sensory signals from the eyes are processed by an ascending hierarchy of at least 32 known computational modules in the lateral geniculate and in the visual, parietal, and temporal cortical regions [Felleman and van Essen91]. The part of the image processing hierarchy that ascends into the temporal cortex extracts geometric features, entities, and attributes. It groups these into higher-level objects and relationships, assigns them to classes, and gives them names. The part of the image processing hierarchy that ascends into the parietal cortex extracts temporal patterns and events, tracks objects through time and space, and computes state variables such as position, orientation, velocity, and acceleration. These two hierarchies work together to segment and detect entities and events, compute their attributes and state, remember them, recall them, classify them into categories, and recognize them as having been experienced before—or if not, label them as being novel and unusual. The parietal cortex on the right side of the brain generates labeled images and maps of places and things. The parietal cortex on the left generates symbolic lists of attributes and relationships [Carter98].

Similar hierarchies exist among the computational modules that process information from the ears, the tactile system, and the proprioceptive systems. Data from the ears are processed in the brain stem and inferior colliculus to detect frequency components and detect phase differences between the ears that identify the direction of incoming sound. Higher levels of acoustic processing detect patterns and analyze the meaning of sounds. This takes place in the temporal lobes. Acoustic processing of music is typically accomplished in the right temporal lobe, while acoustic processing of language is accomplished in the left temporal lobe, near the visual parietal region that processes written text. This section of the brain is sometimes called *Wernicke's area* and has long been known to be involved in language understanding.

At each level in the brain, axon pathways convey information back and forth between the sensory hierarchy and the behavior hierarchy. Control loops are closed at each level of the sensory-motor hierarchy. In the lower spinal cord, stretch and tension sensors are tightly coupled with motor neurons. In the midbrain, sensory data related to balance and coordination are tightly coupled with motor commands. In the primary sensory-motor cortex and basal ganglia, neurons that process tactile sensory information are tightly coupled with neurons concerned with generating motor behavior for each area of the body (e.g., fingers, hands, arms, legs, feet, face, lips, and torso). Wernicke's area involved in language understanding is tightly coupled with Broca's area, involved in generating meaningful and syntactically correct speech [Peele61].

Information also flows from behavior generation to sensory processing. At each level of the sensory processing hierarchy, expectations based on planned behavior are sent to sensory processing modules to aid in focusing attention and interpreting sensory input. At each level, connections are made to the limbic system, where values, priorities, and motives are computed and plans are evaluated. This general hierarchical structure exists both in the cortex and the underlying subcortical regions. There are many connections between cortical and subcortical regions at each level of the sensory-motor hierarchy. These involve the thalamus, basal ganglia, limbic system, midbrain, cerebellum, and spinal cord. Embedded in this general hierarchical structure are specialized processors such as the superior colliculus (for visual gaze control), inferior colliculus (for sound localization), and vestibular nuclei (for computing estimates of "self motion" and "up") [Grossman67, Peele61].

In addition, there exists another largely separate computing structure, the *autonomic nervous system*, for maintaining body functions such as blood pressure, temperature, and blood sugar and oxygen levels. The goal of the autonomic nervous system is to optimize the operating parameters for breathing and heart muscles, blood flow circulation control valves, digestive organ secretions, and intestinal contractions during activities such as sleeping, eating, walking, running, jumping, fighting, fleeing, resting, mating, vomiting, and defecating.

A SHOPPING CENTER SCENARIO

An example of the richness of behaviors that are produced routinely by computing structures in the human brain can be seen from a shopping center scenario described in Albus[81]. We reprint portions of it here for the convenience of the reader.

Consider a simple 10-minute behavioral task of visiting a shopping center on the way home from work for the purpose of buying a record. A detailed examination of what is required to execute this task will illustrate the enormous complexity of the computations required and the wide range of knowledge necessary for performing such a task. We begin by defining a task or behavior vocabulary for each level of the behavior-generating hierarchy. Each task consists of an activity and a goal. Each entry in the task vocabulary is defined by a verb, an object, and some modifier parameters. The task vocabulary at the highest level must include the task $\langle$PICK UP $(x, y)\rangle$, where x is the name of the object to be picked up (in this case a record) and y is a list of parameters such as "on the way home from work" which identifies the time slot during which the task should be done.

The task vocabulary at the next lower level must contain at least the set $\{\langle$GO TO (shopping center)$\rangle$, $\langle$PARK (car)$\rangle$, $\langle$FIND (record shop)$\rangle$, $\langle$BUY (record)$\rangle$, $\langle$GO TO (car)$\rangle$, $\langle$LEAVE (shopping center)$\rangle\}$. Each of these tasks typically takes about 2 minutes.

To decompose $\langle$FIND (record shop)$\rangle$, the task vocabulary at the next lower level must contain at least the set $\{\langle$GET OUT OF (car)$\rangle$, $\langle$LOCK (door)$\rangle$, $\langle$FIND (entrance to building)$\rangle$, $\langle$WALK (down corridor)$\rangle$, $\langle$SEARCH FOR (corridor containing record shop$\rangle$, $\langle$FIND (entrance to record shop)$\rangle\}$. Each of these tasks may take 10 to 30 seconds.

To decompose $\langle$GET OUT OF CAR$\rangle$, the task vocabulary at the next lower level must contain at least the set $\{\langle$REACH FOR DOOR HANDLE$\rangle$, $\langle$PULL HANDLE$\rangle$, $\langle$PUSH DOOR$\rangle$, $\langle$PUT LEFT FOOT OUT$\rangle$, $\langle$TURN BODY LEFT$\rangle$, $\langle$PUT RIGHT FOOT OUT$\rangle$, $\langle$STAND UP$\rangle$, $\langle$STEP FORWARD$\rangle\}$. Each of these tasks are elemental movements lasting from 3 to 10 seconds.

Elemental movements must be further decomposed by the motor cortex and basal ganglia into sequences of desired limb motions lasting a few hundred milliseconds. Desired trajectories of limbs and fingers are further decomposed by the midbrain and cerebellum into sequences of desired forces, velocities, and positions of muscles lasting a few tens of milliseconds. These are transformed by the spinal motor neurons into strings of impulses to muscles lasting a few milliseconds.

Note that at each level, task commands and goals from the higher level are longer term and more abstract than the task commands and goals to the next lower level. At each lower level, tasks become more specific, short term, and detailed. At each lower level, feedback from the environment provides more specific information for guiding behavior. For example, $\langle$PICK UP (record)$\rangle$ could apply to any record shop. The decomposition of this task into a plan for the next lower level required the name and location of the shopping center where the particular record shop is located. At the next lower level, decomposition of the task $\langle$WALK (down corridor)$\rangle$ requires visual information concerning the position of the walls, the position and trajectory of other persons walking in the same corridor, and the position of obstacles such as benches, potted plants, and posts. At the next lower level, the placing of feet and the motions of the body depend on the position of floors, stairs, doors, and windows. At each lower level, there is increasing dependence on timely feedback from the nearby environment.

At the upper levels, task decomposition requires sophisticated perceptual and reasoning skills. Consider, for example, this firsthand account of the series of events during a particular execution of a specific ⟨PICK UP (record)⟩ task.

On this occasion, I parked my car next to the roofed part of the parking garage with the shopping center building to my back. I got out, locked the car door, turned to the right, and entered the parking garage. Note that this sequence of actions requires a formidable amount of information processing, pattern recognition, and motor control. To make the human body stand up requires computations involving the coordination of sight, balance, and tension control in hundreds of muscle groups throughout the body. Inserting the key into the car door lock requires a three-dimensional visual system and a subtle sense of touch and force control. Walking erect on two legs over rough surfaces, up and down stairs, and through doorways at a fast pace, avoiding both stationary and moving obstacles, is a control problem of the first magnitude. The ability of the vision system to locate and fixate cars, people, corridors, and stairs requires an enormous amount of sophisticated information processing, pattern recognition, and deductive reasoning power.

Upon entering the building, I walked down the entrance corridor. Once in the brightly lit corridor with shiny walls and floors, the visual cues were quite different from those in the dimly lit concrete garage filled with parked cars. In the corridor, lines formed by the intersection of the walls, floor, and ceiling apparently converge at the end of the corridor. They are aligned with the flow of motion in the visual field, which radiates outward from the direction of motion. Any discrepancy between the wall–ceiling or wall–floor lines and the visual flow lines can be interpreted by the vision system as a velocity pointing error.

This particular shopping center consists of a circular building with corridors radiating out like spokes on a wagon wheel. Once I reached the center of the wheel, which contained a restaurant, garden, fountain, and elevator, the problem became: "Which way to turn?" I had been to the record shop once before but did not remember where it was. Thus I made an arbitrary choice to turn right and move clockwise around the center of the shopping mall and begin the task ⟨SEARCH FOR (corridor containing record shop)⟩. All that I remembered about the record shop was that it had an old-fashioned décor. The first corridor I encountered was definitely modern in style, so I tried the second, which had an "old world" appearance. This, of course, required the ability to visually distinguish "old" décor from "modern."

A search of the first corridor turned out to be fruitless. As I returned to the central hub, I realized that a linear search of each spoke of the shopping mall would require an extensive amount of time, especially since there were three levels to each corridor. At this point, I invoked the strategy "If lost, ask directions." I thus began a new task, ⟨FIND (someone to ask)⟩. The question then became

"Which someone?" I decided to ask one of the nearby shopkeepers and set off to find one who was not busy with a customer. This, of course, requires the ability to recognize a nonbusy shopkeeper. It also requires the ability to speak and understand natural language.

I finally found such a person in a tobacco shop. He said that he wasn't certain, but he thought there was a bookstore that also sold records a few stores down the next corridor. Following this suggestion, I finally came to the record store. This accomplished the ⟨FIND (record shop)⟩ task successfully .

I then executed the ⟨BUY (record)⟩ task and began the ⟨GO TO (car)⟩ task. I easily found my way back to the central hub, but now had another right or left decision to make. If I went left and simply retraced my steps, I could surely find my way back, but my previous search had been lengthy, and I probably would be shorter and take less effort to continue around to the right. Thus I continued counterclockwise, attempting to recognize the corridor where I had entered. The first corridor I encountered had a cluttered appearance. I remembered that my entrance corridor had been rather plain. So I rejected this corridor and moved further counterclockwise. The next corridor appeared plainer, and I decided that this probably was the right one. However, after walking halfway down it, I came to a health food store that I did not recall seeing on my way in—but I wasn't certain. My confidence that I was in the right corridor was shaken but not destroyed. Next, there was a meat market with a striking appearance. I did not recall this either, and my confidence that I was going the right direction diminished even further. However, by this point, the doorway was only a few steps away and it did look familiar. So, with misgivings, I pressed onward.

Passing through the door, I encountered the parking garage, which also looked familiar. However, because it had a plain look of concrete, I reasoned that it must look about the same at every door. How could I be sure whether to go back and search some more—a long walk at best—or go on and perhaps get hopelessly lost? I then remembered that I had parked my car at the edge of the roof of the parking garage, and as I walked from my car to the shopping center entrance, the shopping center building had been on my right. Therefore, if I looked up and to my left from where I was now, I should see the corner of the parking garage roof. When I looked, I could see that the roof did not end. Therefore, I definitely was at the wrong entrance.

Now certain of my error, I retraced my steps to the central hub and continued my search counterclockwise. At the next corridor, I saw a sign on the wall that said "Gartenhaus." I remembered seeing that sign on the way in. This landmark gave me great confidence. As I progressed down the Gartenhaus corridor, I noticed a window display with a fur coat that I also remembered. I was now certain that I was on the right path. Going out the door, I looked up to the left again, and sure enough, there was the edge of the parking garage roof. I had found my way back to my car.

This scenario illustrates the complexity of the "simple" tasks that we perform every day. If we performed a similar detailed analysis of other simple behaviors such as getting dressed in the morning, going to school or work, preparing and eating meals, walking through woods, or attending a sporting event, we would see that these simple activities are composed of intricate and complicated activities of manipulation and locomotion that require many subtle and complex cognitive decisions. Ordinary life consists of hundreds of such simple tasks that are organized into daily routines by habits, schedules, and social customs. Daily routines are organized into weekly, monthly, seasonal, and yearly patterns, customs, and rituals.

At each level in the decomposition of our daily routine, plans are formulated in real time using perceived knowledge of objects and events observed in the world. Top-down goals interact with bottom-up feedback to produce sensory-interactive goal-seeking behavior. By this mechanism we can achieve our goals despite uncertainty and unexpected events in the world. For most levels of behavior, the interaction between goals and feedback occurs naturally, without conscious thought or apparent effort. It is only the upper two or three levels that appear to be involved in conscious behavior; the bottom three or four levels operate subconsciously. The bottom layers can be taught by the upper layers, so that tasks that require a great deal of conscious thought when we are learners, become second nature, or subconscious, once we are accomplished performers.

The secret of generating complex sensory-interactive goal-directed behavior lies in structuring computational modules into a hierarchical architecture such that each level decomposes its task into about an order of magnitude more detail before passing it to the level below. This principle allows an arbitrary number of relatively simple computing modules to be arranged in hierarchical structures that are able to produce behavior of arbitrary complexity.

It should be noted, however, that many apparently simple behaviors require many years for humans to learn. For example, a child typically cannot tie his or her shoes without practicing to use his hands and fingers in manipulation tasks for three to five years. Most children with less than eight years of experience in exploring their environment and finding their way from place to place would be unable to avoid getting lost in a typical large shopping mall. The apparent ease with which we solve everyday problems is deceptive. Even the enormous computing power of the human brain cannot deal with such problems without long periods of training and a vast store of knowledge to guide our sensory-motor skills.

The acquisition of skills such as are required for walking, running, riding a bicycle, or ice skating is difficult and time consuming even for the human brain. It is not surprising that is has proven hard to build robots with these capabilities. It requires a rich body of knowledge about the world and how it works. Many sensors and complex sensory processing capabilities are needed. Sophisticated task and coordination skills are necessary. Vast amounts of computational power must be employed.

But the basic mechanisms are computational. There is no need to appeal to spiritual essence, supernatural forces, or quantum effects. Once knowledge is acquired and skills are learned, task execution appears deceptively simple. The skilled performer appears to move without significant effort. The more skilled the performer, the more smooth and effortless the performance. Similarly, with the skilled intellect: Well-trained mathematicians or engineers are fluent in the jargon of their profession. Their conversations and thought processes appear effortless, and the greater their skill, the more effortless their performance.

There are many things about intelligent behavior that remain a mystery, but one thing is perfectly clear—the concept of a goal is central. Intelligent creatures do not wander aimlessly or simply react reflexively to their immediate environment. They act proactively to accomplish their goals of survival and propagation in the most efficient manner possible under given environmental conditions. They react to perturbations and unexpected events with alternative tactics and behaviors so as to accomplish their goals despite difficulties and distractions along the way. Intelligent creatures are goal-directed in performing behaviors such as such as hunting for food and water, building nests, singing songs, grooming, displaying, building webs, constructing dams, digging dens, performing dances, pursuing mates, and avoiding a multitude of dangers, including being eaten by stronger, faster, bigger, and often more intelligent creatures. Intelligent creatures from insects to humans conserve resources and focus attention on entities and events such as food and danger that are relevant to their goals. The more intelligent creatures are, the more resourceful they are in dealing with difficulties and unexpected events. The most intelligent creatures can predict the future. They can make plans that anticipate or preempt problems, and invoke tactics that accommodate difficulties and compensate for unexpected events.

In many cases, success in survival and reproduction requires the ability to communicate and understand what is communicated. Success may require complicated social skills such as the ability to recognize the emotional mood of peers and to outwit rivals through ruse and deception. Successful behavior often involves tactics and strategies to accomplish desired results under uncertain and often hostile conditions. Success may require the ability to use tools for building shelter and hunting food. The probability of success in gene propagation may be enhanced by the development of weapons and the evolution of strategies for defense and conquest.

The evidence for the hierarchical structure of the brain is overwhelming, and the role that goals play in behavior is obvious. Yet many researchers in the behavioral sciences as well as in the field of robotics still do not subscribe to a hierarchical goal-directed model. The influence of behaviorism is still strong. Many behaviorist psychologists continue to reject concepts such as goals, intention, motives, and will as untestable hypotheses that should be discarded along with Freudian theories of the unconscious and metaphysical beliefs in the supernatural [Rachlin70]. Especially controversial is the attribution of intention or motive to subhuman creatures. Yet any intelligent system has a set of goals by its very nature. Goals are explicit in our definition of intelligence. To gain a clear insight into the concepts of goal and state, we turn to control engineering.

CONTROL THEORY

Among the fundamental principles of control theory are that systems have internal states and there exist goal states that the system acts to achieve or maintain. States and goals are central to analyzing control system behavior and are crucial to understanding the control process. The simplest form of goal-seeking device is the servomechanism. The *set point*, or reference input to the servomechanism, is a simple form of goal. Sensors monitoring the state of the world produce feedback that is compared with the goal. If there is any discrepancy between the desired state and the observed state, compensatory action is generated to reduce or eliminate the discrepancy. The control system thus approaches the desired state, or seeks the goal [Truxel55, Warwick96].

From earliest times, control theory has focused on the problem of controlling machinery to achieve a goal state. The ancient Egyptians, Babylonians, and Romans designed control systems to maintain the level of water in tanks. In these cases the desired water level is a goal state, and the current water level is an internal state that can be measured or estimated. Some mechanism for controlling the flow of water is then actuated so as to reduce the difference between the measured state and the goal. This is the essence of *feedback control*.

In the nineteenth century, a formal mathematical theory of feedback control was developed by Maxwell[1868] to analyze the behavior of the flyball governor that had been invented by James Watt to regulate the speed of steam engines. Control theory matured during World War II, spurred by the need for radar-controlled antiaircraft guns. The concept of a goal (i.e., to hit an enemy aircraft) was well understood and the concept of state (i.e., status of the gun and the aircraft) was unambiguous. The relevance of feedback control theory to intelligent behavior was recognized by early workers in the field of cybernetics, including Norbert Weiner[48], Ross Ashby[52,58], and Stafford Beer[95].

Another branch of control theory, *programmable automation*, also began during the nineteenth century with the invention of programmable control systems for looms. In 1804, Joseph Jacquard built a loom capable of achieving a goal of weaving a desired pattern in cloth. Each action of the Jacquard loom was programmed by a punched card to achieve a subgoal by positioning the threads and moving the shuttle in a pattern prescribed by the pattern of holes in the punched card [Malone78]. Programmable automation also matured during the middle of the twentieth century with the invention of the numerically controlled (NC) machine tool. Programmable automation also has a clear concept of a goal—a desired product. The significant difference between feedback control and programmable automation lies in how actions are generated. Programmable automation generates actions by issuing a series of commands that are assumed to be accomplished without error. Feedback control generates actions by establishing a desired state and compensating for errors between the desired state and a measured state. Programmable automation functions by

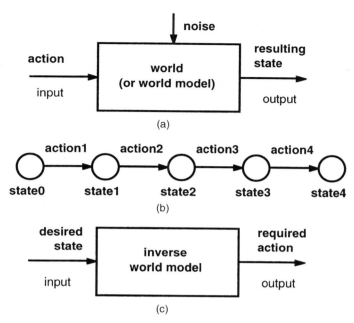

FIGURE 4.1 Basic concepts of control theory. (*a*) The world exists in some state. Action applied to the world (or to a model of the world) results in a state. (*b*) A series of actions may produce a series of states (i.e., a path through state space). (*c*) A desired state applied to an inverse world model may compute the action required to achieve that desired state.

issuing commands, while feedback control functions by compensating for errors [Groover80, Harrington79].

For either feedback control or programmable automation, goals, states, and actions are indispensable components of modern control theory. A *goal* is a desired state of the world that the control system is designed to achieve or maintain. The set point is a fixed goal. A string of goals along a time line is a *reference trajectory*. A *state* is a condition or set of dynamic properties such as position, orientation, velocity, temperature, energy, or momentum of an object or system at a point in time. An *action* is the output of an actuator that exerts force, causes motion, or transmits energy. The part of the world being controlled always exists in some state. Actions applied to the world can cause it to change states.

A notation for describing these basic concepts is illustrated in Figure 4.1. The world (or a model of it) can be treated as a box with an input and output. Figure 4.1*a* illustrates that action applied to the input causes some resulting state at the output. In most cases, the state of the world does not depend solely on the action of the control system itself. There are many other factors, such as the internal dynamics of the physical world (e.g., inertia, energy, gravity), disturbances (e.g., wind, collisions, friction), and the actions of other systems (e.g., machines, people, animals). What cannot be accounted for by the model is typically represented as a noise input. For a control system to be successful, it must act so as to achieve the desired goal despite disturbances. The task is to produce actions that cause the world to achieve or maintain a goal state regardless of the actions of other systems and the internal dynamics of the world. Control theory provides a set of mathematical tools for designing processes that can achieve this result.

Figure 4.1*b* is a state graph that shows how action of the system can cause the world to change states. When in state0, action1 causes the world to change to state1. Action2 causes the world to change to state2, and so on. Figure 4.1*c* illustrates the concept of an inverse world model. If a desired state is input to the inverse world model, the output produces the action required to produce this state. For example, if a particular position of a robot arm is desired, the inverse world model will predict what action is required to achieve that position.

What we have here called a *world model* is typically referred to in the control theory literature as a *system model* or *plant model* [Chen84]. A system model is usually a set of differential equations (for a continuous system) or difference equations (for a discrete system) that predict how a system will respond to a given input. For example, a linearized form of a system model is

$$\frac{d\mathbf{x}}{dt} = \mathbf{A}\mathbf{x} + \mathbf{B}\mathbf{u} + \eta$$

where

$d\mathbf{x}/dt$ = rate of change in the system state

$\mathbf{A}$ = matrix that defines how the system state evolves over time without control action

$\mathbf{x}$ = state of the system

$\mathbf{B}$ = matrix that defines how the control action affects the system state

$\mathbf{u}$ = control variable that causes action

η = noise

The *state of the system* is the information necessary to predict the future behavior of the system. Given a perfect system model and precise knowledge of the present and future control actions, the system model can predict the future states of the system. A particular state is a snapshot of the system at a point in time. Of course, no model can ever precisely capture every aspect of the real world. Thus there will always be differences between what is predicted by the system model and what actually occurs in the world. Thus a noise factor is introduced to account for events that occur in the world that are not represented accurately in the world model [Chen84].

The more general form of a system model is of the form

$$\frac{d\mathbf{x}}{dt} = f(\mathbf{x}, \mathbf{u}, \eta)$$

where f is a function that defines how the system state changes over time in response to both its own internal dynamics and the input control actions. What we call the world model in this book comprises a much richer and more comprehensive body of knowledge than is contained in the system model described above. Our world model contains knowledge of entities, events, attributes, relationships, situations, rules, images, and maps in addition to states and differential equations that are typically all that are included in the classical control system model. Our world model is thus a superset of the classical system model.

There are two basic approaches to control system design: feedback and feedforward. *Feedback* compares the state of the world with the desired goal state and takes action designed to reduce the difference. *Feedforward* uses a model of the world (explicit or implicit) to predict what actions will be required to achieve the desired goal.

Feedback Control

Feedback control produces reactive behavior. Feedback control reacts to sensory observations with actions designed to correct errors between observed and desired states. The strategy of feedback control is:

1. Sense the world.
2. Process the sensed data to estimate the state of the world.
3. Compare the estimated state of the world with the desired state and compute the difference, or error, between estimated and desired states.
4. Compute the feedback compensation required to reduce the error.
5. Output the control actions to the actuators and go back to step 1.

It should be noted that in simple cases, the state estimation referred to in step 2 may be degenerate. For example, in many PID (position, integral, differential) control systems, the sensor directly measures the state of the world to be controlled. In this case the data sensed can be compared directly with the desired state to compute the error. A typical feedback control system is illustrated in Figure 4.2. The desired state $\mathbf{xd}(t)$ is the goal. The actual state $\mathbf{x}(t)$ is measured by sensors to produce sensed data $\mathbf{y}(t)$. Sensory processing computes an estimated state $\mathbf{x}(t)$ from the data sensed. The estimated state is compared with the desired state, and an error $\mathbf{xd}(t) - \mathbf{x}(t)$ is computed. This

96

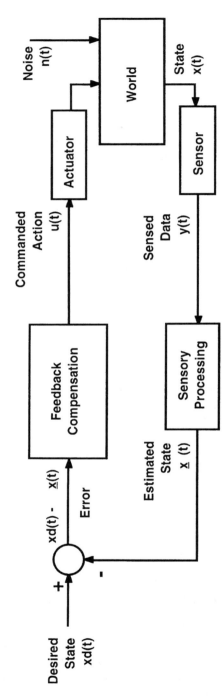

FIGURE 4.2 Feedback control. The desired state is input and compared with the estimated state. An error is computed and feedback compensation computes a commanded action designed to reduce the error.

error is submitted to a feedback compensation control law that computes a commanded action to reduce the error.

Feedback control requires an error between the estimated state and the desired state to generate action. Thus feedback control can never exactly achieve the goal state. It can only approximate the goal more and more closely by increasing the gain of the error compensation or by integrating the error over time. However, increasing the gain and/or integrating the error can cause instability, particularly in systems with time delay. Delay in a high-gain feedback loop will typically produce oscillations that persist or grow until the system drives itself into saturation.

Unfortunately, there is always some delay in any feedback control loop. For example, there may be delay between when a control action is commanded and when a measurable change occurs in the world. There may also be delay between when an event occurs in the world and when it is sensed. Finally, there typically are computational delays between when sensed data appear at the sensor output and when a command for action is sent to the actuator. The sum of all these loop delays can be modeled as a single lumped delay, as shown in Figure 4.3. Here the estimated state is shown as lagging the world state by the delay time Δt. Thus, if the estimated state $\mathbf{x}(t - \Delta t)$ is compared directly with the desired state $\mathbf{xd}(t)$, instability may result.

The likelihood that a feedback control loop will be stable is improved when the loop delay is minimized. The effect of unavoidable delay can be mitigated by predictive filtering that anticipates the future state of the world. Figure 4.3 illustrates a predictive filter inserted into the loop to compensate for the delay. A predictive filter uses the estimated state $\mathbf{x}(t - \Delta t)$, and a world model to predict how the world state will evolve as a result of internal dynamics plus the commanded action $\mathbf{u}(t - \Delta t)$. The predicted state is represented in Figure 4.3 as $\mathbf{x}^*(t)$.

FEEDFORWARD CONTROL

Feedforward control produces anticipatory behavior. Feedforward uses a model of the world to predict what actions are required to achieve the goal state. The feedforward control strategy is:

1. Build a model of the world that predicts how the world will react to control actions.
2. Invert the world model so that the inverse world model predicts what control action is required to produce a desired next state.
3. Input the desired state to the inverse world model and compute the control action required.
4. Output the required control action to the actuators as a feedforward command and go to step 1.

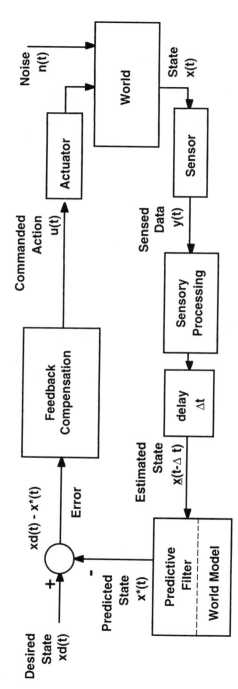

FIGURE 4.3 Feedback delay can be mitigated through the use of predictive filtering.

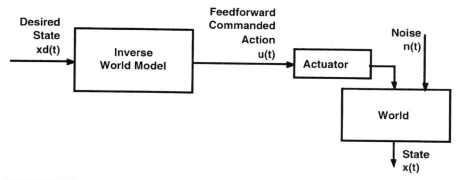

FIGURE 4.4 Feedforward control uses an inverse model of the world to generate feedforward commanded actions that cause the world state to follow the desired state.

In many cases, the inverse world model is computed only once and steps 1 and 2 are performed off-line. In this case, step 4 loops back only to step 3. However, in a dynamic environment where the world is changing rapidly in unpredictable ways, the world model must be updated frequently. In this case, steps 1 and 2 must be included in the real-time loop.

Feedforward control is illustrated in Figure 4.4. Feedforward control has an advantage over feedback control in that it anticipates the future and takes action before errors develop. However, feedforward control has a number of disadvantages. One is the problem of modeling the complexity of the real world. In general, the world can never be modeled exactly. Furthermore, there are many factors such as wind, weather, and unknown objects, and agents such as machines, animals, and people that are impossible to model precisely and are typically treated as noise. Thus the model can never predict exactly how the world will behave in the future, and hence the system cannot compute precisely what action is required to produce a desired result.

Perhaps even greater problems are associated with inverting the world model. A world model predicts how the world will react to an action. But feedforward control requires an inverse model to compute what action is required to achieve a desired state of the world. For example, assume that we have a kinematic model of a digging machine that enables us to predict what shape hole a particular commanded action will produce. This model cannot tell us what digging action is required to produce a hole with a particular desired shape. For that we need to invert the kinematic model.

For relatively simple systems, the world model can often be described by a set of linear differential equations expressed in matrix form such that the matrix can be inverted. However, for more complex systems, the world model may be highly nonlinear and a mathematical procedure for computing an inverse may not exist. The world model may also have redundant degrees of freedom or contain symbolic information about objects and events. The world model may include entities, attributes, images, maps, and relationships expressed in terms of names, lists, pointers, or iconic information in images and maps. In such cases, it may be extremely difficult or impossible to compute an inverse world model.

Attempts to circumvent the need to invert the world model have led a number of researchers to develop to a variety of techniques that generate inverse models directly. These include system identification theory, adaptive filter theory, and neural networks [Narendra86, Miller et al.90, Sutton and Barto98]. Widrow [95] has developed methods that can generate inverse models even for non-minimum-phase plants where there are substantial delays between input and output. These methods can also be applied to plants that are unstable as well as stable, nonlinear as well as linear, and multiple-input/multiple-output as well as single-input/single-output.

Programmable automation achieves feedforward control through a combination of machine design and implicit inverse modeling that takes place in the brain of a human programmer. Programmable machines such as numerically controlled machine tools and looms are designed with a library of actions designed to produce specific results. For example, machine tools are built strong and precise so that they do exactly what they are commanded to do. A RS274 command $\langle GOTO\ x,y,z \rangle$ causes the machine tool to travel to the specified goal point along a straight line at a speed defined by a previous feed-rate command. A human programmer can then use his or her brain to do the inverse modeling required to generate a NC program consisting of a string of commands that will produce the desired part shape. Similarly, a NC loom has a library of commands, each of which produces a given pattern of thread positions for each pass of the shuttle. The loom has a mechanism to interpret each of these commands to produce the desired thread positions. The programmer then uses an implicit world model to generate a string of commands that produce a desired pattern in the woven cloth. Software tools such as programming languages and postprocessors have been developed to facilitate this process.

Feedforward control is widely used for chemical plants and steel mills. Plant models are developed for each plant through careful analysis of the process to be controlled, and explicit inverse models are embedded in software or hardware. Neural nets, fuzzy logic, and genetic algorithms may be used in developing the models [Marlin95]. These types of feedforward control systems are typically maintained and optimized by manual heuristic tweaking of parameters.

Combining Feedback with Feedforward Control

The best properties of both feedback and feedforward control can be achieved by combining feedback and feedforward in a single system [Passino and Antsaklis89]. The principal benefit of feedforward control is that it can anticipate what control actions will be needed for the system to move toward the goal without waiting for feedback. This is important for processes that have long time delays. For example, long thermal delays are common in controlling the temperature of chemical reactions. To achieve the necessary thermal profile may require turning on heaters long before any thermal error is observed. Feedforward control can anticipate the correct action before any error is observed. Knowledge of when to apply anticipatory action requires an inverse model of the chemical reaction. Of course, feedforward control is only as good as the inverse model. If the inverse model is incomplete, incorrect, or out of date, the feedforward control actions may not produce the desired results.

The principal benefit of feedback control is that it can compensate for errors. With feedback control, much less depends on the accuracy of the system model. However, feedback control acts only after an error has been detected. Therefore, feedback control tends to have a persistent error and always lags behind the desired state. If the compensation is modified to minimize these difficulties, the system may become unstable.

Working together, feedforward and feedback control can achieve the best features of both. Feedforward anticipates what actions will be necessary to produce the desired results. Feedback corrects for errors caused by noise or inaccuracies in the inverse model. An example of this approach is shown in Figure 4.5. Note that there is both a world model and an inverse world model in this figure. A forward world model in the feedback loop enables the system to predict the current state based on past measurements. This can be used to eliminate delays in the feedback loop. An inverse world model in the feedforward path enables the system to predict what control action is required to produce the desired result.

PLANNING

Typically, inverse models are represented in the form of differential equations that can accurately predict only the immediate future. To generate anticipatory action leading toward long-term goals in the distant future, planning is required. Planning is a process that selects, discovers, or generates a plan. Planning requires predicting what is likely to result from alternative courses of action. Planning also requires evaluating predicted results, and selecting the best course of action for execution. Typically, planning uses a world model to simulate the future from a starting state at $t = 0$ to a goal state on a planning horizon, $t = t_{ph}$. A plan can be defined in at least three different ways: (1) by a path (or set of paths) through state space from a starting state to a goal state, (2) by a series of actions (or activities) that cause a system to achieve or maintain a goal state, or (3) by both a series of actions and a path of resulting states.

The intermediate states between the starting state and the goal state can be described as *subgoals*. The intermediate actions that lead from one subgoal to the next can be described as *subtasks*. A string of subgoals that define a planned path is sometimes called a *reference trajectory*. A series of commands

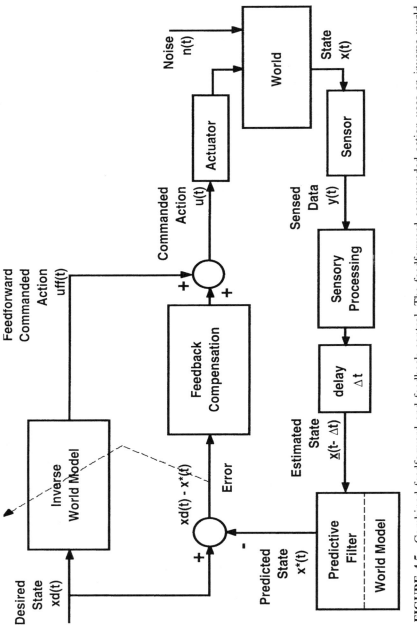

FIGURE 4.5 Combined feedforward and feedback control. The feedforward commanded action uses an inverse world model to compute a feedforward command. The world state is sensed and processed to compute an estimated and predicted state which compared with the desired state. The feedback error is compensated and added to the commanded action. The feedback error can also be used as a learning input to modify the world model or its inverse [Widrow95]. This is indicated by the dashed arrow.

or statements that produce actions that cause the system to transition from one subgoal to the next along the path to the goal state is sometimes called a *program* or *procedure*.

A plan for continuous motion may be represented by a spline or curve through state space defined by a differential equation. A plan for discrete actions may be represented by a state graph, a Petri net, or a Gantt chart. A state graph can be represented by a state table, making it mathematically equivalent to a set of rules in an expert system. A discrete plan can also be represented procedurally as a computer program that controls an agent or mechanical device in generating behavior that leads from a starting state to a goal state. In any case, a plan defines a trajectory through state space. A plan may specify what each agent is responsible for doing and a schedule of when each action should be done. The schedule may specify the timing of events and requirements for coordination between the actions of two or more agents or devices. A plan may also include a list of required resources, such as fuel or material, or a set of conditions, constraints, and priorities that must be satisfied.

A set of subgoals connected by a set of actions can be represented by a state graph that can be executed by finite-state automata. Figure 4.6a illustrates a plan expressed as a state graph where nodes represent desired states (or subgoals) and edges represent actions that cause state changes (or subtasks). Figure 4.6b illustrates a plan expressed as a state table or set of IF/THEN statements. For any state graph, a state table can be constructed such that each edge in the state graph corresponds to a line in the state table, and each node in the state graph corresponds to a state in the state table. The state table form has the advantage that the IF/THEN statements can be executed directly as case statements in a computer program. This form suggests how knowledge expressed in the form of expert system rules can be translated directly into plans, and vice versa.

Planning is the process of generating a plan. In general, planning involves a search over the space of possible futures to generate or select a series of actions and resulting states (or a series of desired states and required actions) that lead from a starting state to a goal state. There are two fundamental approaches to planning. The first is to:

1. Hypothesize a series of actions that lead from a starting state to a goal state.
2. Use a world model to predict the results.
3. Use value judgment to evaluate both actions and predicted results.
4. Choose the action series producing the best evaluation as a plan.

The second is to:

1. Hypothesize a series of states that lead from a starting state to a goal state.
2. Use an inverse model to compute the required series of actions.

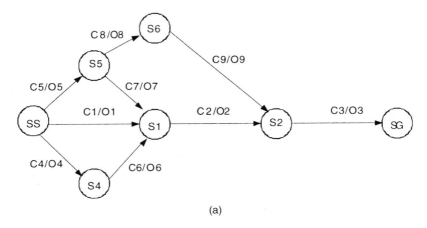

(a)

IF		THEN	
State	Feedback	Next State	Output
SS	C1	S1	O1
SS	C5	S5	O5
SS	C4	S4	O4
S1	C2	S2	O2
S5	C7	S1	O7
S5	C8	S6	O8
S6	C9	S2	O9
S2	C3	SG	O3

(b)

FIGURE 4.6 Two forms of plan representations: (*a*) example of a state graph representation of a plan that leads from starting state SS to goal state SG; (*b*) state table that is the dual of the state graph in part (*a*).

3. Use value judgment to evaluate both actions and results.
4. Choose the states producing the best evaluation as a plan.

Both approaches involve a world model or inverse to predict the future. Both involve a search procedure to explore the space of possible future actions and results. Both require value judgment to evaluate the cost, benefit, risk, and payoff of each hypothesized plan. Both result in a series of actions and subgoals.

It should be noted that all the steps in planning need not be done in real time. A hypothesized series of actions or states may be precomputed and simply retrieved from memory. A set of entire plans may even be built into hardware or read-only memory and simply be triggered when desired. The generation of plans by search over the space of all possible actions and predicted results can easily consume infinite time and computing resources. However, a search

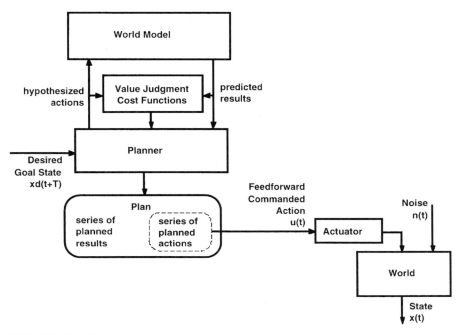

FIGURE 4.7 Planning and feedforward control. A planner generates hypothesized actions that the world model uses to predict results. The cost, risk, and benefit of the actions and predicted results are computed by a value judgment process that evaluates tentative plans for the planner. The tentative plan with the best evaluation is chosen as the plan to be executed. The plan consists of a series of planned actions and planned results. The planned actions are sent to actuators as feedforward commanded actions.

through a library of containing a few precomputed plans can often be a trivial task. Very effective real-time planning can often be achieved by:

1. Using a hypothesized series of actions and/or states (however generated) to predict the future.
2. Evaluating how desirable the predicted results are.
3. Selecting the behavioral hypothesis with the best predicted results.

Planning versus Feedforward Control

Figure 4.7 illustrates that planning is closely related to feedforward control. A planner accepts a desired goal state and generates a series of actions required to achieve that goal. This is the same input/output function as an inverse world model. Thus planning can be viewed as a means of inverting the world model. Both planning and feedforward control depend on the precision of the world model (or its inverse). At best, any world model is only an approximate representation of the real world. Thus, neither planning nor feedforward control can do better than to approximate the series of actions needed to reach the goal. To the extent that the approximation is in error, the actual results will deviate from the results predicted. In a dynamic unpredictable environment, the state of the world may change at any time. When this occurs, the world model must be updated and replanning must be invoked. Any delay between when an unexpected change occurs in the world and when a new plan is ready for execution may produce an error between is observed and what was planned.

One approach to dealing with a dynamic and unpredictable world is to replan frequently. Replanning requires that the world model be updated and a new plan be selected or generated. It can, of course, be computationally expensive to replan frequently enough to deal with high frequency variations in the world. A second approach is to use feedback from sensors to modify planned actions so as to compensate for differences between desired states in the plan and sensed states in the world. This approach makes it possible to accomplish goals even though the world model used in planning may be imprecise or incomplete. Feedback from sensors is processed into predicted states of the world and compared against desired states in the plan. Compensation can then be computed to reduce the error. This is illustrated in Figure 4.8. In this figure, the series of desired results (or subgoals) in the plan becomes the reference trajectory, which is compared with the predicted state of the world. At each point in time t, the error between desired state $\mathbf{xd}(t)$ and predicted state $\mathbf{x}^*(t)$ can be used to compute feedback compensation, which is added to the feedforward action $\mathbf{uff}(t)$ to produce the commanded action $\mathbf{u}(t)$ to the actuator.

It is possible to combine these two approaches. An engineering trade-off can be made between frequent replanning and feedback compensation. A portion of the available computing resources can be committed to replanning and the remainder committed to feedback compensation. Replanning requires updating the world model, selecting or generating one or more hypothesized plans, evaluating each hypothesis, and selecting the best. Feedback compensation requires processing sensory data to update the estimated and predicted state of the world, computing the error between desired and predicted states, and computing feedback compensation. Typically, feedback compensation requires less computation and hence can be performed more quickly than planning. Thus feedback compensation is typically performed more frequently than replanning. During the interval required for replanning, several feedback compensation cycles can be executed to provide high bandwidth response to unexpected events. A more complete discussion of the integration of planning with control can be found in Meystel[82,98] and Passino and Antsaklis[89].

THE DIFFICULTIES OF SEARCH

The process of searching for a plan to reach a goal can be a computationally intensive and time-consuming process. The number of possible plans that can

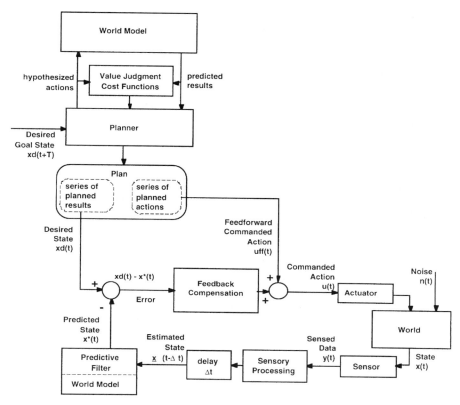

FIGURE 4.8 Combined feedforward and feedback control with planning. The planner generates a plan consisting of both a series of planned results that serve as a reference trajectory, and a series of planned actions that serve as feedforward commands. The current world state is sensed and processed to compute an estimated and predicted state which compared with the current desired state. The feedback error is compensated and added to the commanded action.

be hypothesized is the number of possible actions at each step in the plan raised to the power of the number of steps. Even in the case of relatively simple finite-state systems, the space of possible plans can easily become too large to search in a practical amount of time. For example, the number of possible plans for the game of chess is estimated to be on the order of 10^{120}. This is more than the number of microseconds since the big bang! Planning for real-world tasks involves a much richer space of possible behaviors than chess. Thus unless some heuristic method is employed to reduce the search space to a reasonable size, planning would be a practical impossibility.

A *heuristic method* is a rule of thumb that often works (but is not guaranteed.) A planning heuristic is a process that usually finds an acceptable plan, but may fail. A heuristic planner may or may not find the best plan, or any plan at all, even if one exists. But a well-designed heuristic usually finds an acceptable plan within a feasible period of time.

The simplest and most often used method for reducing the search space is the following:

1. Store a library of plans that are known to work under given circumstances.
2. When the appropriate circumstance occurs, select the corresponding plan from the plan library and execute it.

This is called *case-based planning* or *rule-based planning*. Case-based planning is widely used in manufacturing. For example, for each part to be machined, a plan in the form of a part program is developed by a human planner and expressed procedurally as a series of statements or instructions in a programming language such as RS274. Thus, for each part, there is a part program (or plan) stored in a library indexed under the name of the part. When the time comes for a part to be machined, the appropriate part program is withdrawn from the library, loaded on a machine tool controller, and executed. This methods works because manufacturing is a highly constrained environment. There are a finite number of parts and a limited number of machines. The machines are stiff and accurate and the material to be machined is uniform and predictable. Thus the stored plans produce precisely the sequence of tool movements required to produce the desired parts reliably.

However, in less constrained environments, where the results of actions are less predictable, the universe of possible states and actions is much larger. Case-based planning may require an impracticably large library of plans. For example, it is not feasible to build a library of all possible chessboard configurations and store the best move for each configuration. It is not feasible to build a library of all possible ways for a bird to fly through the woods, and even less feasible to store all possible ways a bird could fly through all possible woods. It is not possible to store a library of all possible grammatical sentences that might be appropriate for every possible situation that might occur in the life of an average speaker. Many real-world situations require plans that can adapt to unforeseen events so as to remain valid over minutes, hours, days, months, or even years. The number of possible sequences of low-level actions that could be generated over such planning periods is infinite. There is no way that all possible reference trajectories could be computed in advance. Thus, for most real-world situations, case-based planning is not a feasible approach for planning behavior.

A second heuristic for reducing the search space is to develop a library of plans wherein each plan contain a large number of conditional branches. This can enable any given plan to accommodate a wide variety of conditions. This may be how instinctive behaviors are implemented in many creatures. A variation on this heuristic is to use partially developed plans, called *schema* or *scripts*, that give general form to behavior that is appropriate to common situations, but leave open a number of parameters to be supplied by sensing the environment. An example of this type of heuristic is the development of process plans by a human programmer for machining operations. For each type of part feature to be machined, there is a basic schema for the machining operations needed to make it. The specific dimensions, choice of tools, depth of cuts, and feed rates are parameters that are provided by the programmer to meet the specified dimensions, tolerances, and material properties of the part. More complex examples can be found in a script for attending a restaurant, or a schema for interpreting a visual image [Schank and Abelson77, Arbib92,95].

A third heuristic is to develop efficient algorithms for exploring those paths through the search space that appear most likely to achieve success [Nilsson80, Fikes and Nilsson71]. A vast literature of planning algorithms has been developed, including A* search, dynamic programming, expert systems, and game theory [Meystel2000]. Many of these have been used for shop scheduling operations in manufacturing, path planning in robotics, and mission planning for military and space operations.

A fourth heuristic is to partition the planning process into hierarchical layers so that the range and resolution of the search space at each level of the planning hierarchy can be limited [Sacerdoti75, Meystel94, Albus79,91, Albus et al.87]. This reduces the search space of possible plans at each level while maintaining good performance through stepwise refinement at multiple levels. High-level plans look far into the future but contain little detail. Low-level plans are highly detailed but address only the immediate future. Intermediate-level planners decompose the first steps in the next-higher-level plan into more detailed but shorter-range plans for the next lower level.

The four heuristic methods described above can be combined in various ways to enable planners to generate acceptable plans within an acceptable period of time. In particular, hierarchical planning can be combined with all of the other heuristic methods to improve planning performance.

HIERARCHICAL MULTIRESOLUTIONAL PLANNING

Hierarchical multiresolutional planning reduces the search space in five ways: (1) it limits the number of states that need to be considered at each level, (2) it limits the number of actions that need to be considered at each level, (3) it limits the planning time horizon at each level, (4) it limits the resolution along the time axis at each level, and (5) it limits the number of entities, agents, events, and situations that need to be considered. At each level, the resolution of state space can be limited by grouping points in the space into discrete states. This effectively tessellates the state space into a limited number of regions. The number of possible actions can be limited by grouping actions into discrete classes that can be mapped onto a limited vocabulary of verbs. Parameters can be used to modify actions within a verb class. For example, all possible actions that take a system from one place to another can be grouped into a single action verb ⟨GO_TO⟩ modified by a set of parameters such as the coordinates of the goal point, desired arrival time, and desired velocity at the goal point. At each level, resolution along the time axis can be limited by dividing the time line into a discrete set of time increments. The number of steps in a plan can be limited by limiting the planning horizon at each level.

Also at each level, the number of entities, agents, events, and situations can be limited and be mapped onto a limited vocabulary of nouns. Attributes can be used to specify characteristics of persons, places, and things within a noun class. Pointers can be used to specify relationships between entities and events. Fuzzy logic, neural nets, and /or computing with words can be applied at each level [Zadeh94, Cleveland and Meystel90].

Hierarchical planning is often mistakenly associated with the concept of centralized planning. Nothing could be further from the truth. Hierarchical planning systematically distributes the planning process over a large number of local planners, each of which generates local plans for a limited number of objects within a limited region of space and time. Each lower-level plan is nested within one or two resolution elements of the next higher-level plan. Therefore, each lower-level plan refines the spatial and temporal precision of the next-higher-level plan, but only within a small region of state space. At each lower level, the planning horizon decreases as the resolution increases. Conversely, at each higher level, the planning horizon increases while resolution decreases. The result is a hierarchy of distributed planners and localized plans. High-level plans for overall system behavior are long range with low resolution, and low-level plans for local subsystem behavior are short range with high resolution.

Without hierarchical distributed planning, real-time planning for behavior that must be both long-term and precise would be impossible. However, with hierarchical planning, any desired precision in space and time can be achieved by reducing the range and increasing the resolution of planners at each successively lower level. Range in time and space increases geometrically with each higher level, while precision in time and space increases geometrically with each lower level. The range and resolution in space and time at each level can be selected so as to make planning computationally tractable at each level. The number of levels in the planning hierarchy can be chosen to achieve any desired scope and planning horizon at the top level with any desired resolution in space and time at the bottom level. Meystel[94] demonstrated that there exists a minimum of computational complexity that can be achieved for multilevel planning by proper choice of the ratio between resolution at adjacent levels [Khazen and Meystel98, Albus et al.97].

The use of hierarchical organizations for planning and control in complex systems is an old concept. It has been used by military, religious, and government organizations for millenia. Higher levels in the hierarchy deal with problems that are wider in scope and longer in time with less detail. Lower levels deal with problems that are more restricted in scope and time with more detail. At each level, the product of range and resolution is roughly constant. Thus the command and control problems at each level remain within the capacity of the human mind to cope.

Hierarchical planning is particularly effective for ground vehicle path planning using maps with different range and resolution representations of terrain features. High-level goals can be represented on a low-resolution map of a large region. A high-level path plan can be generated and expressed in terms of waypoints that are relatively far apart in distance and coarsely defined in position. The first waypoint in the high-level path plan then defines a goal region for the next lower-level-path planner on a higher-resolution map of a smaller area. A lower-level path can then be generated and expressed in terms of waypoints that are closer together in distance and more precisely defined in position. This process can be repeated at successively lower levels with waypoints closer together and more precisely defined on higher-resolution maps at each lower level. At some level, a priori maps are no longer available but must be generated from stereo or LADAR images. At this level, path plans extend only a few seconds into the future and are defined by waypoints that are only meters apart. Eventually, at the lowest level, plans consist of coordinated reference trajectories for propulsion, braking, and steering actuators.

PLANNING RESEARCH

Planning is a topic that has occupied the attention of researchers in a number of fields including operations research, artificial intelligence, game theory, and intelligent control. During the 1940s, von Neumann and others developed the theory of games [von Neumann and Morgenstern44]. Game theory has been used to determine the best course of action for many applications, including casino games, war games, antiaircraft missile guidance, and air-to-air combat maneuvers [Lucas et al.77]. In the 1950s, operations research began with the analysis of queues, graph theory, and methods for optimization of schedules [Bellman57, Ashby58]. Also during the 1950s, planning was addressed by researchers in artificial intelligence through the study of strategies for playing games such as checkers and chess [Shannon50, Samuel59]. Doran and Michie[66] applied graph-theoretic mechanism for path planning. Howden[68] introduced the "sofa movers problem," treating the geometric issues of motion planning for objects in three-dimensional space. The A* algorithm was introduced by Hart et al.[68] for searching state graphs. During the 1970s, STRIPS was developed by Fikes et al.[72] as a path planning system for the SRI Shakey robot. The concept of search was extended by Lozano-Perez and Wesley[79] to the problem of obstacle avoidance. Sacerdotti[75] introduced the notion of hierarchical planning.

During the 1980s, Lozano-Perez[81] applied the concept of *configuration space* to planning for robot manipulation. Khatib[86] applied concepts from field theory to problems of path planning in an environment filled with obstacles. Julliere et al.[83] developed a mobile robot with planning via tessellated space. Chavez and Meystel[84] introduced a concept of searching in a space with nonuniform traversability. Latombe[91] published a comprehensive textbook on robot path planning that outlines most of the theories and experiences accumulated over two decades in a variety of applications. Brooks and others have explored behaviorist alternatives to planning and explicit world modeling [Brooks99].

Traditionally, artificial intelligence research in path planning has not addressed the issue of dynamics. This has been considered the prerogative of control theory. Planning was treated as a problem of finding either a reference trajectory or a set of subgoals, or both. Following the reference trajectory or achieving the subgoals in a dynamic environment was left as a control problem. Fu[69], Saridis[77], and their students initiated research in control systems that incorporated planning and recognition [Saridis and Valvanis87]. This work eventually grew into the new field of intelligent control [Saridis and Meystel85]. Intelligent control blends operations research, artificial intelligence, and control theory. Albus integrated the concepts of task decomposition and hierarchical planning into intelligent control for both biological and artificial intelligent systems [Albus79,81,91].

EMOTIONAL BASIS OF BEHAVIOR

To this point we have discussed how behavior is generated from goals and how plans can be formulated based on predictions of how the world works. We have examined how knowledge of the world can be represented and used to generate behavior that is likely to be successful in achieving goals. Several questions remain, such as: Where do goals come from? Why is one goal selected rather than another? What is the basis for selecting one plan and not another? These questions require the concept of cost and benefit. In operations research, cost and benefit are embodied in an objective function [Bellman57]. In control theory, they are expressed as a cost function [Bryson and Ho75]. In biological systems, cost/benefit is evaluated by the emotions [Levine and Leven92].

Emotions are clearly an important aspect of generating and controlling behavior. Researchers in psychology have long recognized the effect of emotions on behavior. Yet the role of emotions has been largely ignored by the cognitive and computational sciences communities. This is perhaps because there exist strong metaphysical objections to the notion that emotions can be explained in computational terms. Although controversial and deeply disturbing in some quarters, the idea that computers might exhibit elements of artificial intelligence, has at least become established within the realm of scientific discourse. However, the idea that emotional feelings could be subjected to computational analysis is considered by many to lie outside the limits of respectable science.

Many agree with Penrose[89] that "feeling" is surely something that a computer could never have.

Yet almost all theories of the brain attribute emotion to the brain's limbic system, which is presumably an evolutionary old part of the brain involved in the survival of the individual and species [MacLean52,73]. It is also generally conceded that many emotional response patterns are hard-wired in the brain's circuitry and that the particular stimulus conditions that activate emotional responses are mostly learned by association through classical conditioning. Thus, the emotions would seem to be a relatively rudimentary aspect of brain functioning and thus relatively simple to model by computational methods.

Some of the areas in the brain usually included in the limbic system are the hippocampal formation, septum, cingulate cortex, anterior thalamus, mammillary bodies, orbital frontal cortex, amygdala, hypothalamus, and parts of the basal ganglia. The mammillary bodies contain areas that are related to hunger and thirst. The hypothalamus is involved in sexual drives. In primitive brains, the limbic system consists primarily of neurons that recognize smell, taste, nausea, and pain. The most basic of behavioral decision making is to eat what smells or tastes good and to reject what smells or tastes bad; to avoid what causes pain or produces nausea and to pursue what brings relief [Peele61].

Among the areas of the limbic system that have been studied most systematically are the hippocampus and the amygdala. The hippocampus is clearly involved in what has been called *declarative memory* (i.e., the ability to consciously recall experiences from the past) [Squire et al.93]. The amygdala appears to be implicated primarily in computation of fear and rage and the generation of rewarding and punishing features of stimuli, particularly the role of fear in conditioning. It is significant that the hippocampus and amygdala and many other limbic regions receive inputs from each of the major sensory systems and from higher-order association areas of the cortex. The sensory inputs arise from both the thalamic and cortical levels. These various inputs allow a variety of levels of information representation (from raw sensory features processed in the thalamus to whole objects processed in sensory cortex to complex scenes or contexts processed in the hippocampus) to cause the amygdala to activate emotional reactions. Inputs to the amygdala come from most, if not all input modalities, but more from secondary and tertiary sensory areas than from primary areas. Output from the amygdala projects to a variety of brainstem systems involved in controlling emotional responses, such as species-typical behavioral (including facial) responses, autonomic nervous system responses, and endocrine responses [LeDoux and Fellous98]. There are many connections between the limbic system and the frontal lobes, and considerable evidence that cognition plays an important role in generating emotions. Ortony et al.[98] developed a list of 22 distinct emotion types associated with representative emotion words. This is shown in Table 4.1.

Emotions also have a strong effect on cognitive analysis. The emotions are intimately involved with deciding what is good and bad, what is desirable and undesirable, what is attractive and repulsive, what should be approached and

TABLE 4.1   Twenty-two Distinct Emotion Types with Representative Emotion Words[a]

Well-Being Emotions
(arising from the appraisal of events evaluated in terms of goals and interests)

Pleased about a desirable event (Happy)	Displeased about an undesirable event (Unhappy)

Fortunes-of-Others Emotions
(well-being subtypes)

Pleased about an event desirable for other (Happy-for)	Displeased about an event desirable for other (Resentment)
Pleased about an event undesirable for other (Gloating)	Displeased about an event undesirable for other (Pity)

Prospect-Based Emotions
(well-being subtypes)

Pleased about a prospective desirable event (Hope)	Displeased about a prospective undesirable event (Fear)
Pleased about a confirmed desirable event (Satisfaction)	Displeased about a confirmed undesirable event (Fear confirmed)
Pleased about a disconfirmed undesirable event (Relief)	Displeased about a disconfirmed desirable event (Disappointment)

Attribution Emotions
(arising from the appraisal of actions evaluated in terms of standards and values)

Approving of one's own praiseworthy action (Pride)	Disapproving of one's own blameworthy action (Shame)
Approving of other's praiseworthy action (Admiration)	Disapproving of other's blameworthy action (Reproach)

Well-Being/Attribute Emotions
(compounds)

Approving of one's own praiseworthy action and pleased about the related desirable event (Gratification)	Disapproving of one's own blameworthy action displeased about the related undesirable event (Remorse)
Approving of someone else's praiseworthy action and pleased about the related desirable event (Gratitude)	Disapproving of some else's blameworthy action and displeased about the related undesirable event (Anger)

Attraction Emotions
(arising from the appraisal of objects evaluated in terms of tastes and preferences)

Attracted by an appealing object (Liking)	Repelled by an unappealing object (Dislike)

Source: Ortony et al.[98]. [Reprinted by permission of Oxford University Press.]
[a]Events are evaluated in terms of goals, actions of agents in terms of standards and values, and the appealingness of objects in terms of tastes and preferences.

avoided, what is cause for fear or hope, what should be loved or hated. The emotions are also associated with deciding what is important and trivial, what is worthy of attention, what can be safely ignored, what should be remembered, and what can be forgotten. These are all attributes that can be computed for, and assigned to, entities and events. There seems to be little or nothing about any of these attributes that requires properties of mind that are uniquely human, and there seems to be nothing about any of them that lies outside the bounds of computational theory.

There are also numerous connections between the limbic system and the autonomic nervous system and endocrine system. These connections would appear to provide adequate explanation for the "feelings" associated with emotional responses. There is no need to appeal to mysterious or uniquely human properties of consciousness, intuition, or spirituality to account for emotional "gut" feelings. There are direct neural connections from computational modules that produce emotional state variables to actuator muscles in the gut as well as to glands that secrete adrenaline, endorphins, and other mood-altering drugs into the bloodstream. These neuronal connections are present in virtually all mammalian brains and are not unique in humans. These connections are perfectly capable of producing muscle contractions and secretions of chemicals that cause feelings of joy, dread, happiness, depression, nausea, sinking feelings in the stomach, tingling in the spine, clamminess in the palms, and raising of hair on the back of the neck.

Recently, with the advent of functional magnetic resonance imaging (fMRI), the computational role of emotion in behavior has begun to be appreciated. fMRI has the capacity to produce three-dimensional maps of the regions of the brain that are active during various activities, such as perceiving, reflecting, thinking, planning, and acting. Brain imaging has progressed to the point where distinctions can readily be observed between areas of the brain involved in activities such as recognizing objects and thinking about objects, between perceiving written words versus spoken words, or between thinking about what to write versus what to speak. fMRI can also distinguish between areas in the brain that are (1) involved in generating plans, (2) involved in evaluating plans, and (3) involved with selecting which of various alternative plans to execute. Two-way communications between these three areas suggests how the loop between generating, evaluating, and selecting plans might work. For example, there are strong connections that loop between: (1) dorsolateral prefrontal cortex, which formulates high-level long-range plans for the future; (2) the amygdala, which computes emotional feelings of fear; the ventromedial prefrontal cortex, which generates emotional feelings of well-being; and (3) the orbitofrontal cortex, which decides which plan to select [Carter98]. A simplified block diagram of this loop is shown in Figure 4.9.

Yet despite the primitive origins and relatively simple mechanisms of the brain structures that give rise to emotional feelings, there remains something about emotions that causes most people to reject computational explanations. The mainstream of psychology rejects even the possibility that humans and

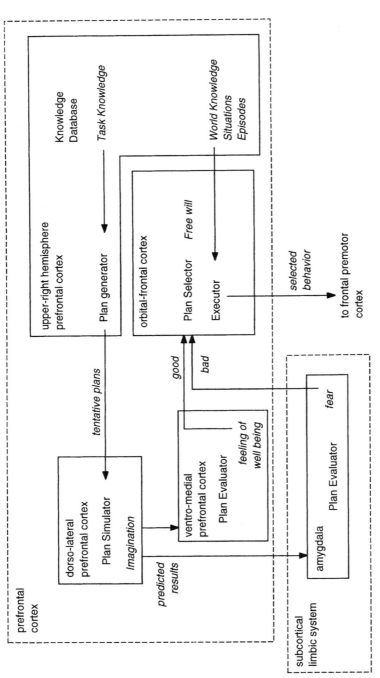

FIGURE 4.9 Simplified block diagram of high-level planning. A plan generator in the upper-right hemisphere of prefrontal cortex uses task knowledge to hypothesize tentative plans. These are submitted to a plan simulator in the dorsolateral prefrontal cortex. This simulator imagines predicted results of the tentative plans and submits them to plan evaluators in the ventromedial prefrontal cortex and the amygdala in the subcortical limbic system. The amygdala computes a fear (i.e., a "bad" evaluation). The ventromedial prefrontal cortex computes well-being (i.e., a "good" evaluation). A plan selector in the upper-right hemisphere of prefrontal cortex then selects a planned behavior based on these evaluations. This selection process is what is commonly called *free will*. The plan selected is then combined with knowledge of situations and episodes in the world and sent to the frontal premotor cortex for further decomposition into action.

animals share any "real" emotional experiences, even though humans and animals share the same basic neural structures that give rise to emotional experiences. The formal debate goes back a century, to William James. It was James who posed the famous question: "Do we run from the bear because we are afraid, or are we afraid because we run?" James concluded that we are afraid because we run [James1890].

Viewed from the prospective of control theory, this interpretation makes absolutely no sense. Creatures run for a thousand reasons, only a few of which involve fear. And it is obvious to anyone who has ever experienced fear that it produces an almost irresistible urge to run. One wonders how such an astute observer as James, who placed such great faith in introspection, could conclude that we are afraid because we run. Yet a century later, psychologists are still reluctant to attribute the behavior of running away to an internal state variable called fear.

Perhaps it is because admitting an internal state of fear that triggers flight would open the door to the possibility that animals might experience the same feeling of fear that humans experience. This, of course, would lead to the possibility that humans and animals might share emotional experiences. Apparently, this is a hypothesis that many scientists are not willing to consider. For example, modern psychologists such as LeDoux and Fellous readily admit that all animals, regardless of their state of evolutionary development, have the ability to detect and escape from danger. But they conclude from this widespread distribution of fear-avoidance behaviors in the animal kingdom that "it is unlikely that the subjective experience of fear is at the heart of this ability" [LeDoux and Fellous98]. LeDoux and Fellous support their argument by asserting that emotional feelings occur only in brains that have the capacity for consciousness. But this reasoning is completely circular. The conclusion that a subjective experience of fear is not responsible for fear-avoidance behavior in animals is inevitable if one begins with the assumption that the subjective experience of fear is unique to humans and therefore could not exist in animals. Presumably, LeDoux and Fellous would not argue that feelings of hunger, thirst, and pain are unique to humans. Why, then, are feelings of fear? Or for that matter, feelings of love?

This is not to argue that emotional feelings in humans are not demonstrably more refined and profound than those in animals. However, it appears plausible to us that if emotions are products of evolutionarily primitive brain structures, then humans and lower forms should share at least the most primitive forms of emotional experiences such as fear. From a control theory viewpoint, it seems quite likely that emotional feelings such as fear lie at the heart of behavior, and do so in all species. It also seems perfectly reasonable that emotional state variables such as fear, hate, and love can be generated by computational mechanisms and used for generating behavior in artificial intelligent systems.

Although popular belief defines computers as incapable of exhibiting and experiencing any form of emotion, it is clear that computer models have been useful in modeling certain aspects of emotional processes. For example, Gross-berg and Schmajuk[87] developed a computational model of conditioned affective states based on the notion that conditioned reinforcement involves pairs of antagonistic neural processes, such as fear and relief. This model predicts various effects related to the acquisition and extinction of conditioned excitation and inhibition that have been observed experimentally. The model suggests a mechanism by which neurons, or artificial neural nets, can produce reinforcing values (either positive or negative) dependent on the previous activity of the model.

It is also clear that computers are quite capable of calculating objective functions and computing cost, benefit, risk, and payoff. Most planning algorithms evaluate goodness or badness of planned actions. Most path planning algorithms compute the cost and benefit of traversing alternative trajectories to a goal. Most task planners evaluate the cost and benefit of sequences of actions that lead to a goal payoff. The entire field of operations research is based on concepts of cost minimization. Neural net modeling is based on concepts of reinforcement (both positive and negative) that can be computed by numerical methods. Computers routinely place economic value on objects. Ordinary grocery store checkout registers and inventory control computers assign monetary value to objects. Nonmonetary values such as attractiveness or repulsiveness are simply additional attributes that can be assigned to entities as readily as attributes of cost, benefit, size, shape, mass, temperature, and color. Love and hate are attributes that can be assigned to objects, animals, and people. Hope and fear are attributes that can be assigned to anticipated future events or situations. Thus, there seems to be no reason that computational modules cannot be used to build the equivalent of a limbic system for artificial brains.

MOTIVES

Motivation is yet another psychological concept that is central to intelligent behavior but largely unexplored by computational theories of cognition. Motivation is the internal force that is hypothesized to produce actions on the basis of a momentary balance between internal needs and external demands of the environment. Motives are instrumental in selecting goals. Motivated behavior is goal directed. A goal may be associated with a primary motivation such as hunger, thirst, or sexual urge. In this case, the goal is to reduce the primary motivation. Primary motivations are sometimes called *drives*.

Hull's drive reduction theory [Hull43] proposes that events that threaten survival give rise to internal drive states—and behaviors that reduce the drive are rewarding. For example, Hull argues that lack of food is motivational because it causes an increase in the hunger drive, and the consumption of food is rewarding because it leads to a reduction in the hunger drive. Hull also postulates that learning can lead to acquired drives (called *secondary motivation*). Hull's drive reduction theory has fallen out of favor in recent years due to experiments that have shown that electrical stimulation can be a direct source of motivation. Artificial stimulation of certain areas in the brain can cause behaviors in the absence of appropriate stimuli or homeostatic cues [Mowrer60].

Yet motivations, drives, and goals are simply patterns of neuron firings (i.e., state variables) that can implement task commands modified by parameters such as priority, aggressiveness, and urgency. High-level task commands are generated in the prefrontal cortex and are decomposed into actions that can be modified by sensory feedback at successively lower levels of the sensory-motor system. It is clear that as the brain evolves from primitive forms to more complex, it has added levels to the computational hierarchy. Each new layer adds about an order of magnitude to the temporal duration of plans that can be constructed. Each new layer adds about an order of magnitude to the scope of forethought and imagination. Each new layer increases by about an order of magnitude the sophistication of plans that a species can make and the complexity of thoughts that can be supported. The addition of the prefrontal cortex to the brain of *Homo sapiens* multiplies the sophistication of the behavior of humans over their closest cousins on the evolutionary tree by at least an order of magnitude.

This is an effect that can also be seen in the development of behavior in children. Piaget[76] and his followers have demonstrated clearly that there are periods of development in the behavior of children as they mature. At a very early age, the infant is virtually a reflex machine. Early months are dedicated to development of muscle coordination, hand–eye coordination, and eye tracking skills. As the child matures, his or her motor skills develop more sophistication. Crawling, walking, and running become possible. Perceptual skills necessary to support these motor skills develop correspondingly. The development of the brain in a child corresponds roughly to the development of the human brain over evolutionary history. As the child's brain becomes more mature, it adds layers with longer temporal horizon. This enables the child to imagine longer-range plans. It enables the brain to build a better world model that is better able support long-term planning.

CONSCIOUSNESS AND EMOTION

Perhaps we should not end this chapter without a word about consciousness. There is little agreement among cognitive scientists as to what consciousness is. We therefore cite a dictionary definition.

Consciousness: being aware of oneself and one's relationship to the world; perceiving, apprehending, or noticing with a degree of controlled observation; capable of or marked by thought, will, design, or perception

Many people insist that consciousness is something uniquely human that distinguishes humans from all other forms of life. Many consider it beyond the realm of possibility that consciousness could ever exist in a computer.

Cartesian dualism is still a strong influence in science today. There is a widespread belief (actually, a faith)* that the conscious mind (soul, will, or karma) is something other than the product of computation in the brain, regardless of how complex and powerful. There is both popular and scientific resistance to the notion that the mind can be explained as a computational process running on the computing machinery of the brain. Because if it could, then there is no reason to doubt that a conscious mind will someday emerge in a computer [Kurzweil99].

In some ways, consciousness and emotion are the last refuge of dualism. Both are poorly understood and mysterious. Both are intimately linked with our innermost feelings and self-image. Both have deep religious connotations. Both are widely believed to lie beyond scientific explanation. Yet history is filled with instances where physical mechanisms have been demonstrated to account fully and successfully for what had previously been considered to be outside the domain of science. Many examples can be given in astronomy, chemistry, physics, biology, meteorology, and medicine.

The scientific method operates by hypothesizing the simplest possible physical explanation and testing to see if that hypothesis is adequate to account for all the phenomena observed. In the case of emotions, the simplest hypothesis is that emotions are state variables produced by a value judgment system that evaluates objects, events, situations, and thoughts. The values of these emotional state variables provide input to a wide variety of decision processes involved in selecting behavior and making plans. Emotional state variables provide input to control processes that focus attention, establish levels of alertness, prepare systems for action, and control behavior. Emotional state variables also produce reward and punishment signals that control learning. Emotional state variables (such as fear, hope, pain, relief, attraction, repulsion, love, hate) can be assigned to objects, events, and situations in the world model. This enables the sensory-motor system to place value on entities and events and to generate appropriate behavior relative to objects and situations in the world. Emotional state variables can be combined with motivations (such as hunger, thirst, and sex drive) to influence behavioral decisions and control internal parameters such as heart rate, breathing, blood pressure, artery and vein flow valve settings, chemical excretion rates, and sensitivity to pain. The "gut" feelings that result from human emotional experiences may readily be explained by connections between the limbic system and the autonomic nervous system. The subcortical connections between primitive limbic structures where subconscious evaluations are made and prefrontal cortical

---

*The apostle Paul defines *faith* as "the assurance of things hoped for, a conviction of things not seen" (Hebrews, Chapter 11, verse 1).

regions where conscious thinking takes place could account for the mysterious phenomenon of intuition [Carter98].

In the case of consciousness, the simplest hypothesis is that awareness emerges in any system that has an internal world model with sufficient sophistication to represent the richness and dynamics of the physical world in which it exists. Self-awareness then follows naturally in any system that includes a self-object in its representation of the world. Any system that can build and maintain a world model that represents objects, events, situations, and relationships in the world should easily be able to include itself as one of the objects. A self-aware system can keep track of itself. It can know its own position, velocity, and orientation in the world relative to other objects in situations and relationships. It can compute self attributes and assign all manner of properties (both real and imagined) to its self-object. It can be aware of how it "feels" about things. Events or objects that are labeled with particular emotional state variables may cause the system to prepare for action, to flee from danger, to attack an enemy, or to defend territory, friends, or loved ones. A self-aware system will be aware of pain state variables which signal that some part of it is being damaged by stress or physical trauma. It will feel the pleasure of reward for success and the pain of punishment for failure. It will experience the thrill of victory and the agony of defeat. This level of awareness of the world and of one's internal self is what most people define as consciousness.

Any system that can plan can use its world model (including its self-object) to simulate and predict the results of hypothesized actions. The ability to simulate possible future states of the world is commonly called *imagination*. Planning consists of imagining the future and choosing the most attractive course of action. Thus, if a system can plan, it can imagine, and vice versa. If a system is aware of itself, it can imagine itself in many kinds of situations and relationships. It can have imaginary friends. It can imagine relationships with angels, demons, spirits, and gods. It can imagine its place in a universe filled with galaxies, atoms, black holes, and quarks. It can imagine death and the afterlife. It can formulate myths to explain its origins and destiny. It can organize rituals and social customs to shape the imagination of others.

It seems clear that it is possible to build machines that at least appear to possess emotions. We certainly will be able to build machines that act as if they are afraid, angry, happy, sad, or hopeful. Once a system can assign emotional state variables to observed objects in the world, it can act afraid by running away from objects labeled with a high-fear state variable. An intelligent machine can be programmed so that a fear state variable would cause an increase in engine speed or hydraulic pressure. A machine could even be programmed to verbalize that it is afraid and assume a defensive or submissive posture when its fear state variable is elevated. Who could say if such a machine is "really" feeling fear?

Similarly, it seems clear that it will be possible to build intelligent machines that at least appear to be conscious. If we build a machine with a rich and dynamic model of the world, it could focus attention and use sensors to maintain its world model. If we program it to use its knowledge of the world to plan behavior and control actions that are successful in accomplishing its goals, it will behave as if it is intelligent. If it behaves toward objects and events in the world in a way that indicates that it is aware of them and of their significance, if it can acquire knowledge by interpreting language correctly, and if it can converse about a wide variety of subjects, it will behave as if it is intelligent. If it also appears to be aware of the thoughts and feelings of others, and if it exhibits a sense of humor and an appreciation of beauty, at what point do we admit that the machine is conscious?

We are probably a long way from building such machines, so this question won't have to be answered for many years, perhaps many decades. For now we can only say that no one really knows. Perhaps no one will ever know. We are only beginning to understand the mechanisms of mind. There may be many things that are simply unknowable. We ultimately may be forced to return to Hume's conclusion that no one can know for sure even whether the external world exists. Perhaps we will always be limited to Kant's conclusion that although we can never really know for sure, we can build a good enough model of the world to behave successfully in the real world—and that is good enough. For those who insist that only humans can experience consciousness or emotions, then by definition machines can never experience consciousness or emotions. However, this is not a fruitful scientific hypothesis. We prefer to hypothesize that machines not only can, but must, have at least some elements of both consciousness and emotions if they are to behave in an intelligent goal-directed manner.

To this point, we have presented a brief outline for a computational theory of intelligence. Next, we turn our attention to the task of constructing a reference model architecture that can be used to design and build intelligent systems.

5   A Reference Model Architecture

Interior perspective of atrium, New Tokyo City Hall complex, from *The Art of Architectural Illustration*, Resource World Publications, Inc. [By permission of Nihon Sekkei, Inc., architects; Hiroyuki Takahashi, artist.]

In this chapter we introduce a reference model architecture to express our computational model of intelligence. A reference architecture is a sufficiently specific formulation that hypotheses can be constructed, tested, and either validated or disproved. Once validated, a reference model architecture can form the basis for an engineering methodology whereby intelligent systems can be designed and built to meet specified requirements.

The reference model that we have chosen is RCS (Real-Time Control System). The RCS architecture represents the culmination of over 200 person-years of research and development in intelligent control theory and practice over a period of more than two decades. The work has been done at the National Bureau of Standards (NBS) [renamed the National Institute of Standards and Technology (NIST) in 1988] and the following industry and university laboratories: Martin-Marietta Denver Aerospace, NASA Goddard Space Flight Center, Martin-Marietta Baltimore, Advanced Technology Research, Transitions Research Corporation, Helpmate Robotics, Servus Robots Inc., General Dynamics Robotic Systems, General Dynamics Electric Boat, Ohio State University, Drexel University, University of Maryland, University of New Mexico, and the European Space Agency. RCS was originally developed by Barbera for sensory-interactive robotics [Barbera et al.79,84]. It has been applied to a wide variety of projects, including computer-integrated manufacturing systems, open-architecture controllers for machine tools, industrial robots, multiple autonomous undersea vehicles, experimental controllers for nuclear submarines, space station telerobotic systems, telerobotics for aircraft maintenance, postal service stamp distribution and general mail facilities, controllers for laser, plasma, and water-jet cutting machines, vision-guided highway vehicles, automated mining machinery, robot cranes, and unmanned ground vehicles.

There are many other architectures that we could have chosen. For example, SOAR (State, Operator, and Result) is an architecture designed to serve as a model for both human and artificial cognition. SOAR implements the problem-solving act of applying an operator to a state and producing a result. It relies primarily on search methods, production rules, and hierarchical goal decomposition for solving problems. SOAR is based on Newell, Shaw, and Simon's work [Newell et al.58] on human problem solving. SOAR employs AI concepts such as problem solving, planning, learning, knowledge representation, and interaction with the external world. SOAR deals with many of the critical elements of intelligent systems, such as states, goals, plans, agents, behaviors, and representations of knowledge. A two-volume compendium of papers on SOAR has been published [Rosenbloom et al.93].

The reason that we did not choose SOAR is because it does not incorporate the fundamental concepts of real-time control, image understanding, and map-based reasoning as an integral part of its architecture. SOAR is vulnerable to the criticisms of Dreyfus and Searle. Essentially all knowledge is represented in terms of symbolic data structures and production rules. There is little emphasis on symbol grounding. No significant use is made of iconic representations, geometric reasoning, map manipulation, time or frequency analysis, closed-loop feedback control, or parallel processing in the image domain. Despite these shortcomings, SOAR has recently been used successfully for flying simulated aircraft and has been incorporated into the MODSAF combat simulator at Fort Knox, Kentucky. We believe that SOAR could be integrated into the upper levels of RCS to provide reasoning and problem-solving capabilities, but it is not a good candidate for representing the lower-level demands of real-time interaction with a dynamic world.

Subsumption is another architecture that we could have chosen. Subsumption is the behaviorist architecture originally conceived by Rodney Brooks [86,99]. Subsumption has had a profound influence on many of the architectures developed by academic and NASA researchers for robotic systems. Behaviorist architectures emphasize a direct path from sensing to acting. They deliberately minimize or bypass the issue of internal representations of the external world. In the tradition of J. J. Gibson[79], Brooks considers the world to be its own model that can be sensed when necessary to enable behavioral decision making. Behaviorist architectures are primarily reactive and typically do not include mechanisms for planning or problem solving that anticipate and avoid future difficulties and optimize future results. They emphasize stimulus–response mechanisms and largely ignore concepts such as goals, plans, and symbolic reasoning. The principal contribution of behaviorist architectures is that they have demonstrated how complex behaviors can be generated by simple reactive systems operating in a complex environment. They stand in sharp contrast to good old-fashioned artificial intelligence architectures like SOAR that apply complex reasoning mechanisms to relatively simple environments such as the blocks world and discrete board games. Behaviorist architectures have been used for many robotic systems. For example, Bekey has applied a version of Subsumption to controllers for robot hands and helicopters. Maes[89], Mataric[92], and many others have applied it to a variety of mobile lab vehicles. Payton[86] applied it to undersea vehicles.

Our reason for not choosing a behaviorist architecture is the lack of a rich internal model of the world. Information from different sensors is nowhere fused into a single best estimate of the state of the world. Instead, sensors are typically connected directly to action. Each sensor-actuator system generates its own estimate of what behavior is most appropriate, and arbitration heuristics are used to select a single "best" behavior for execution. Behaviorist architectures are often not goal-directed and typically do not generate plans for future behavior.

A number of hybrid architectures have been developed to compensate for the shortcomings of the architectural extremes of SOAR and Subsumption. For example, AuRA is a hybrid architecture designed by Arkin that incorporates both planned and reactive behavior in a goal-directed schema-based system [Arkin90,98]. AuRA combines a high-level deliberative hierarchical planner based on traditional AI techniques with a low-level reactive controller based on schema theory [Arbib92]. AuRA has been used successfully for a wide variety of mobile robot projects in the laboratory and in simulation. SAUSAGES is a hybrid architecture developed jointly by researchers at Carnegie Mellon University and Lockheed-Martin Corporation for the Demo II autonomous ground vehicle program [Gowdy97]. SAUSAGES (System for Autonomous Specification, Acquisition, Generation, and Execution of Schemata) is designed to be a bridge between the worlds of planning and perception. It combines arbitration concepts from Subsumption with map-based planning and control.

MRHA (Multiple Resource Host Architecture) for the Mobile Detection, Assessment, and Response System (MDARS) program is a specification for the design of a computer control system for a fleet of mobile security robots that patrol the interiors of buildings such as warehouses [Everett et al.98]. MRHA has a supervisor, an operator station, and a number of controller/dispatchers that plan paths and issue commands to multiple mobile robots. Atlantis is a three-level hybrid architecture designed at JPL for the Mars Rover project [Gat92]. Atlantis was used on the *Sojourner* vehicle that landed on Mars.

Many agent architectures have been developed and there are many different kinds of agents [Maes90,94]. Some agents are relatively simple computational structures consisting of a few finite-state machines that generate interesting behaviors through interactions with other agents. More complex agents have been designed that can perform complicated tasks such as scheduling factory operations, monitoring Internet traffic, conducting electronic commerce, implementing computer games, and performing web searches [Lieberman et al.99]. Many agents are purely software objects that do not embody physical sensors and actuators. Agents may cooperate or compete with each other in the process of generating group behavior. Agent architectures are particularly well suited for modeling activities such as market behavior, social interactions, or political processes. Often, agent architectures are not directed by any unifying goal. Groups of agents are expected to produce interesting or useful results through behavior that emerges as a result of each agent pursuing its own goals.

Other significant architectures include the Task Control Architecture developed by Simmons at Carnegie Mellon University [Simmons94]. TCA has been demonstrated on a wide variety of applications, including a six-legged planetary rover, several indoor mobile robots, a tile inspection robot, an autonomous excavator, and a prototype autonomous spacecraft. TCA has also been used in work on virtual reality and teleoperation. TCA has many features in common with RCS, but is does not include as rich a world model or as complete a system for sensory processing.

Meystel[87,91] developed an architecture at Drexel University for autonomous control of a dune buggy vehicle that is very similar to RCS in its use of multiresolution maps and its hierarchical approach to path planning and goal decomposition [Chavez and Meystel84, Isik and Meystel88, Albus et al.93]. Other architectures include the EAVE Autonomous Underwater Vehicle Architecture developed at the University of New Hampshire by Blidberg[86], the Pilots Associate architecture developed by Lockheed Aerospace, the Submarine Operational Automation System developed by General Electric for the next-generation nuclear submarine, and the PIC architecture developed at the University of Pennsylvania for undersea vehicles [Roeckel et al.99].

There are also a number of architectures developed for integrated manufacturing systems. These include the Computer-Integrated Manufacturing Application Framework Specification [Eng96] developed by the SEMATEC consortium and the National Information Infrastructure Protocol developed by the NIIP consortium. These are designed to facilitate integration of manufacturing systems software and data using standard communications protocols such as CORBA (Common Object Request Broker Architecture) [Object Management Group95]. CIMOSA (Computer-Integrated Manufacturing Open System Architecture) was developed by the European ESPRIT project [Jorysz and Vernadat90]. A reference model architecture for industrial process control was developed by Williams[89]. GERAM (Generalized Enterprise Reference Architecture and Methodology) was developed by the IFAC (International Federation of Automatic Control) Task Force on Architectures for Manufacturing [Bernus et al.98]. The Reference Model for Manufacturing and Planning and Control was developed by Biemans[89] at Phillips. The MSI (Manufacturing Systems Integration) architecture was developed by the Manufacturing Systems Integration Division at NIST [Senehi94b]. Most of these manufacturing architectures are expressed as high-level representations of the manufacturing enterprise. They do not address the lower levels of real-time interaction with the physical world, which are critical to understanding intelligent systems from the perspective of the mind and brain. For example, the MSI architecture was designed specifically to integrate the upper levels of factory, shop, and workstation scheduling with intelligent machines and processes using RCS for real-time control [Senehi et al.94a, Senehi and Kramer98].

RCS is a hybrid architecture in that it combines deliberative with reactive components. However, RCS differs from other hybrid architectures in that it does not relegate planning to the upper levels and reactive behavior to the lower levels of a hierarchical control system. RCS combines deliberative with reactive components at every hierarchical level. Every level of the RCS hierarchy contains a planner and reactive executor. Planners at higher levels in RCS have longer-term planning horizons with lower-resolution planning space. Planners at lower levels have shorter-term planning horizons with higher-resolution planning space. Reactive feedback loops at higher levels have lower bandwidth response, while reactive feedback loops at lower levels have higher bandwidth response.

RCS is an agent architecture in that it is composed of computational modules of the type that are often termed *agents*. For example, RCS can control groups of vehicles or machines that cooperate or compete in ways that are reminiscent of autonomous agents. However, RCS differs from most agent architectures in that it organizes its "agents" into a hierarchy of operational units within which each agent possesses specific skills and abilities and may be assigned particular duties and responsibilities. RCS also differs in that the overall system is designed to accomplish specified tasks with clearly identified global goals. Within RCS operational units, agents make plans, decompose tasks and goals, and react with sensory feedback so that the overall system functions as a unified goal-seeking entity. Within RCS, agents work together to accomplish specified objectives. Behavior does not simply emerge but is purposefully directed toward achieving desired results.

In summary, the reasons we have chosen RCS as a reference model architecture are:

1. RCS combines many concepts from artificial intelligence with control theory.
2. RCS supports a rich dynamic world model at many different levels of resolution of space and time.
3. RCS combines deliberative with reactive behavior at many levels of resolution.
4. RCS integrates prior knowledge with current observations at many levels.
5. RCS was specifically designed to model the functional architecture of the human brain.
6. RCS provides a framework that addresses the entire range of human perception, knowledge, and behavior in both time and space.
7. RCS is a mature system with a number of software engineering tools and software libraries that are available to potential users.

RCS partitions the control problem into four basic elements: behavior generation, world modeling, sensory processing, and value judgment. RCS clusters these elements into computational nodes that have responsibility for specific subsystems and arranges these nodes in hierarchical layers such that each layer has characteristic functionality and timing. Each layer provides a rich and dynamic world model and a sensory processing hierarchy to keep the world model up to date. Each layer provides a mechanism for integration of deliberative (planning) and reactive (feedback) control. The RCS reference model architecture has a systematic regularity and recursive structure expressed in a canonical form that provides a basis for an engineering methodology.

BACKGROUND

RCS has evolved through variety of versions over a number of years as understanding of the complexity and sophistication of intelligent behavior has increased [Quintero and Barbera93, Albus97]. The first version of RCS (RCS-1) was designed for sensory-interactive robotics in the mid-1970s [Barbera et al.79, Albus et al.81]. In RCS-1, the emphasis was on combining commands with sensory feedback so as to compute the proper response to

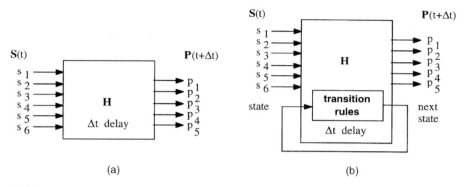

FIGURE 5.1 CMAC module. (*a*) The CMAC implements an **H** function that maps an input vector $\mathbf{S}(t) = (s_1, s_2, s_3, s_4, s_5, s_6)$ into an output vector $\mathbf{P}(t + \Delta t) = (p_1, p_2, p_3, p_4, p_5)$. (*b*) The CMAC module can become a finite-state automaton if there exists a feedback path from the output to the input that represents the state of the CMAC module, and if the *H* function contains a set of transition rules that compute a state transition function.

combinations of desired goals and observed states. Among the earliest applications was the control of a robot arm with a structured light vision system in real-time visual pursuit tasks [Nagel et al.79, VanderBrug et al.79]. RCS-1 was heavily influenced by biological models such as the Marr–Albus model of the cerebellum [Marr69, Albus71,72], and the Cerebellar Model Articulation Controller (CMAC) [Albus75a,b]. The Marr–Albus model is one of the most widely accepted models of the cerebellum used by neurophysiologists today. CMAC is a neural network abstraction of the Marr–Albus cerebellar model.

CMAC is a two-layered neural net that can implement a single-valued nonlinear function of the form

$$\mathbf{P}(t + \Delta t) = \mathbf{H}(\mathbf{S}(t))$$

where $\mathbf{S}(t)$ is a vector of input variables at time t, $\mathbf{P}(t + \Delta t)$ is a vector of output variables at time $t + \Delta t$, and **H** is a matrix of single-valued multivarient functions. CMAC can learn nonlinear functions much faster than back propagation and can be implemented in real-time systems for real-time control.

CMAC can be represented as a computational module *H* that transforms an input vector **S** into an output vector **P** as shown in Figure 5.1*a*. If the **H** function keeps track of the state and has a set of state transition rules that maps the current state and input into a next state, the **H** module is a finite-state automaton. Thus CMAC can be transformed into a finite-state automaton by taking one or more of the outputs and looping them back to become inputs as shown in Figure 5.1*b*.

If the input **S** is a combination of a command vector from a higher level plus a feedback vector from sensors, and the output **P** is a command to a lower-level module plus a next state, the CMAC can decompose a task into a string of subtasks. CMAC thus became the conceptual building block of RCS-1, as shown in Figure 5.2a. A hierarchy of these building blocks can be used to implement a hierarchy of behaviors such as observed by Tinbergen[51] and others in simple creatures such as fish, birds, and insects.

RCS-1 was implemented as a set of state machines arranged in a hierarchy of control levels [Barbera et al.84]. At each level the input command effectively selects a state table that implements a behavior driven by feedback in stimulus–response fashion [Nagel et al.79]. RCS-1 predates, and in many respects anticipates, the Subsumption architecture developed by Brooks[86] and the hybrid architecture of Arkin[86]. The principal difference between RCS-1 and Subsumption is that RCS selects behaviors based on goals expressed in commands, whereas Subsumption selects behaviors based on heuristic arbitration algorithms.

The next generation, RCS-2, was developed during the early 1980s by Barbera et al.[84], Kent and Albus[84], Albus[81a], Albus et al.[82], and many others for intelligent control of the NIST Automated Manufacturing Research Facility (AMRF) [Simpson et al.82, Furlani et al.83, Wavering and Fiala87, Fiala and Wavering87, Haynes et al.84, Kilmer et al.84, McCain85, Murphy et al.88, Norcross88, Barkmeyer et al.96, Rippey et al.83, Bloom et al.84]. The basic building block of RCS-2 is shown in Figure 5.2b. The **H** function is a finite-state machine state-table executor. The new feature of RCS-2 was the addition of a G function consisting of a number of sensory processing algorithms, including structured light and blob analysis image processing algorithms [Kent84].

RCS-2 defined an eight-level hierarchy consisting of servo, coordinate transform, E-move, task, workstation, cell, shop, and facility levels of control [Albus82]. Only the first six levels were actually built. Two of the five AMRF workstations fully implemented five levels of RCS-2. One was a horizontal machine tool workstation consisting of a machine tool, a robot with a vision system and quick-change grippers, and automated buffer table [Barbera et al.79,84, Wavering and Fiala87, Scott and Strouse84]. The second was a cleaning and deburring workstation with two robots, a number of interchangable deburring tools tools and buffing wheels, and a washing machine [McCain85, McCain et al.85, Murphy et al.88, Proctor et al.89, Lumia88]. The overall architectural design of the AMRF is shown in Figure 5.3.

RCS-2 was fully documented in a user's reference manual (Leake and Kilmer88) and has been commercialized by Advanced Technology Research (RCS-2000).* RCS-2 was used by Barbera and Fitzgerald at Martin-Marietta Baltimore to design the control system for the Army TMAP (Tactical Mobile Autonomous Platform) vehicle. RCS-2 was used by NBS and Martin-Marietta for the Army Field Material Handling Robot (FMR) [McCain et al.86]. A

---

*Commercial products identified in this book are not intended to imply recommendation or endorsement, nor is it intended to imply that the products identified are necessarily the best available for the purpose.

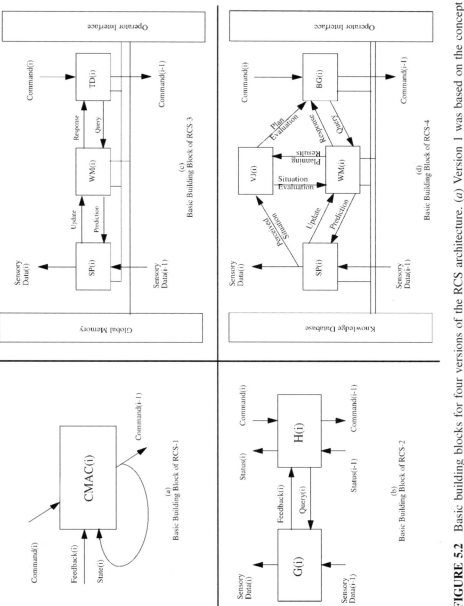

FIGURE 5.2 Basic building blocks for four versions of the RCS architecture. (*a*) Version 1 was based on the concept of a completely reactive neural net feedback loop. (*b*) Version 2 introduced sensory processing into the reactive loop. (*c*) Version 3 introduced the concept of an explicit internal world model. (*d*) Version 4 introduced the concept of an explicit value judgment system.

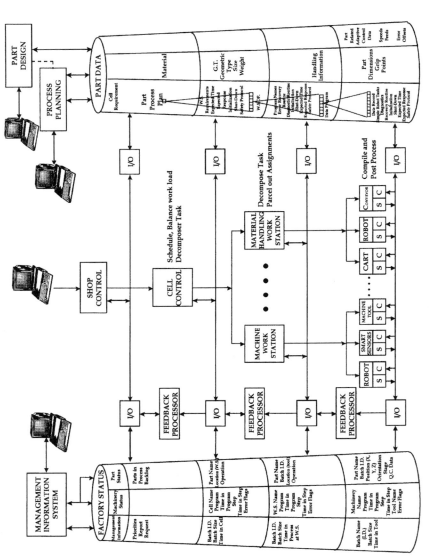

FIGURE 5.3 Architecture for the NIST Automated Manufacturing Research Facility (AMRF). On the right is a hierarchical database containing process plans and control programs that define how to manufacture parts. On the left is a hierarchical database containing information about the state of the parts and machines in the AMRF. In the middle are the controllers that compute plans, sequence commands, measure results, and compute actions to compensate for errors between plans and results. The arrows represent a communications system that moves information throughout the architecture. [From Simpson et al.82.]

NASREM: NASA/NBS STANDARD REFERENCE MODEL

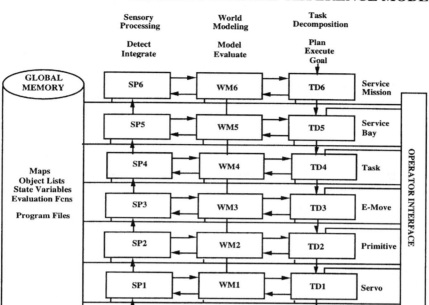

FIGURE 5.4 NASREM (RCS-3) control system architecture. [From Albus et al.87.]

RCS-2 controller for a semiautonomous HMMWV (Highly Mobile Multi-purpose Wheeled Vehicle) was developed for the Army TEAM and Demo I programs [Szabo et al.90,92, Murphy et al.93].

RCS-3 was developed for the NASA/NBS Standard Reference Model Telerobot Control System Architecture (NASREM) [Albus et al.87, Lumia94]. NASREM was developed for the NASA's Space Station Flight Telerobotic Servicer. The basic building block of RCS-3 is shown in Figure 5.2c. A block diagram of NASREM is shown in Figure 5.4. The significant new features introduced in RCS-3 are the world model and the operator interface. The inclusion of the world model provides the basis for task planning and for model-based sensory processing. This led to refinement of task decomposition processes such that each task decomposition (TD) module has a job assignor and a set of planners and executors (i.e., a planner and executor for each subsystem reporting to the TD module). The job assignor, planners, and executors in each TD module correspond roughly to the Saridis three-level control hierarchy [Saridis85]. The Drexel planner–navigator–pilot controller has many features in common with RCS-3 [Isik and Meystel88]. RCS-3 also served as the foundation for the UTAP (Universal Telerobotic Architecture Project) architecture developed jointly for the Air Force by NIST and the NASA Jet Propulsion Lab [Lumia et al.95].

RCS-3 was used for control of an autonomous vehicle vision-guided lane following at speeds of up to 80 kilometers per hour on freeways [Schneiderman et al.94] and vehicle following in stop-and-go driving in local traffic [Schneiderman and Nashman94b]. RCS-3 was also used by the U.S. Bureau of Mines as a control system architecture for automated mining systems. A comprehensive mining scenario was developed starting with a map of the region to be mined, the machines to be controlled, and the mining procedures to be applied. Based on this scenario, an RCS control system with simulation and animation was designed, built, and demonstrated. This same RCS control system was then used to control the operation of an actual mining machine instrumented with sensors [Huang et al.91,92b]. As a result, RCS-3 was adopted by the U.S. Bureau of Mines as a reference model architecture for automated coal mining systems [Horst94, Horst and Barbera94].

RCS-3 was implemented as a candidate control system for the next-generation nuclear submarine on a project for DARPA. The RCS-3 controller was used to drive a maneuvering and engineering support systems simulator for a 637 class nuclear submarine. The maneuvering system required an automatic steering, trim, speed, and depth control system. The RCS system demonstrated the ability to execute a lengthy and complex simulated mission involving transit of the Bering Straits under ice. Ice-avoidance sonar signals were integrated into a local map using a CMAC neural network memory model. Steering and depth control algorithms were developed that enabled the sub to make trim and ballast adjustments to avoid hitting either the bottom or the ice while detecting and compensating for random salinity changes under the ice. The RCS engineering support system demonstrated the ability to respond to a lubrication oil fire by reconfiguring ventilation systems, rising in depth to snorkel level, and engaging the diesel engines for emergency propulsion [Huang et al.92a,93].

A RCS control system was developed for a U.S. Postal Service Automated Stamp Distribution Center. This system demonstrated the ability to route packages through a series of carousels, conveyors, and storage bins, to maintain precise inventory control, provide security, and generate maintenance diagnostics in the case of system failure. The distribution center was designed and tested first in simulation and then implemented as a full-scale working system. The system contained over 220 actuators, 300 sensors, and 10 operator workstations [Advanced Technology and Research Corporation93].

The RCS-3 reference model is currently being used as the basis for an open-architecture enhanced machine controller (EMC) for machine tools, robots, and coordinate measuring machines [Proctor and Albus97]. The EMC combines NASREM with the Specification for an Open System Architecture Standard (SOSAS) developed under the Next Generation Controller program sponsored by the Air Force and National Center for Manufacturing Sciences. EMC functional modules and application programming interfaces (APIs) have been defined, and an EMC controller has been installed and has been operational for three years on a four axis machining center in a General Motors production prototype shop in Pontiac, Michigan. The ultimate goal of this effort is to define a set of standard application programming interfaces for open-architecture industrial machine controllers [Proctor and Michaloski93, Stouffer et al.93, Michaloski00, Proctor et al.96].

RCS-4 was designed for the NBS/DARPA Multiple Autonomous Undersea Vehicle (MAUV) project [Albus88, Herman and Albus88, Herman et al.91]. The basic building block is shown in Figure 5.2*d*. The principal new feature in RCS-4 is the explicit representation of value judgment (VJ). VJ processes provide the RCS-4 control system with the kind of cost functions that are generated by the limbic system in the biological brain. VJ processes compute the value of objects, materials, territory, situations, events, and outcomes. VJ processes compute cost, benefit, and risk of planned actions. VJ processes compute value state variables that define the importance of goals and the value of objects or territory. This enables the behavior-generating processes to decide what objects or regions should be attacked, defended, assisted, attended to, remembered, or otherwise acted upon, and how much cost or risk should be accepted in pursuing these behaviors. Value judgment evaluation functions are an essential part of any form of planning or learning.

In RCS-4, evaluation functions are made explicit and specific VJ processes are defined for computing them. The incorporation of value judgment into RCS-4 was heavily influenced by the concepts developed by Pugh[77] in *The Biological Origin of Human Values*. The structure and function of VJ processes in RCS-4 are elaborated further in Albus[91,93] and Lucas et al.[77].

The most recent versions of RCS are 4-D/RCS and ISAM. 4-D/RCS integrates RCS with the four-dimensional approach to dynamic machine vision developed by Dickmanns et al.[94] at Universitat der Bundeswehr Munchen for the VaMoRs vehicle. 4-D/RCS is being developed as a reference model architecture for the design, engineering, integration, and test of experimental unmanned ground vehicles for the Department of Defense Demo III program [Albus99b]. ISAM (Intelligent System Architecture for Manufacturing) is being developed at NIST to meet the future needs of U.S. industry for measurements and standards for intelligent manufacturing systems [Scott96, Rippey and Falco97, Senehi and Kramer98].

THE RCS REFERENCE MODEL ARCHITECTURE

Architecture: the structure of components, their relationships, and principles of design, including the assignment of functions to subsystems and the specification of the interfaces between subsystems

Reference model architecture: an architecture in which the entire collection of functions, entities, events, relationships, and information flow involved in interactions between and within subsystems are defined and modeled

A RCS reference model architecture for an intelligent system has the following properties:

1. It defines the functional elements, interfaces, and information flow within and between intelligent systems.
2. It specifies the informational units and data models for both static (long-term) and dynamic (short-term) representations of knowledge necessary to describe the environment and the intelligent systems operating within it.
3. It specifies processes by which goals are selected, plans are generated, tasks are decomposed, subtasks are scheduled, and feedback is incorporated into control so that both deliberative and reactive behaviors can be combined and coordinated in a single integrated system.
4. It specifies processes by which signals from sensors are transformed into knowledge of situations and relationships.
5. It specifies symbolic and iconic representations for knowledge of objects, events, relationships, and situations in space and time, including semantic, pragmatic, and causal relationships in data structures that can support reasoning, decision making, and control.
6. It specifies how knowledge can be acquired (learned), stored (remembered), and retrieved (recalled).
7. It specifies how values can be represented and used to compute cost, benefit, risk, and uncertainty for evaluating plans for the future and assessing results of past behavioral choices.
8. It specifies the timing of processes and temporal relationships between functional elements.

Functional elements: the fundamental computational processes from which the system is composed

The functional elements of a RCS reference model architecture are sensory processing, world modeling, value judgment, and behavior generation.

Sensory processing: a set of processes by which sensory data interacts with prior knowledge to detect and recognize useful information about the world

In the RCS reference architecture, sensory processing accepts signals from sensors that measure properties of the external world or conditions internal to the system itself. Sensory processing scales, windows, and filters data, computes observed attributes, and compares them with predictions from internal models. Correlation and variance between sensed observations and internally generated expectations are used to update internal models so that the internal models can track and predict the behavior of objects and events in the external world. Correlation and variance are also used to detect events and recognize entities and situations. Sensory processing computes attributes of entities and events such as distance, shape, orientation, color, texture, and dynamical motion. Sensory processing focuses attention and clusters (or groups) recognized entities and detected events into higher-order entities and events.

In most cases, sensors cannot directly measure the state of the world. Sensors typically can only measure phenomena that depend on the state of the world. Thus sensory processing must infer the state of the world from input signals and a priori knowledge. Sensor signals are typically corrupted by noise. Sensor signals are also often confounded by control actions that cause the sensors to move through the world. The set of functions that describe how sensory signals depend on the state of the world, the control action, and sensor noise is called a *measurement model*. A measurement model is typically of the form

$$\mathbf{y} = \mathbf{G}(\mathbf{x}, \mathbf{u}, \eta)$$

where

$\mathbf{y}$ = signals from sensors

$\mathbf{G}$ = function that relates sensor output to world state, control action, and noise

$\mathbf{x}$ = state of the world

$\mathbf{u}$ = control action

η = sensor noise

A linearized form of the measurement model is typically of the form

$$\mathbf{y} = \mathbf{Cx} + \mathbf{Du} + \eta$$

where $\mathbf{C}$ is a matrix that defines how sensor signals depend on the world state and $\mathbf{D}$ is a matrix that defines how sensor signals depend on the control action. Sensor signals are frequently ambiguous (i.e., there may be many states of the world that might produce the same sensory signals). This means that the sensory processing system must be supported by a world modeling system that makes hypotheses about the state of the world and generates predictions that can be compared with observations by sensory processing.

World modeling: a functional process that constructs, maintains, and uses a world model knowledge database (KD) in support of behavior generation and sensory processing

World modeling performs four principal functions:

1. It predicts (possibly with several hypotheses) sensory observations based on the estimated state of the world. Predicted signals can be used by sensory processing to configure filters, masks, windows, and schema for correlation, model matching, recursive estimation, and focusing attention.

2. It generates and maintains a best estimate of the state of the world that can be used for controlling current actions and planning future behavior. This best estimate resides in a knowledge database describing the state and attributes of objects, events, classes, agents, situations, and relationships. This knowledge database has both iconic and symbolic structures and both short- and long-term components.

3. It acts as a database server in response to queries for information stored in the knowledge database.

4. It simulates results of possible future plans based on the estimated state of the world and planned actions. Simulated results are evaluated by the value judgment system to select the "best" plan for execution.

World model: an internal representation of the world

The world model is the intelligent system's best estimate of the world. The world model may include models of portions of the environment, as well as models of objects and agents. It also includes a system model that represents the internal state of the intelligent system itself. The world model is stored in a dynamic distributed knowledge database that is maintained by world modeling processes. Knowledge stored in the world model is distributed among computational nodes in the RCS reference architecture. The world model in each node contains knowledge of the world with range and resolution that is appropriate for control functions in the behavior generation process in that node. The RCS concept of a world model is closely related to the control theory concept of a system model.

System model: a set of differential equations (for a continuous system) or difference equations (for a discrete system) that predict how a system will respond to a given input

A system model is typically of the form

$$\frac{d\mathbf{x}}{dt} = \mathbf{F}(\mathbf{x}, \mathbf{u}, \zeta)$$

where

$d\mathbf{x}/dt$ = rate of change in the system state

$\mathbf{F}$ = function that defines how the system state changes over time in response to internal dynamics and control actions

$\mathbf{x}$ = state of the system

$\mathbf{u}$ = control action

ζ = error in the system model

A linearized form of the system model above is of the form

$$\frac{d\mathbf{x}}{dt} = \mathbf{Ax} + \mathbf{Bu} + \zeta$$

where $\mathbf{A}$ is a matrix that defines how the system state evolves over time without control action and $\mathbf{B}$ is a matrix that defines how the control action effects the system state.

Learning may enable the world model to acquire a system model. This type of learning is often called *system identification*. Learning of world model parameters may be implemented by neural nets, adaptive filtering, or system identification techniques.

Knowledge database: the data structures and the static and dynamic information that collectively form the world model

The knowledge database is a store of information about the world in the form of structural and dynamic models, state variables, attributes and values, entities and events, rules and equations, task knowledge, images, and maps. The knowledge database has three parts:

1. An internal representation of immediate experience consisting of sensor signals and current values of observed, estimated, and predicted images, entities, events, attributes, and relationships.
2. A short-term memory containing iconic and symbolic representations of entities, events, and relationships that are the subject of current attention. The list of items that are the subject of current attention is defined top-down by task commands, requirements, and expectations. The attention list is defined bottom-up by items that are surprising or recognized as dangerous.
3. A long-term memory containing symbolic representations of all the generic and specific objects, events, and rules that are known to the intelligent system. In some systems, long-term memory may also contain information necessary to generate iconic representations.

Value judgment: a process that: (a) computes cost, risk, and benefit of actions and plans, (b) estimates the importance and value of objects, events, and situations, (c) assesses the reliability of information, and (d) calculates the rewarding or punishing effects of perceived states and events

Knowing the expected cost, risk, and benefit of plans and actions is crucial to making good behavioral choices. Value judgment evaluates perceived and planned actions, events, objects, relationships, and situations, thereby enabling behavior generation to select goals and set priorities. Tentative plans evaluated as more beneficial, less risky, or less costly will be selected for execution over those evaluated as less beneficial, more risky, or more costly. Objects evaluated as attractive or valuable will be pursued or defended. Objects evaluated as repulsive or feared will be avoided or attacked.

Value judgment computes what is important. Entities and events evaluated as important become the focus of attention. This enables cameras and other sensors to be directed toward objects and places in the world judged to be important. Evaluation of what is important is crucial for deciding what entities, events, and situations to store in memory. What is evaluated as important can be remembered by transfer from short-term to long-term memory. What is unimportant can be safely ignored and forgotten.

Value judgment assesses the reliability of information and judges whether current sensory experience or world model information is more believable and a more reliable guide for behavior. Information judged reliable will be trusted, and confidence assigned to predictions based on that information. Confidence and believability factors can be attached to various interpretations of entities and events. Value judgment also computes what is rewarding and punishing. This can be used directly for feedback control of action as well as for learning. An intelligent system will learn to repeat actions leading to events and situations evaluated as good or rewarding, and will learn to avoid those actions resulting in punishment.

Behavior generation: the planning and control of actions intended to achieve or maintain behavioral goals

Behavioral goal: a desired result that a behavior is intended to achieve or maintain

Desired result: a result that value judgment evaluates as desirable or beneficial

Intent: a high-level behavioral action and goal conveyed to behavior generation by a command

Command: a name, a commanded action, and a command goal. Both commanded action and command goal may include parameters

The command name is an identifier. The commanded action may include parameters that specify how, where, when, how much, how fast, and on what. The command goal is a desired state that may include parameters that specify tolerance in space and time on various state variables. Goal parameters may also specify the value of the goal.

Behavior generation accepts commands, and using information from the world model formulates and/or selects plans, and controls action. Behavior generation develops or selects plans by using a priori task knowledge and value judgment functions combined with real-time information provided by sensory processing and world modeling to find the best assignment of tools and resources to agents and to find the best schedule of actions (i.e., the most efficient plan to get from an anticipated starting state to a goal state). Behavior generation controls action by planning, feedforward action, and feedback error compensation. Behavior generation includes the control theory concept of a control law.

Control law: a set of equations that compute control action given current state, desired state, and feedforward action

A control law is typically of the form

$$\mathbf{u} = \mathbf{E}(\mathbf{uff}, \mathbf{xd}, \mathbf{x}^*)$$

where

$\mathbf{u}$ = control action

$\mathbf{E}$ = function that defines how the output depends on goals, plans, and feedback

$\mathbf{uff}$ = feedforward control action (from a plan)

$\mathbf{xd}$ = desired world state (from a plan)

$\mathbf{x}^*$ = predicted world state (from the world model)

A linearized form of a control law is

$$\mathbf{u} = \mathbf{uff} + \mathbf{G}(\mathbf{xd} - \mathbf{x}^*)$$

where G is a matrix that defines the feedback compensation applied to the difference between the desired and predicted state of the world.

The Elementary RCS Module

The elementary RCS_MODULE consists of an augmented finite-state automata, or state machine, that computes mathematical, logical, iconic, or symbolic operations on input vectors to produce output. Each input vector consists of an input command from a supervisor and/or an operator, feedback from the world model, and status from one or more subordinates. Each output vector consists of an output command to one or more subordinates and status to the supervisor and operator display, as illustrated in Figure 5.5. The input/output relationships for a RCS_MODULE can be described as a many-to-one vector mapping from an input space to an output space. This is the type of functional mapping performed by a CMAC or a finite-state machine.

The elementary RCS_MODULE executes cyclically. During each execution cycle, the module reads from its input buffers, performs a preprocess, a deci-

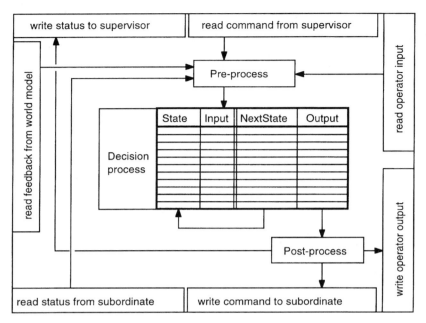

FIGURE 5.5 Elementary RCS_MODULE. Each module has a set of input and output buffers. Input consists of a command from a supervisor and/or an operator, feedback from the world model, and status from one or more subordinates. Output consists of commands to one or more subordinates and status to a supervisor and operator display. The decision process illustrated here is a state table.

sion process, a postprocess, and writes to its output buffers. The pre-process translates the input vector into a format that can be accepted by the decision process. The decision process decides what action should be performed given the current state and input conditions. The decision process functions by executing the appropriate line in a state table. Each line in the state table is essentially an IF/THEN rule. The decision process also invokes a procedure to compute the appropriate action parameters. The postprocess translates the decision process output into a format that can be written to the output buffers. Both pre- and postprocesses may also perform various housekeeping functions, diagnostics, and performance measures. The read–compute–write cycle runs to completion each time it is triggered. A typical RCS_MODULE takes only a few microseconds to execute a cycle. At low levels in the RCS hierarchy, the time interval between one execution trigger and the next is typically periodic, with the period depending on the sample rate required by the process being controlled. For example, to control a typical vehicle, a read–compute–write cycle rate of 100 to 500 per second is adequate. To mimic the performance of a neuronal reflex arc, an RCS_MODULE should sample input and compute output at a rate of 500 to 1000 cycles per second. Servo control for a high-precision machine tool may require rates up to 10,000 cycles per second. Electronic control of radar beams may require even faster rates. However, for most machines and vehicles, 200 hertz (Hz) is fast enough. At higher levels in the RCS control hierarchy, the RCS_MODULE may cycle less frequently. At higher levels, the execution trigger may be event driven.

The elementary RCS_MODULE can be arranged in a hierarchical architecture in which commands and feedback into each module are used to decompose input commands into strings of commands to the next lower level. This was the initial RCS-1 implementation. However, the RCS_MODULE can also be used as a generic functional module with a relationship between input and output that implements a wide variety of processes other than hierarchical command decomposition. Many different mathematical or logical functions can be called from within the preprocessing, decision-processing, or postprocessing stages to implement planning, world model updating, simulation, prediction, windowing or grouping of sensory data, computation of attributes, filtering, recognition, and computation of value judgment costs and benefits. RCS_MODULEs can be arranged in a client–server relationship such that queries to world model modules are serviced and replies are generated on demand; or they can be used to sample and process sensory input, compare observations with world model predictions, compute states and attributes, detect events, recognize objects and situations, and establish and maintain correspondence between the system's world model and the real world. The elementary RCS_MODULE is a standardized functional module that can be interconnected and nested to perform any kind of functional operation on an input vector to produce an output vector. The RCS_MODULE thus provides a generic tool for modeling a brain or designing an intelligent machine system.

Tools for creating RCS_MODULEs are described in a number of papers [Huang et al.2000, Horst2000, Shackelford99] and in a textbook with lab experiments [Gazi et al.2001]. Commercial software tools are available that can be used to generate RCS_MODULEs, and free software libraries and tools can be downloaded from *http://www.isd.mel.nist.gov/projects/rcs_lib*.

NML Communications

Communication between modules in recent applications of the RCS architecture is implemented by a real-time communication process called the Neutral Message Language (NML) [Shackleford99]. NML consists of a set of communications channels and a vocabulary for composing messages that can be transmitted over those channels. NML includes a process that moves information from output buffers of one RCS_MODULE to the appropriate input buffers of one or more other RCS_MODULEs. At the end of each computational cycle, a RCS_MODULE calls NML to move the information stored in its output buffers into an external NML buffer, or mailbox. Once this is done, any RCS_MODULE can call NML to move the information from the NML mailbox into its input buffers. This is illustrated in Figure 5.6. Each path-

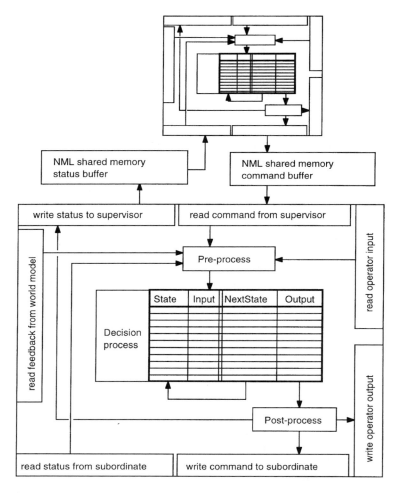

FIGURE 5.6 Pair of NML communications channels between two RCS_MODULEs. In this diagram, the NML mailboxes are implemented as shared memory buffers.

way from a RCS_MODULE output buffer to a RCS_MODULE input buffer is called an NML channel.

The mailboxes defined by NML contain information. The entire set of mailboxes contain all of the information that is currently flowing between RCS_MODULES. The NML mailboxes are a subset of the knowledge database. RCS can thus be described either by a functional block diagram where the functional modules are drawn as boxes with arrows indicating communication pathways, or by a data flow diagram where data flowing between functions is drawn as circles with arrows indicating the functional transformations between input and output data sets. The NML communications channels are defined by a NML configuration file. A configuration file consists of a set of lines containing ASCII characters. There is a line in the configuration file for every NML buffer, a line for every process that writes to that buffer, and a line for every process that reads from that buffer.

Buffer lines specify the buffer name, the buffer type, the host computer in which the buffer resides, the size of the buffer, and whether the data stored in the buffer are being communicated between processors with incompatible data formats and therefore needs to be converted into a neutral format. Buffer lines also specify some additional information that we need not discuss here. A detailed discussion of how to write NML messages and configuration files is contained at *http://www.isd.mel.nist.gov/projects/rcs_lib* along with tools for building NML channels and NML messages. Examples of how to use NML can also be found in *The RCS Handbook* [Gazi et al.2001].

NML supports a variety of communications protocols, including common memory, point-to-point messaging, queuing or overwrite message delivery, and blocking or nonblocking read mechanisms. A typical NML message consists of a unique identifier, the size, and the message body. The message body can contain a data structure such as a C struct or a C++ class. NML provides a mechanism for handling many different kinds of messages. It can support communicating between local modules that reside on the same computer or remote modules that reside on different computers. NML can be configured to provide communication services between modules on a single computer or multiple computers, on singleboard or multiple boards. NML is very portable and has been implemented on a wide variety of operating systems including SunOS, UNIX, Linux, VxWorks, and Windows NT. It can handle messages to be transmitted between different types of computers across a backplane, via a local area network, or across the Internet.

The RCS Computational Node

In the RCS reference architecture, the functional elements: behavior generation, world modeling, sensory processing, and value judgment plus the knowledge database are organized into a RCS_NODE.

RCS_NODE: a part of a RCS system that processes sensory information, computes values, maintains a world model, generates predictions, formulates plans, and executes tasks

A typical RCS_NODE is shown in Figure 5.7. A RCS_NODE is essentially equivalent to Koestler's concept of a *holon* [Koestler67]. Each RCS_NODE looks upward to a higher-level node from which it takes commands, for which it processes sensory information, and to which it reports status. Each RCS_NODE also looks downward to one or more lower-level nodes to which it issues commands and from which it accepts sensory information and status. Each RCS_NODE may also communicate with peer nodes with which it exchanges information.

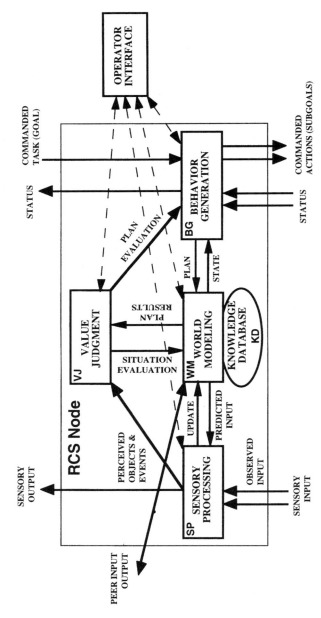

FIGURE 5.7 Internal structure of a RCS_NODE. The functional elements within a RCS_NODE are behavior generation, sensory processing, world modeling, and value judgment. These are supported by a knowledge database and a communication system that interconnects the functional processes and the knowledge database. Each functional element in the node may have an operator interface. The connections to the operator interface enable a human operator to input commands, to override or modify system behavior, to perform various types of teleoperation, to switch control modes (e.g., automatic, teleoperation, single step, pause), and to observe the values of state variables, images, maps, and entity attributes. The operator interface can also be used for programming, debugging, and maintenance.

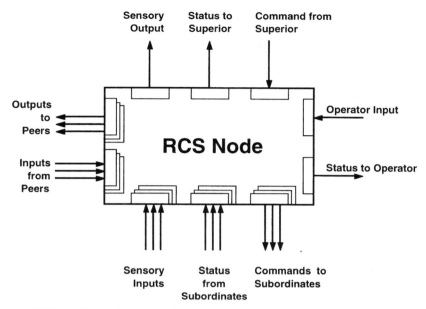

FIGURE 5.8 RCS computational node showing input and output buffers.

Each RCS_NODE contains behavior generating (BG), world modeling (WM), sensory processing (SP), and value judgment (VJ) processes, plus a knowledge database (KD). Any or all of the processes within a node may communicate with an operator interface. Within a RCS_NODE, interconnections between SP, WM, and BG close a reactive feedback control loop between sensory measurements and commanded actions. The interconnections between BG, WM, and VJ enable deliberative planning and reasoning about the costs and benefits of future actions. The interconnections between SP, WM, and VJ enable knowledge acquisition, situation evaluation, and learning. Predicted input generated by the WM from information in the KD is compared in SP with observed input from sensors or lower-level nodes. The difference between observations and predictions can be used to update the WM and implement a recursive estimation process such as Kalman filtering. Information in the world model KD of each node can be exchanged with peer nodes for purposes of synchronization and information sharing.

A RCS_NODE is similar to a RCS_MODULE in that both have a set of input buffers, a set of output buffers, and a set of input and output message formats. This is illustrated in Figure 5.8. Message formats include command frames, status frames, sensory inputs and outputs, peer-to-peer communications, and an operator interface. At any point in time, the set of input buffers define an input vector and the set of output buffers define an output vector.

A RCS_NODE is different from a RCS_MODULE in that a RCS_NODE may be constructed from one or more RCS_MODULEs. The RCS_MODULES within the RCS_NODE are independent and asynchronous except for coordination through flags or semaphores passed in NML messages. NML moves information between asynchronous RCS_MODULEs by providing mailboxes into which messages are posted at times convenient to writers and read at times convenient to readers.

A RCS_MODULE always runs on a single CPU and will often share that CPU with several other RCS_MODULES. In contrast, a RCS_NODE may be distributed over several CPUs. The number of RCS_MODULEs required to implement a RCS_NODE varies depending on the application. In simple cases, it may be possible to implement an entire RCS_NODE within a single RCS_MODULE. For example, simple sensory processing and world modeling functions might be embedded in the preprocess phase of a RCS_MODULE. Behavior generation and value judgment might both be embedded in the decision process single RCS_MODULE. In more complex applications, a number of RCS_MODULEs will be required to implement the functionality of each RCS_NODE.

The processes of BG, SP, WM, and VJ do not necessarily map one-for-one onto RCS_MODULEs. For example, the elements of BG, WM, and VJ that interact to produce planning are tightly coupled and may all be embedded in a single RCS_MODULE rather than distributed over several RCS_MODULEs. Similarly, the elements of SP, WM, and VJ that produce recursive estimation and filtering are tightly coupled and may all be embedded in a single RCS_MODULE. One reason for this choice of mapping of software onto RCS_MODULEs is that communications delays are typically much less between processes within a single RCS_MODULE than between two or more RCS_MODULEs. This is not because NML introduces large communication delays but that communications between two RCS_MODULEs require that NML wait for one RCS_MODULE to finish its compute cycle to produce output and for the next RCS_MODULE to begin its compute cycle to read input. Internal communication within a RCS_MODULE can move data at any time, not just at the beginning and end of a RCS_MODULE execution cycle.

A collection of RCS computational nodes such as illustrated in Figures 5.7 and 5.8 can be used to construct a distributed hierarchical reference model architecture such as that shown in Figure 5.9. The particular architecture in Figure 5.9 is for a machine tool in a manufacturing environment. A similar architecture could be developed for an autonomous land vehicle, a construction machine on a construction site, an undersea vehicle carrying out cooperative operations with other undersea vehicles, and many other applications.

Each RCS_NODE in the architecture in Figure 5.9 acts as an operational unit in an intelligent system. Depending on where a particular RCS_NODE resides in the hierarchy, it might serve as a controller for one or more actuators, a subsystem, an individual machine, a group of machines comprising a man-

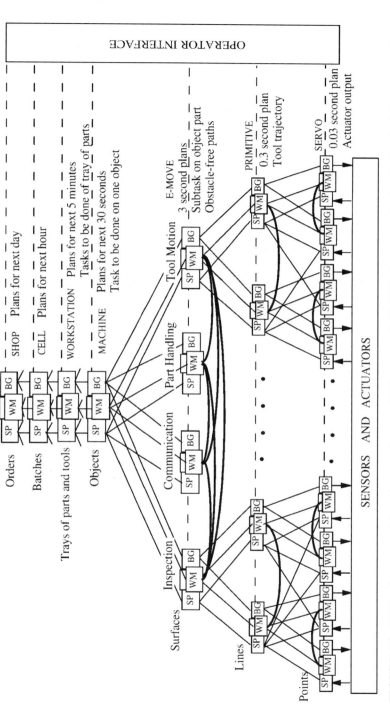

FIGURE 5.9 RCS reference model architecture for a machining center. RCS_NODEs are organized such that the BG processes form a command tree. Information in the world model is shared between WM modules in nodes within the same subtree. KD structures are embedded in WM in each node in this figure. On the right are examples of the functional characteristics of the BG processes at each level. On the left are examples of the type of entities recognized by the SP processes and stored by the WM in the KD knowledge database at each level. Sensory data paths flowing up the hierarchy typically form a graph, not a tree. VJ processes are hidden behind WM processes. An operator interface provides input to, and output from, processes in every node. Note: The entity class hierarchy on the left can be orthogonal to the BG process hierarchy on the right. This is enabled by vertical connections between WM modules. See, for example, Figure 9.4.

ufacturing workstation, a group of workstations comprising a manufacturing cell, or a group of cells comprising a manufacturing shop. The functionality of each RCS_NODE (or RCS_MODULE within a node) can be implemented by a set of software processes or by a person or group of persons.

On the right of Figure 5.9, the typical planning horizon for planners at that level is shown. Note that the planning horizon at each level is approximately an order of magnitude longer than the planning horizon at the level below. This means that the tasks at each level can be decomposed into about 10 subtasks for the next lower level. On the left are entities typical of those formed by grouping hypotheses at each level.

MANAGING THE COMPLEXITY OF AN INTELLIGENT SYSTEM

Intelligent systems are inherently complex. Maintaining a rich representation of the world in the knowledge database and using this representation to plan and execute sophisticated behavior with a high probability of success can require enormous amounts of computing power and memory capacity. Visual images may consist of tens or hundreds of thousands of pixels, each of which may have numerous attributes that change with time, and hence must be recomputed many times per second. For many tasks, such as driving autonomous vehicles, image processing rates of more than 10 times per second are required. Correlating incoming sensory data with predictions based on stored information is a computationally intensive process. An intelligent control system may need to store and retrieve information about hundreds of thousands of entities and events in long-term memory. Sophisticated intelligent control systems can require gigabytes of memory and many gigops of processing power.

Software that can produce reliable intelligent behavior in a wide variety of situations may contain millions of lines of code. Thousands of processes may be required to run simultaneously. Many of these require real-time performance with reaction times measured in milliseconds. The difficulty of writing and debugging large programs with tightly coupled highly interactive code is legendary. If intelligent systems are ever to become an economically practical reality, a method must be found for managing the problem of system complexity.

One method for dealing with system complexity is hierarchical layering with tessellation of detail within each layer. This method for organizing complex systems has been used throughout time by nature and by manufactured systems for effectiveness and efficiency of command and control [Ashby52, Koestler67, Albus79, Meystel82,96, Beer95]. In hierarchical systems, higher-level management units have broader scope and longer time horizons with less concern for detail. Lower-level units have narrower scope and shorter time horizons with more focus on detail. At no level does a unit have to cope with both a broad scope and a high level of detail. At every level in RCS, computational nodes have a limited range and resolution of responsibility and control.

This enables the design of systems of arbitrary complexity without computational overload in any node at any level. Hierarchical layering and tessellation of detail is used in the RCS reference model architecture as a methodology for managing computational complexity in intelligent systems [Maximov and Meystel92, Meystel94, Albus and Meystel95,96, Albus et al.97].

In RCS, behavior-generating processes at the upper levels choose strategic goals and formulate long-range plans. Priorities and values defined at the upper levels influence the selection of goals and the prioritization of tasks throughout the lower levels of the hierarchy. At intermediate levels, tasks and goals with priorities are received from the level above, and subtasks with subgoals and attention parameters are output to the level below. Throughout the entire RCS hierarchy, behavior-generating processes successively refine longer-range plans into shorter-term tasks with greater detail. At each level, higher-level tasks and goals are decomposed into strings of lower-level, more narrow, and more detailed subtasks and subgoals for subordinates.

At each level, behavior-generating processes make plans of roughly the same number of steps. At higher levels, the space of planning options is larger and world modeling simulations are more complex, but more time is available between replanning intervals for the planner process to search for an acceptable plan. At lower levels, planning is more constrained and plans are often selected from libraries of well-learned and practiced procedures. The effect of each hierarchical level is thus to refine the tessellation of details of tasks geometrically and limit their scope so as to keep computational loads within manageable limits. Thus the amount of computing resources needed for behavior generation in any RCS_NODE remains within bounds. At all levels of the BG hierarchy, details of execution are left to subordinates. At the bottom, drive signals to actuators produce action on the world.

At the bottom of the sensory processing hierarchy, sensors measure phenomena in the world. Output from sensors provide input to sensory processing functions that focus attention, filter, and group (or cluster) signals and attributes into features, entities, and events. Sensory processing functions at lower levels process data over local neighborhoods and short time intervals. At all levels, sensory information from multiple sources are integrated over time and space so as to increase the scope geometrically and encapsulate the detail of entities and events observed in the world. At higher levels, data are integrated over long time intervals and large spatial regions. Thus the amount of computing resources needed for sensory processing in a RCS_NODE at any level remains within reasonable limits.

At all levels, task commands are decomposed into subtasks and subgoals. Sensory input is filtered, integrated over space and time, and compared with expectations. At all levels, plans are generated or selected and executed. Entities and events are detected and information is evaluated. At each level, state variables, entities, events, and maps are maintained to the resolution in space and time that is appropriate to that level. Throughout the hierarchy, as resolution is increased, range is decreased, and vice versa, for both spatial and temporal dimensions. This produces a ratio of range to resolution that remains relatively constant at all levels in the hierarchy.

Within nodes at every level, feedback loops are closed to provide reactive behavior. At lower levels, feedback is of high bandwidth and servo loops provide fast response to sensed conditions. At higher levels, feedback is integrated over longer time periods to detect long-term trends and determine long-range behavioral decisions. At each level in the RCS hierarchy, sensory processing functions detect errors between what is expected and what is observed. Small errors are used to update estimated attributes of entities and events in the world model. Large errors indicate the need to hypothesize new entities or objects to account for discrepancies between predicted and observed sensory inputs.

Also at each level, behavior-generation processes detect errors between desired and estimated states of the world. Control laws are designed so that feedback compensation can correct small errors between desired and estimated states. Large or persistent errors are indicative of problems that are beyond the capacity of feedback compensation within a single RCS node. Errors that exceed limits can trigger emergency actions and/or initiate replanning. In RCS, errors are detected at lower levels first where they can be addressed most quickly. The lower-level sensory processing functions are the first to detect states and events that indicate problems or emergency conditions, such as limits being exceeded on position, velocity, acceleration, vibration, pressure, force, current, voltage, or temperature. The lower-level behavior-generation processes are also the first to act to correct or compensate for errors between desired and estimated states.

Only if low-level reactive control laws are incapable of correcting the differences between expectations and observations will errors filter up to higher levels where planners look further into the future and develop alternative tactics and strategies to solve problems and find alternative routes to high-level goals. Thus longer-range or harder-to-solve problems are addressed at higher levels. Information in the knowledge database at lower levels is typically short term and fine grained. Knowledge at the higher levels is broader in scope and more general in nature.

Timing

Time is a critical factor in sensing and control. The most obvious thing about time is that is flows in one direction and there is a unique point called the present that divides the past from the future. For each RCS_NODE, the origin of the time line is the present (where $t = 0$). All actions take place and all sensory observations occur at the instant $t = 0$. Everything to the right of $t = 0$ is in the future. Everything to the left is in the past. If we quantize time with a ticking clock, the present exists during the zeroth tick of the clock that separates the future from the past.

Figure 5.10 is a diagram of the temporal relationships involved in representing the time line. At each level of the RCS reference model, time is represented

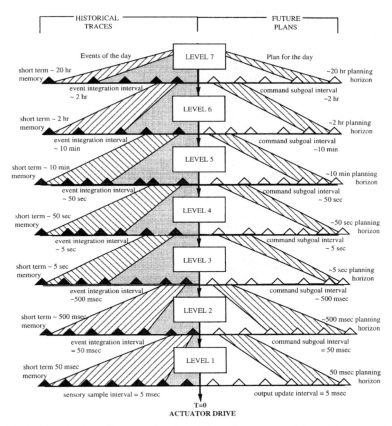

FIGURE 5.10 Timing diagram for the RCS reference model architecture showing the temporal range and resolution for planning and short-term memory at different hierarchical levels.

at a different scale. The triangles on the time line represent significant points in time. At the lowest level, time is represented at the highest resolution. At each successively higher level, the planning horizon expands by an order of magnitude and the resolution along the time horizon decreases by an order of magnitude. At low levels, time is measured in milliseconds and seconds. At high levels, time is measured in minutes, hours, or days. At all levels, $t = 0$ separates the future from the past. At all levels, future plans scroll left to be executed at $t = 0$, and then become past events that recede into the past as a historical trace. At low levels, time appears to flow rapidly. At higher levels, time appears to move more slowly.

At each level, input tasks are decomposed into plans that contain sequences of subtasks that are output as commands to the next lower level. To the right of $t = 0$, the triangles along the time line represent goals and subgoals. Lines between triangles represent task activity that is required to achieve subgoals. Inputs to each level are commands that specify the current and next task in the higher-level plan. Each level has a planner that generates a set of planned subtasks and subgoals out to its own planning horizon. The plan to the current goal is shown as a light gray area. The plan beyond the current goal to the planning horizon is shown as a hatched area. At lower levels, plans are highly detailed but extend only a brief period into the future. At higher levels, plans are less detailed but extend further into the future. The number of subtasks in a plan is approximately the same at all levels.

Each level has a short-term memory that maintains a historical trace of about 10 events that are significant to that level. At each level, events held in short-term memory are grouped (clustered or integrated) onto patterns that are detected and labeled as higher-level events. To the left of $t = 0$, the triangles mark the beginning and ending of detected events. The lines between triangles represent the historical intervals during which events occur. Events may include the successful accomplishment of planned goals, or the failure to achieve goals. Events may represent unexpected happenings, surprises, or emergencies. At the lowest level, events represent single samples of sensor outputs. At higher levels, events represent patterns or clusters of lower-level events (or subevents). The detection of an event may signal the occurrence of a state or situation in the world. In Figure 5.9, the clustering of a set of subevents into a higher-level event is shown as a gray area. The set of subevents remaining in short-term memory that belong to an event detected previously are shown as a hatched area. Low-level events have high-frequency components but extend only a brief moment into the past. At higher levels, events have lower-frequency components but cover a longer historical period. The number of events stored in short-term memory is approximately the same at all levels.

Note that Figure 5.10 shows symmetry between future plans and past events. The RCS reference model maintains a historical trace in short-term memory with duration approximately equal to the planning horizon at each level. Thus the number of events in short-term memory is approximately the same as the number of subtasks in a plan for each level. Figure 5.10 indicates specific timing for planning horizons and short-term memory. These timing intervals are examples only. For any particular application, timing will depend on the dynamics of the system being controlled. Typically, the numerical values of timing parameters are determined by the dynamics of the physical system being controlled. To prevent aliasing, events must be sampled at a rate that is at least twice the highest frequency in the input signal. This is the theoretical limit set by the Nyquist criterion. In practice, the sampling rate is typically set at 10 times the highest frequency in the input. Sensor signals are typically filtered before sampling to assure that this criterion is met. Similarly, outputs to actuators must be updated fast enough to control the highest-frequency modes of the system being controlled.

It is important that systems with nested control loops maintain separation between the loop bandwidth of the inner and outer loops. Otherwise, the inner and outer control loops may interact in very undesirable ways to produce instabilities. In Figure 5.10, the bandwidth of each successively higher-level loop is defined to be approximately an order of magnitude lower than the level that it controls. For example, at the output from level 1, feedback loops are closed and commands are sent to actuators every 5 milliseconds (ms). At the output from level 2 (input to level 1), commands are sent to groups of actuators every 50 ms. At the output from level 3 (input to level 2), new commands are sent to primitive-level subsystems approximately every 0.5 second (s). At the output from level 4 (input to level 3), new commands are sent to elemental move-level subsystems on average every 5 s. At the output from level 5 (input to level 4), new commands are sent to machines on average every 50 s. At the output from level 6 (input to level 5), new commands are sent to groups of machines in workstations about every 10 minutes. At output from level 7 (input to level 6), new commands are sent to cells of workstations about every 2 hours. At the input to level 7, new commands to the shop are received about every 24 hours. These specific times are for illustration only. For different environments, timing requirements may vary.

This hierarchical structure does not necessarily prevent higher levels from responding quickly to new information in the world model. At each level, any executor can respond within one compute cycle trigger interval. However, temporal integration of events in the sensory processing hierarchy limits the bandwidth of the control loop that is closed through each level and thus prevent higher levels from fluctuating rapidly in response to high-frequency components in sensory feedback.

The particular grouping of actuators into subsystems and systems is application dependent. For example in manufacturing, level 1 might control actuators on machine tools, robots, and material handling systems. Level 2 might coordinate actuators to define desired tool motion or robot gripper position. Level 3 might develop obstacle free paths and coordinated motions for arms, hands, and tools. Level 4 might coordinate activities between subsystems of a machine tool or robot. Level 5 might plan and control coordinated activities between a robot, machine tool, and material delivery system within a manufacturing workstation. Level 6 might plan and control coordinated activities of a group of workstations in a group technology cell. Level 7 might plan and control activities of cells in a shop.

In the case of unmanned ground vehicles, level 1 might control actuators for throttle, steering, braking, gear shift, and camera pan/tilt head. Level 2 might coordinate driving subsystems and camera-pointing subsystems. Level 3 might coordinate driving and camera for obstacle avoidance. Level 4 might coordinate attention, locomotion, and communications activities for an entire vehicle. Level 5 might coordinate the activities of two or more vehicles in coordinated section tasks. Level 6 might coordinate two or more sections in scout platoon tactics. Level 7 might coordinate scout platoon tactics with tank company tactics in a mechanized armor battalion.

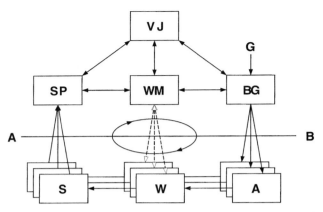

FIGURE 5.11 Elementary loop of functioning consisting of two domains, controller and controlled system. The line AB is a divider between these domains. It denotes the fact that the modules above the line belong to the controller system while the modules below the line belong to the controlled system. [From Albus and Meystel96.]

Elementary Loop of Functioning

A control loop through a typical RCS_NODE is shown in Figure 5.11. This is what Meystel calls an *elementary loop of functioning* [Meystel and Albus2001]. The bandwidth of this control loop is determined by the sampling rate of the sensors, the temporal filter properties of the SP and WM processes, and the computation update frequency of the execution module within the BG process. A task command specifying a goal G arrives at a BG element of a generic control node. The function of the BG process is to transform this task command into a set of signals for a set of actuators {A}. Each actuator operates on part of the world {W}, producing effects that are monitored by a set of sensors {S}. The signals from the sensors are processed and integrated within sensory processing (SP), which provides the information necessary for world modeling (WM) to update the representation of the world in the knowledge database. The dashed arrows between the WM and the world {W} indicate a virtual correspondence between the real world and its representation in the world model. The knowledge maintained by WM is the system's best estimate of the state of the real world W. The virtual correspondence is maintained via a noisy channel through the sensors {S} by filtering and recursive estimation processes in the SP and WM processes. To the extent that the correspondence between W and its representation in WM is an accurate (or at least adequate) reflection of reality, the behavior generated by BG is likely to be successful in achieving the goal G. To the extent that the correspondence between W and WM is incorrect, behavior is likely to be unsuccessful.

   In Figure 5.12, a second control level is shown. The upper, or outer loop, has lower resolution in both time and space. The outer loop has a lower bandwidth,

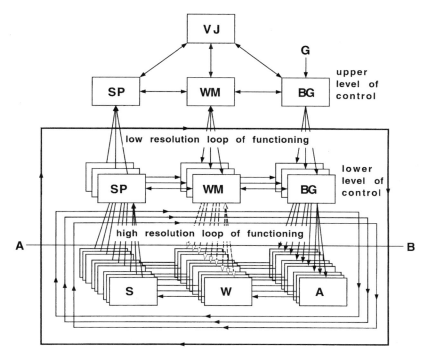

FIGURE 5.12 Elementary loop with two control levels. VJ processes are not shown in the lower level of control in this diagram. [From Albus and Meystel96.]

a longer planning horizon, a larger span of control over subordinate agents, and typically deals with a larger range of space and time. The inner loop has a higher loop bandwidth, a shorter planning horizon, a smaller span of control, and deals with a smaller spatial range with higher resolution. Figure 5.12 is essentially a redrawing of one of the primitive-level nodes in the architecture shown in Figure 5.9. Figures 5.11 and 5.12 emphasize the closed-loop nature of the RCS architecture. Control loops are typically closed through every node in the architecture of Figure 5.9.

RCS is a complex and multifaceted architecture. Figure 5.13 illustrates RCS from three different perspectives. On the left is the organizational hierarchy. This illustrates how each node in the RCS reference model is an organizational unit that receives commands from a superior unit and sends commands to one or more subordinate units. A second perspective, shown in the middle of Figure 5.13, illustrates the computational hierarchy, or chain of command, supported by the world modeling and sensory processing functions. This perspective emphasizes how a control loop is closed through each node. It illustrates how each BG module is supported by local WM and VJ processes plus a local KD that are maintained locally and serviced by SP functions.

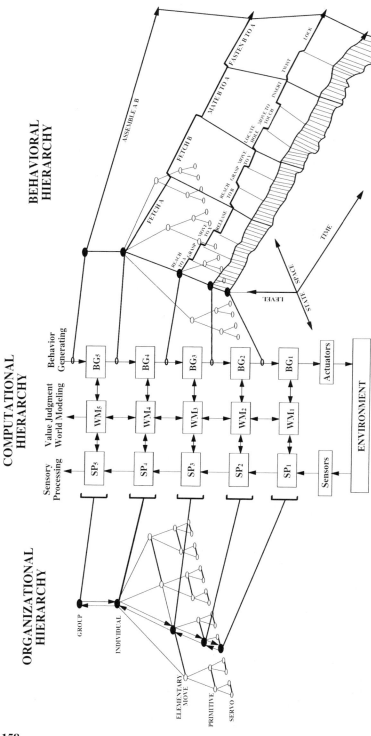

FIGURE 5.13 Three aspects of the RCS reference model. On the left is an organizational hierarchy showing the command and control relationships between RCS_NODEs. In the center is a computational hierarchy showing the horizontal flow of information between SP, WM, and BG processes within the RCS_NODES. On the right is a behavioral hierarchy showing simultaneous trajectories of BG task activity along the time line. [From Albus81.]

158

Tight coupling between BG and WM supports planning. Tight coupling between WM and SP supports recursive filtering, recognition, and world model maintenance functions. A third aspect is the evolution of behavior along the time line shown on the right of Figure 5.13. Commands and state variables in messages flowing between BG, WM, SP, and VJ modules within and between units can be plotted versus time to reveal the past history and future plans of each unit. This view emphasizes the temporal nature of the planning, control, sensory processing, and world modeling processes in the RCS architecture.

IN DEFENSE OF HIERARCHIES

Hierarchical architectures are controversial and unpopular with many researchers in the intelligent systems community. Many argue that hierarchical systems are rigid, inefficient, autocratic, and stifling to creativity. We believe that such arguments are not supported by the facts. To begin with, hierarchical organizations are not intrinsically rigid. Hierarchies can be reconfigured dynamically during the interval between any two computation cycles. There can be frequent realignment of relationships between supervisors and subordinates. Because of its deterministic set of parent–child relationships, a RCS hierarchy can reorganize itself as quickly, if not more quickly, than any other type of organization. All that is required is a procedure for reconfiguration of relationships between subordinates to superiors in response to changing situations.

Second, hierarchical systems are not intrinsically inefficient. Otherwise, they would long since have disappeared from military and corporate organizations. It is a historical fact that winning armies are invariably organized hierarchically, as are economically successful large corporations. Flat organizations may be attractive for a variety of reasons, but they are notoriously inefficient and unwieldy when they grow large. Of course, hierarchical systems can also be inefficient, particularly if they concentrate all planning in a central node or prohibit exchange of information between organizational units at the same level, thus requiring all communication between peers to flow through the supervisor. However, this is an arbitrary and unnecessary restriction that typically occurs only in academic straw-man examples. In real-world hierarchies (as in RCS), information flows easily between peer units at the same level. Typically, the horizontal flow of information within and between nodes at the same level is much larger than the flow of commands and status reports up and down the chain of command. The horizontal flow of world modeling and planning information may even exceed the vertical flow of sensory information up the SP hierarchy. Thus, as long as planning and decision making are appropriately distributed and peer-to-peer communication is facilitated, high levels of efficiency are commonly achieved. The prohibition of information flow between peers is characteristic of a tree, not a hierarchy. A hierarchy is technically a structure in which there is a partial ordering of nodes in a chain of command, but there also exists the possibility of information exchange between peers, or even between nodes at different levels outside of the supervisor–subordinate relationship. Confusion between a hierarchy and a tree is often the source of the fallacy that hierarchies are inefficient.

Some arguments about efficiency of hierarchical systems are based on issues related to central planning versus local decision making. This is a red herring. Hierarchical control is not the same as centralized planning and does not imply a central planner—quite to the contrary. Hierarchical planning requires that decisions be distributed among organizational units at many different levels throughout the entire system. In RCS, local planners allocate resources and schedule activities for immediate subordinates at every level. Each organizational unit has its own planning process that refines high-level commands into lower-level plans. Planning and decision making are distributed throughout the hierarchy as plans become more detailed and specific at each successively lower level. In RCS, hierarchical planning is neither centralized nor monolithic. Within each local planner, a variety of planning processes can be implemented, including bidding for jobs by autonomous agents. Other styles of planning can also be implemented based on cost/benefit analyses that take into consideration contention for resources between peers that cannot be addressed by completely independent agent bidding.

Hierarchical control does imply a well-defined chain of command, wherein high-level goals and priorities are established at the top. Hierarchical control also implies that there exists a command and control structure designed to coordinate the activities of the entire organization in pursuit of high-level goals. This produces more, not less, efficiency.

Arguments against hierarchical control based on the relative merits of democracy versus autocracy are ideological and political in nature and largely irrelevant to issues of effectiveness and efficiency in control systems. It is true that organizational units within a control hierarchy are not completely autonomous or free to pursue their own best interests. It is true that hierarchical control may place unwelcome or unfair demands and disproportionate requirements for effort and sacrifice on some agents within the hierarchy and not on others. But efficiency and effectiveness are typically specified in terms of the success or failure of the organization as a whole in achieving its goals, not in terms of short-term benefits to individual components. A creature in which each leg or arm or muscle group could decide independently of the whole what is best for itself alone could not function effectively as an organism in the real world. Success of the organization as a whole requires that the entire organization function as a unit, with individual subsystems working together for the good of the whole.

Finally, we come to the argument that hierarchical organization stifles creativity. This is highly questionable. As far as we are aware, there are no scientific data to suggest that creativity is more likely in flat organizations than in hierarchical structures. However, there is a strong theoretical argument that creativity is more (not less) likely to occur in a hierarchical architecture. Creativity can be defined as the ability to create new behaviors or procedures that never before existed or never before were applied to a given situation. Creativity therefore requires the ability to search the space of future possibilities for a series of actions that have never before been employed. Searching possible futures for an advantageous series of actions is, of course, a definition of planning. This implies that a system organization that is more efficient in planning is likely to be more creative. If we recall that one rational for applying hierarchical control to intelligent systems was to improve the efficiency of the planning process, it follows that creativity should be higher in a hierarchical system than in a flat organization. We therefore argue that a hierarchical organization does not stifle creativity, but rather, enhances it.

We turn next to a more detailed discussion of how to build intelligent control systems using the RCS reference model architecture, beginning with behavior generation.

6 Behavior Generation

Horde of warriors, cave painting from Teruel Grotto of the Val del Charco del Agua, Amarga, Spain. [After Cabre; by permission from Museum of Man, Paris.]

We now begin elaboration of the RCS reference model architecture, starting with the internal structure of the RCS_NODE. In Chapter 5 the relationships were illustrated between the functional elements of behavior generation, world modeling, sensory processing, value judgment, and the knowledge database within the RCS_NODE (see Figure 5.7). In this chapter we examine the internal structure of behavior generation (BG) at a level of detail that can enable software engineering methods to be applied.

BG is a set of processes that provides planning and control functions for the RCS_NODE. A BG process corresponds to an operational unit in an organizational hierarchy such as a bureaucracy or chain of command. Within each BG unit there is a planner and a set of executors. A BG unit accepts tasks in the form of commands from a superior BG unit. The BG planner selects or generates plans. The BG executors issue commands to subordinates to perform behavior designed to accomplish those plans.

Task: an activity and goal assigned to a BG process by an executor in a higher-level BG process

   A task is evoked by an input command to a BG process. A task defines the work to be done by a BG process. A task consists of a task activity and a task goal.
   Figure 6.1 illustrates a BG unit that accepts task command input from an executor in a higher-level BG unit. Within the BG unit there is a task decomposition planner that decomposes each commanded task into set of a tentative plans for subordinate BG units. These tentative plans are submitted to a WM simulator/predictor which generates expected results. Value judgment computes the cost and benefit of each tentative plan and its expected results and reports this evaluation back to the task decomposition planner. The planner then selects the best set of tentative plans to give to executors for execution. For each subordinate BG unit, there is an executor that compares each step in its plan with feedback from the knowledge database. Each executor then outputs subtask commands to a subordinate BG unit and monitors its behavior. The KD is kept up to date by processed sensory information. The SP, WM, and VJ processes that supply the KD with the information needed by the BG unit for planning and control are also shown in Figure 6.1. The BG unit in Figure 6.1 is essentially equivalent to the AI feedback planning system described by Passino and Antsaklis[89].
   The task decomposition planner in Figure 6.1 can be decomposed further into a job assignor, a set of schedulers, and a plan selector, as shown in Figure 6.2. Thus a BG process contains four subprocesses:

   1. Job assignor (JA)
   2. Set of schedulers (SC)
   3. Plan selector (PS)
   4. Set of executors (EX)

The JA, SC, PS, and EX subprocesses are arranged within BG operational units as shown in Figure 6.2.

Job assignor (JA): a subprocess of BG that decomposes input tasks into job assignments for subordinate BG units

   The job assignor (JA) performs four functions:

   1. JA accepts input task commands from an executor in a higher-level BG unit.

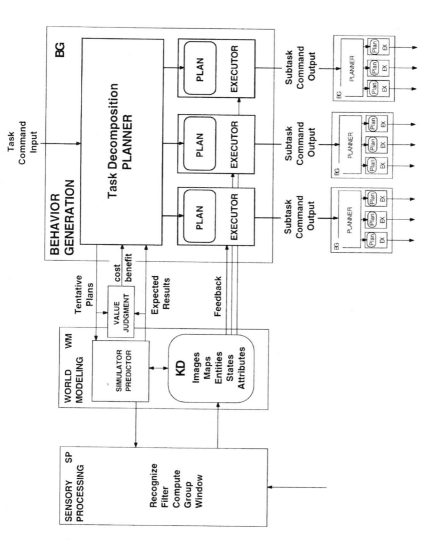

FIGURE 6.1 Interfaces between BG, WM, SP, and VJ. The task decomposition planner within the behavior generation (BG) process selects or generates tentative plans. These are sent to the WM simulator/predictor to generate expected results. The value judgment process computes cost and benefits of tentative plans and expected results. The planner selects the tentative plan with the best cost/benefit ratio and places it in plan buffers within the executors. Each executor cycles through its plan, compares feedback with planned results, and issues subtask command output to lower-level BG modules.

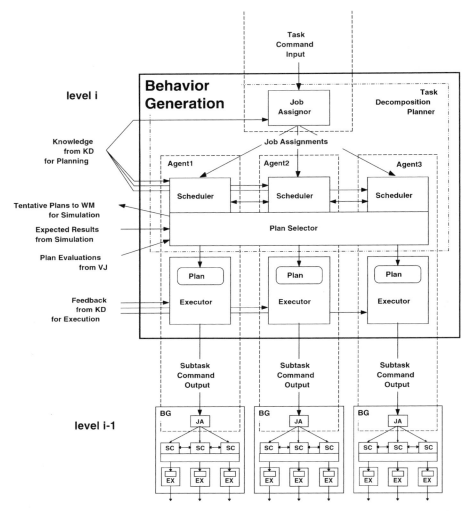

FIGURE 6.2 Internal structure of BG units. The task decomposition planner contains a job assignor (JA) and a scheduler (SC) for each of the subordinate BG units. The JA and SC processes generate tentative plans that are submitted to WM for simulation and to VJ for evaluation. The plan selector (PS) selects the best of the tentative plans to be sent to the appropriate executor. The three agents are peers at level i and supervisors at level $i - 1$. The JA subprocess at level i is part of a supervisor agent from level $i + 1$.

2. JA decomposes each commanded input task into a set of jobs which will be assigned to schedulers for subordinate BG units. This is a spatial decomposition.

3. JA transforms each job assignment into a coordinate frame of reference which is appropriate for the subordinate BG unit.

4. JA allocates resources to the subordinate BG units to enable them to accomplish their assigned jobs. (At lower levels, the assignment of resources to BG units may be fixed by the system design. At higher levels, resources such as tools, materials, and subordinate BG units may be reassigned to a BG unit depending on the situation.)

Job: an activity and goal assigned by the Job Assignor to a Scheduler within a BG process

A job is evoked by a command from a job assignor to a scheduler subprocess. A job defines the work to be done by an agent. A job consists of a job activity and a job goal. The difference between a job and a task is that a task is an assignment to be performed by a BG unit at the ith level, whereas a job is an assignment to be performed by a single agent within a BG unit at the ith level. A task is an input to a JA subprocess. A job is an output from a JA subprocess.

Scheduler (SC): a subprocess within BG that accepts a job assignment and computes a schedule for its subordinate BG unit

There is a SC subprocess for each subordinate BG unit. Each SC accepts its job assignment from the JA and computes a sequence of planned actions and resulting states, from a job starting state to a job goal state. This is a temporal decomposition. The SC subprocesses within the BG unit typically communicate with each other, and may negotiate with each other to resolve conflicts and coordinate their respective job plans. SC processes may define plan synchronization flags that EX processes can use to synchronize activities between subordinate BG units.

Plan selector (PS): a subprocess within BG that works with a WM plan simulator and VJ plan evaluator to select the best overall plan (i.e., job assignment, resource allocation, and coordinated schedule) for its subordinate BG units

The plan selector subprocess submits tentative plans generated by the JA and SC subprocesses to the world model for simulation. The world model simulator generates a prediction of what would probably result if the tentative plan were executed. The results of simulation are then sent to value judgment processes for evaluation. The value judgment evaluates the cost, benefit, and risk of the plan and its predicted result, and returns its evaluation to the plan selector. The plan selector then selects the best of the tentative plans to be placed in plan buffers for execution. This is a decision process. Many tentative plans may be evaluated during the selection of a "best" plan.

The JA and SC subprocesses within a BG unit comprise a management team. The JA subprocess acts as the supervisor of the BG unit. The SC subprocesses act as planners for the peer agents within the BG organizational unit. This management team can implement different management styles by different distribution of duties and responsibilities between the JA and SC subprocesses. For example, an autocratic or micromanagement style can be achieved if the JA subprocess assumes most or all of the responsibility for both job assignment and scheduling, leaving very little discretion to the SC subprocesses. On the other hand, a collegial management style can be achieved if the SC subprocesses negotiate with the JA for job assignments and assume complete responsibility for scheduling their respective jobs. In either extreme, or some combination of the two styles, the JA and SC subprocesses work together to create tentative plans that allocate resources and responsibilities to subordinate BG units to accomplish the commanded task.

It is the responsibility of the plan selector to assure that the plan buffer always contains an acceptable plan. This guarantees that execution is never starved for input. Whenever a new tentative plan is evaluated as better than the current plan in the buffer, the plan selector may replace the plan in the buffer with the better plan. This can assure that the plan buffer always contains the best plan found to date. The plan buffer decouples planning from execution. Plan execution can respond immediately to sensory feedback, whereas planning proceeds asynchronously at its own pace. As long as the planner generates new plans at least as fast as the executors carry them out, execution will flow smoothly. If the planner cannot keep up with real-time events, the system may have to pause periodically, or revert to purely reactive behavior, until a new plan is completed.

For applications where the world environment is dynamic, the planner can be computing a new plan continuously while the executor is executing the current plan. As soon as a newer and better plan is ready, the PS can replace the old plan in the plan buffer, and the executor can continue without interruption. In extremely dynamic and unpredictable environments, the planner may need to generate a new plan by the time the current plan is only 10 percent completed.

Executor (EX): a subprocess within BG that executes its portion of the selected plan by generating subtasks for its subordinate BG process

The EX subprocess coordinates actions between itself and other EX subprocesses. It handles errors between planned results and the evolution of the world state reported by the world model. There is an EX subprocess for each subordinate BG unit. Each EX subprocess closes a feedback control loop within its own BG unit. This is a reactive process. Each EX subprocess:

1. Computes feedforward output commands based on planned actions
2. Computes errors between the current planned subgoal (i.e., desired state) and the observed (or predicted) state of the world

3. Computes error compensation and modifies output command parameters so as to correct errors

4. Outputs commands consisting of feedforward and feedback compensation to its subordinate BG unit

5. Detects when subgoals are complete and increments its subtask sequencer to the next action and subgoal in the plan buffer

6. Reacts to emergency conditions

For both continuous and discrete systems, EX subprocesses produce strings of output subtask commands that become input task commands to JA subprocesses in BG units at the next-lower level.

ORGANIZATIONAL UNITS VERSUS AGENTS

Note the distinction in Figure 6.2 between the BG organizational unit and the agents that belong to the units.

Agent: an entity that plans and executes jobs (possibly in coordination with other agents)

Each agent contains three subprocesses (Figure 6.2):

1. A scheduler (SC) at level i
2. An executor (EX) at level i
3. A job assignor (JA) at level $i - 1$

The RCS convention defines agents as separate from the SP, WM, VJ, and KD processes that support them within a node. In most of the artificial intelligence literature, the definition of an agent includes the SP, WM, VJ, and KD functions within the agent itself. The RCS convention is preferred here because the algorithms and software development tools required for building SP, WM, and VJ processes and KD data structures are significantly different from those required for BG processes.

It is important to maintain a clear distinction between organizational units and the agents that belong to them. Otherwise, system design and discussions between system designers about the respective roles and responsibilities of agents and organizational units can become very confused and confusing. The RCS BG process is defined from the perspective of organizational units. Examples include military units such as division, battalion, company, platoon, and squad. Agent architectures popular in the current literature are typically organized from the perspective of individual agents [Maes90,94, Lieberman et al.99]. Examples include personnel with rank such as major, captain, lieutenant, sergeant, and private.

The RCS architecture embeds a hierarchy of agents within a hierarchy of organizational units. For example, in Figure 6.2 it can be seen that each organizational unit consists of agents with two different ranks: (1) a commander agent of higher rank, and (2) subordinate agents of lower rank. Conversely, each intelligent agent belongs to two organizational units at two different levels: (1) a lower-level unit in which the agent is a commander; and (2) a higher-level unit in which the agent is a peer with other agents of similar rank that are subordinate to a commander of higher rank. RCS makes this distinction clear by defining tasks as assignments to operational units, and jobs as assignments to agents belonging to those units.

In Figure 6.2, a BG unit at level (i) consists of a JA subprocess that performs the role of unit supervisor and one or more SC and EX subprocesses that fill the role of subordinate staff in the unit. Within the dashed lines, the SC, EX, and JA subprocesses are arranged as agents with SCs and EXs acting as peer subordinates in the BG unit at level (i) and the corresponding JAs acting as supervisors of BG units at level ($i - 1$). Within each BG process, the JA subprocess of a supervisor agent works with the SC subprocess of peer agents to develop a plan. The EX subprocess of each agent works with the JA subprocess at the next lower level to command the subordinate BG unit. The EX thus acts as a server to the lower-level BG unit. It interfaces between upper- and lower-level planners and provides synchronization by establishing start times of subtasks. Thus each BG unit can plan and coordinate the actions of subordinate BG units. The set of agents within each BG unit can work as a team to accomplish tasks assigned to the BG unit of which they are a part. Coupling between job assignment and scheduling functions produce teamwork between supervisor and subordinate agents in planning and execution of BG tasks.

Each agent within a BG operational unit may compete or cooperate with other agents within the same BG unit. Within a BG unit, a SC subprocess may negotiate with the JA subprocess of the supervisor agent and compete against peer agent SC subprocesses for job assignments and resource allocations. The SC subprocesses of an agent may also work together with other SC subprocesses to plan cooperative behavior. SC subprocesses may work with their supervisor agent to formulate tentative job assignments and schedules for plans. The executor subprocess of each agent then executes its part of the selected plan by outputting task commands to the JA subprocess in a BG unit at the next lower level.

Whether RCS is viewed as a hierarchy of organizational units, or a hierarchy of agents, it must be understood that this hierarchy is a nested multiresolutional structure in which communication flows both vertically between supervisors and subordinates and horizontally between peers. Mathematically, RCS is not a tree. It is a layered multiresolutional lattice that is richly interconnected by horizontal communications within each layer.

The structure shown in Figure 6.2 makes clear a number of different ways that dividing lines between levels in the RCS hierarchy could be established.

1. The boundary between BG units at different levels establishes a dividing line between organizational units. Information flows across this boundary in the form of task commands.
2. The boundary between JA and SC subprocesses established a dividing line between agents. Information flows across this boundary in the form of job assignments.
3. The boundary between the PS and EX subprocesses establishes a dividing line between tasks and subtasks. Information flows across this boundary in the form of plans.

The overall result is that tasks are input to a BG unit at one rate, and subtasks are output to lower-level BG units at about an order-of-magnitude-higher rate. Similarly, jobs are assigned to agents in a BG unit at one rate, and to agents in BG units at the next lower level at an order-of-magnitude-higher rate.

TASK COMMANDS

Input to each BG process from its supervisor consists of a task command.

Task command: a command to a BG process to perform a task

A task command can be represented as an imperative sentence or set of sentences in a message language that directs a BG process to do a task, and specifies the task goal, the task parameters, and the object(s) upon which the task is to be performed. A task command has a specific interface defined by a task command frame.

Task command frame: a data structure containing the information necessary for commanding a BG process to do a task

A task command frame may include:

1. *Task name* (from the vocabulary of tasks the receiving BG process can perform)
2. *Task identifier* (unique for each commanded task)
3. *Task goal* (a desired state to be achieved or maintained by the task)
4. *Task goal time* (time at which the goal should be achieved or until which it should be maintained)
5. *Task object(s)* (on which the task is to be performed)
6. *Task parameters* (such as speed, force, priority, constraints, tolerance on goal position, tolerance on goal time, tolerance on path deviations, coordination requirements, and level of aggressiveness)
7. *Next task name, goal, goal time, objects, and parameters*

Item 7 in the task command frame enables the BG planning process always to plan to its planning horizon even when the planning horizon extends beyond the current task goal. A task command frame may be represented by a C struct or a C++ class.

Task goal: a desired result or state of the world that a BG process is commanded to achieve or maintain

A task may be one of two types:

1. An activity performed by an operational unit (i.e., a BG process) to achieve a goal state
2. An activity performed by a BG process to maintain a goal state

For type 1 tasks, achieving the task goal terminates the task. For example, a type 1 task may be ⟨to drive to a location⟩, ⟨to look for a target⟩, ⟨to assemble a product⟩, or ⟨to shop for a list of items⟩. The task is done when the goal is achieved. For type 2 tasks, the objective is to maintain the task goal state until a different task is commanded. For example, a type 2 task may be ⟨to maintain a temperature⟩, ⟨to maintain a velocity⟩, or ⟨to maintain an altitude⟩. The task is done when a new task is commanded.

A task implies action. In natural language, action is denoted by verbs. In the definition of task above, the infinitive form of the verb is used as a task name (e.g., task name = ⟨to drive to a location⟩). The application of the verb ⟨do⟩ to the task name transforms the task verb from its infinitive to its active form. It puts the task verb into the form of a command. For example, application of the verb ⟨do⟩ to the task ⟨to look for a target⟩ creates a command ⟨look for target⟩, where ⟨look for⟩ is the action verb, and ⟨target⟩ is the object to which the action is to be applied. For every task there can be assigned an action verb that is a command to perform the task. The set of tasks that a BG process is capable of performing therefore defines a vocabulary of action verbs or commands that the BG process can accept.

A task such as ⟨drive to a location⟩ may be performed by a number of agents (e.g., drivers) on many objects (e.g., cars, trucks, or tanks). A task such as assembling an engine may be performed by a number of agents (e.g., assembly line workers) on a number of engine parts (e.g., pistons, bearings, gaskets, head, oil pan). At the input of the BG process, a task command such as ⟨assemble engine⟩ is encoded as a single command to do a single activity to achieve a single goal. The JA subprocess at the input of the BG process performs the job assignment function of decomposing the task into a set of jobs for a set of agents (e.g., a team of assembly line workers). For each agent, a SC subprocess schedules a sequence of subtasks, which are executed by the EX subprocess within each agent so as to accomplish the commanded task.

TASK KNOWLEDGE

Task knowledge is what enables a BG process to perform tasks. Task knowledge includes skills and abilities possessed by the BG unit and its agents. For example, task knowledge may include knowing how to machine a part, build an assembly, or inspect a feature. It may include driving skills for steering around an obstacle, following a road, or navigating from one map location to another. Task knowledge may include the ability to bake a cake, do the laundry, change a diaper, or play the piano. It may include knowledge of how to move without being observed from one location to another. It may include knowledge of how to cross a creek or gully, to encode a message, or to load, aim, and fire a weapon.

Task knowledge may also include a list of equipment, materials, and information required to perform a task. It may include a set of conditions required to begin or continue a task, a set of constraints on the operations that must be performed, a set of priorities, and a mode of operation. Task knowledge may include error correction procedures, control laws, and rules that describe how to respond to failure conditions.

For any task to be accomplished successfully, knowledge of how to do the task must exist. Task knowledge must either be available a priori, or the system must have a procedure for discovering it. A priori task knowledge may be provided in the form of schemata or algorithms designed by a programmer. It may be discovered by heuristic search over the space of possible behaviors. It may be acquired through learning or inherited through instinct. Task knowledge is invoked by task commands that provide parameters such as priority, mode, speed, force, or timing constraints. Task knowledge can be represented in a task frame.

Task frame: a data structure specifying all the knowledge necessary for accomplishing a task

A task frame is essentially a recipe consisting of a task name, a goal, a set of parameters, a list of materials, tools, and procedures, and a set of instructions on how to accomplish a task. A task frame may include:

1. *Task name* (index into the library of tasks the system can perform). The task name is a pointer or an address in a database where the task frame can be found.
2. *Task identifier* (unique id for each task command). The task identifier provides a means for keeping track of tasks in a queue.
3. *Task goal* (a desired state to be achieved or maintained by the task). The task goal is the desired result of executing the task.
4. *Task goal time* (time at which the task goal should be achieved, or until which the goal state should be maintained).

5. *Task objects* (on which the task is to be performed). Examples of task objects include parts to be machined, features to be inspected, tools to be used, targets to be attacked, objects to be observed, sectors to be reconnoitered, vehicles to be driven, weapons or cameras to be pointed.

6. *Task parameters* (that specify, or modulate, how the task should be performed). Examples of task parameters are speed, force, priority, constraints, tolerance on goal position, tolerance on goal time, tolerance on path, coordination requirements, and level of aggressiveness.

7. *Agents* (that are responsible for executing the task). Agents are the subsystems and actuators that carry out the task.

8. *Task requirements* (tools needed, resources required, conditions that must obtain, information needed). Tools may include instruments, sensors, and actuators. Resources may include fuel and materials. Conditions may include temperature, pressure, weather, visibility, soil conditions, daylight or darkness. Information needed may include the state and type of parts, tools, and equipment, the state of a manufacturing process, or a description of an event or situation in the world.

9. *Task constraints* (upon the performance of the task). Task constraints may include speed limits, force limits, position limits, timing requirements, visibility requirements, tolerance, geographical boundaries, or requirements for cooperation with others.

10. *Task procedures* (plans for accomplishing the task, or procedures for generating plans). Plans may be prepared in advance and stored in a library, or they may be computed on-line in real time. Task procedures may be simple strings of things to do or may specify contingencies for what to do under various kinds of circumstances.

11. *Control laws and error correction procedures* (defining what action should be taken for various combinations of commands and feedback conditions). These typically are developed during system design but may be developed through learning from experience.

Some of the slots in the task frame are filled by information from the command frame. Others are properties of the task itself and what is known about how to perform it. Still others are parameters that are supplied by the WM.

Each BG process has a set of tasks that it can perform. This defines a vocabulary of task names and a library of task frames. The task vocabulary defines the set of task commands that the BG process can accept. When a BG process is given a command to do a task from the task vocabulary, it uses the knowledge in the corresponding task frame to accomplish the task. If a BG process is given a command to do a task for which there is no task frame, for which the resources specified in the task frame are not available, or for which required conditions are not true, it responds to its supervisor with an error message = ⟨can't do⟩. Task frames may reside in the BG process or may be stored in the knowledge database within the same RCS node.

TASK DECOMPOSITION

Task decomposition: a process by which a task given to a BG process at one level (i) is decomposed into a set of sequences of subtasks to be given to a set of subordinate BG processes at the next lower level ($i - 1$)

Task decomposition consists of five generic functions:

1. A *job assignment* function is performed by a JA subprocess whereby (a) a task is divided into jobs to be performed by agents, (b) resources are allocated to agents, and (c) the coordinate frame in which the jobs are described is specified.
2. *Scheduling* functions are performed by SC subprocesses whereby a job for each agent is decomposed into a sequence of subtasks (possibly coordinated with other sequences of subtasks for other agents). Working together, JA and SC functions generate tentative plans.
3. *Plan simulation* functions are performed by a WM process whereby the results of tentative plans are predicted.
4. *Evaluation* functions are performed on the results of tentative plan simulations by a VJ process using cost/benefit analysis.
5. A *plan selection* function is performed by a PS subprocess in selecting the "best" of the alternative plans for execution.

The result of task decomposition is a plan consisting of a set of planned subgoals connected by planned actions. The objective of task decomposition is to maximize the likelihood of success, maximize the benefit, and minimize the cost and risk of achieving the commanded task goal. Planned actions can be used by the EX subprocess to generate feedforward commands. Planned subgoals can be used by the EX to generate desired states. A string of desired states is sometimes called a *reference trajectory*. The desired states can be compared with observed states, and differences can be treated as error signals that are submitted to a control law to compute feedback compensation. Feedforward and feedback compensation can be combined to produce sequences of subtask commands to BG processes at the next lower level in the control hierarchy.

Thus each task command input to a BG process generates a behavior described by a set of sequences of subtask commands output to a set of subordinate BG processes. The JA subprocess within a BG process accepts task commands with goals and priorities. The JA, SC, and PS subprocesses develop or select a plan for each commanded task by using a priori task knowledge, heuristics, and value judgment functions combined with real-time information and simulation capabilities provided by world modeling functions. The planning process generates the best assignment of tools and resources to agents and finds the best schedule of actions (i.e., the most efficient plan to get from an anticipated starting state to a goal state). EX subprocesses control action by stepping through the plan. At each control cycle, the EX subprocesses can both issue feedforward commands and compare current state feedback with the desired state specified by the plan. The EX subprocesses merge feedforward actions with feedback error compensation to generate task commands to the next-lower-level BG processes.

PLANS AND PLANNING

Task decomposition is a planning process that produces a task plan.

Plan: a set of subtasks and subgoals that are designed to accomplish a task or job

A plan defines one or more paths through state space from an anticipated starting state to a goal state [McDermott85, Georgeff84, Stefik81a,b, Shoppers87]. Typically, each path consists of a string (or set of strings) of actions and a string (or set of strings) of resulting states. The actions and resulting states are subtasks. In general, a plan is defined by a state graph (or state table) that might have a number of branching conditions that represent alternative actions that may be triggered by situations or events detected by sensors. For example, a plan might be to go to work if it is raining or go to the beach if it is sunny. A plan may contain many optional actions, depending on sensory feedback. A plan may be to turn right or left, or to stop or go, depending on sensing a switch, or the value of a state variable, or the color of a traffic light. A plan might contain guarded moves or emergency actions that would be triggered by out-of-range conditions.

A plan designed for a set of agents in a BG process is a task plan. A plan designed for a single agent is a job plan. The distinction between a task plan and a job plan is illustrated in Figure 6.3. This shows the decomposition of a task into a task plan consisting of three job plans for three agents. Each job plan is represented as a state graph where subgoals are nodes and subtasks are edges that lead from one subgoal to the next. The task plan illustrated in Figure 6.3 consists of three job plans defined by linear lists of subtasks and subgoals. Coordination between agents can be achieved by making state transitions in the job plan of one agent conditional upon states or transitions in job plans of other agents, or by making state transitions of coordinated agents dependent on states of the world reported in the world model.

Schedule: the timing or sequencing specifications for a plan

A schedule can be represented as a sequence of activities that are indexed by time or triggered by events. A schedule may specify that activities or goal events should occur within certain time constraints [McDermott82]. For example, a train schedule specifies the clock time when a train is due to arrive or

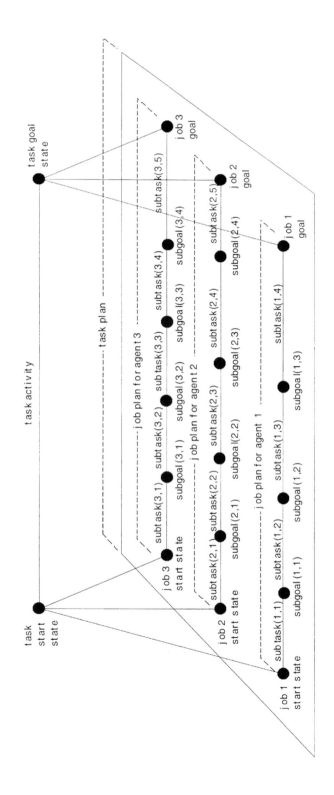

FIGURE 6.3 Task plan for a BG with three agents. Here a single task is decomposed into three parallel job plans for three agents. Each job consists of a string of subtasks and subgoals. The entire set of subtasks and subgoals is a task plan for accomplishing the task.

depart various stations along its route. A student class schedule specifies when classes begin and end. On the other hand, a schedule may simply specify the order in which activities should occur regardless of the time. For example, the sequence of program steps in a robot assembly task typically do not depend on clock time but on events that occur as the assembly task proceeds. If a plan is represented as a state graph or state table, conditions for transition from one subtask activity to the next can be specified in terms of time, or events, or both [Hopcroft and Ullman79].

Planning: a process of generating and/or selecting a plan

A block diagram of a typical task planning process is shown in Figure 6.4. This planning process can be repeated until the library of stored plans is exhausted or until the entire universe of possible plans is searched. Whenever a better plan is discovered, it can be substituted for the plan that already exists in the plan buffer. This assures that the plan buffer always contains the best plan found to date. In a system with only a few binary sensors, the planner may have a library of plans for every combination of command and feedback state. If so, the planner can simply select a plan from the library to meet the condition. Alternatively, the plan library may contain many plans for every possible set of conditions. In this case, each of the potential plans can be submitted by the plan selector to the world model for simulation and to value judgment for evaluation. In some cases, the plan library may contain a schema that must be filled in by JA and SC subprocesses before it becomes a plan. Planning from scratch by heuristic search through the space of all possible future actions and resulting states can be very time consuming and is rarely used in practical applications except for route planning, where the search space is reduced to a two-dimensional map. Real-time planning in configuration space may be successful if the size of the space is limited in dimension, range, and resolution and a planning delay is acceptable.

In general, a plan is a state graph, which is equivalent to a state table, which can be implemented as a procedural computer program. In many cases, particularly for machine tools or industrial robots, plans are embedded in programs that are developed off-line and selected in real time during production by name or by a simple set of rules. VJ algorithms for evaluating plans may take into account priority and mode parameters specified in the task command. For example, if a commanded task is high priority, a plan to achieve the task may be selected even though it evaluates as risky or costly, whereas if a commanded task priority is low, a costly plan to do the task may be rejected in favor of a plan that skips the task. If the commanded mode parameter is set to aggressive, plans that are fast but risky may be selected over plans that are slow but safe. If the commanded mode parameter is cautious, plans that are safe but slow may be preferred.

There are many different planning algorithms [Tate85, Sacerdoti73, Nilsson80, Wilensky83]. The planning procedure illustrated in Figure 6.4 need not

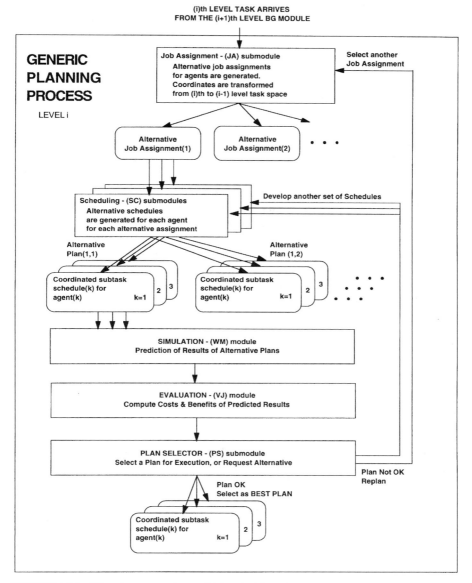

FIGURE 6.4 Planning operations within a RCS node. Boxes with square corners represent computational functions. Boxes with rounded corners represent data structures.

always occur in the order shown. For example, job assignment and scheduling may be reordered or intermingled and iterated to generate alternative assignments of jobs or resources and alternative schedules. Also, simulation and evaluation may be performed on each step in a tentative plan as it is being constructed, or on the entire tentative plan after it is completed.

There are also many different methods that can be used for planning [Rosenschein81, Adams et al.85, Allen and Koomen83]. These include rule-based look-up of precomputed plans, instantiation of scripts or schema, dynamic programming, game theory, heuristic search in state space or configuration space, map-based path planning and navigation, or cost minimization procedures involving differential geometry and dynamical mechanics. Different planning techniques may be used at different levels in the RCS hierarchy. For example, at the servo level, planning may involve simple linear, trapezoidal, or spline interpolation between goal points. At the primitive level, planning may be done by solving systems of differential equations involving force, inertia, and friction. At the subsystem level, plans may be developed by plotting paths on a map containing obstacles or by generating trajectories through configuration space. At higher levels, planning algorithms may involve expert system rules, means–ends analysis, theorem proving, graph search, dynamic programming, evolutionary programming, genetic algorithms, or game theory. Planning may be done automatically, manually, or using computer-interactive tools.

There are two fundamental types of planning: task planning and path planning. Both require finding a good path through state space from a starting state to a goal state. However, path planning typically involves only a two- or three-dimensional search space (although path planning in configuration space for a robot manipulator may involve search in a six-dimensional space.) On the other hand, task planning typically involves an N-dimensional search space, where N is the number of state variables of all the objects and agents involved in the task. For complex tasks, N can be a very large number and the size of the search space becomes completely intractable. Thus, while real-time search strategies for path planning are feasible, task planning typically must be done off-line. Plans for complex tasks such as making a Samurai sword, fabricating a Swiss watch, or building a Gothic cathedral are typically developed by generations of highly skilled craftsmen making incremental improvements in complicated processes over periods of decades or centuries. Task plans for processes such as manufacturing steel or fabricating semiconductors could never be derived from scratch simply by searching the space of all possible actions.

Thus in many practical applications, planning is not done in real time. For example, current practice in manufacturing is to develop process plans (i.e., task plans for making a product) off-line, using computer-aided design, simulation, and planning tools. Process plans typically are formulated well in advance and are optimized long before production begins using computer-interactive modeling tools and experimental test hardware. On the other hand, production schedules typically cannot be formulated in advance because there are too many real-time variables that are unknown until the last minute. Production schedules often become obsolete during production, due to unforeseen problems, and must be modified on-line. In practice, this typically requires human intervention by experienced shop foremen and machinists to perform real-time replanning manually using ad hoc methods.

In the military, planning is also done off-line whenever possible. High-level plans often are prepared months or years in advance by teams of military experts. Strategies and tactics may be developed over decades through experience on the battlefield and experimentation in war games. Tactical mission planning is typically performed as far in advance as possible using maps, charts, and intelligence reports. Plans may be rehearsed by walk-through, by moving models on a sand table or map, or by training exercises. Of course, once the battle begins, plans often go awry and must be modified to accommodate changing events. Human intervention typically is required to compensate for unexpected losses, to take advantage of unexpectedly favorable conditions, or to cope with unanticipated situations. Real-time modifications in plans are typically done by ad hoc methods that depend on the training, experience, and intuition of skilled human beings.

In many practical situations, planning is a distributed process that is carried out by many different planners working in different places at different times. Often, plans are generated off-line by organizational units that are not responsible for executing them. This may work well in predictable environments where plans can be developed and stored until they are used. But in unpredictable environments, where plans must be modified and improvised in real-time, planning and execution must be tightly coupled.

The RCS BG process is designed to support both real-time and off-line planning. Versions I and II of RCS use off-line planning exclusively. Plans are expressed as state tables, stored in a library, and selected in real time by task name [Barbera et al.84]. In these versions of RCS, the planner function consists of nothing more than indexing into a library of state-table plans to find a state-table name that matches the task command name. The EX process then performs task decomposition by sequencing through the state-table plans based on feedback from sensors and status from lower-level EX processes. At each level, a new command from the supervisor selects a new state table from the library. EX then cycles through the state table to generate the appropriate sequence of task commands to subordinates. Version II of RCS is still used for implementing programmable logic control (PLC) and numerical control (NC), which are widely used in manufacturing automation for production lines, machine tools, and industrial robots [Proctor et al.96].

More recent versions of RCS support real-time path planning based on search through a map using cost/benefit analysis [Albus98]. This enables vehicle systems to deal with situations such as planning a cross-country route on a battlefield through a field of obstacles where the attributes of the terrain and the nature of the obstacles are not known until they are observed by sensors. In these applications, off-line planning may still be done, but it is only the beginning. A priori plans must be modified in real time to cope with dynamically changing situations. Real-time replanning must be done quickly so as to minimize the delay between sensing and acting. The question then is: How quickly can a plan be generated?

If the planner (JA, SC, WM, VJ, and PS processes) can generate a new plan by the time the old plan is completely executed, control actuation can be smooth and continuous. The EX subprocesses can step through the plan and control the lower-level BG units while the planner generates a new plan. If the planner is faster and can generate a new plan more frequently, the EX can still execute the beginning of each plan until a new plan replaces it. If the planner is faster still and can generate a new plan before the first step of the old plan is completed, the role of the EX subprocess becomes minimal. Under these conditions, EX simply relays the first step in its plan to the subordinate BG, establishes start times so as to coordinate subsystem behaviors as required and deals with emergencies when necessary.

In all cases, planning within the RCS architecture is distributed at many hierarchical levels with different planning horizons at different levels and many different planning algorithms. At high levels, planning horizons may extend weeks, months, or even years into the future and deal with distances of thousands of kilometers [Albus98]. At lower levels, planning horizons may be only a few minutes, seconds, or milliseconds and deal only with immediate surroundings. At each hierarchical level, planning functions compute plans that extend from the anticipated starting state out to a planning horizon characteristic of that level. On average, planning horizons shrink by about an order of magnitude at each lower level. Thus the number of subtasks between the starting state and the planning horizon at each level remains relatively constant (on the order of 10 steps per plan). This causes the temporal resolution of subtasks to increase about an order of magnitude at each lower level. Thus the number of levels required in a RCS hierarchy is approximately equal to the logarithm (base 10) of the ratio between the planning horizons at the highest and lowest levels.

COMMANDS AND PLANS

In this section we give a simple example of how RCS generates commands and plans at multiple levels. For this example we define a task command frame that consists of:

1. The name of an action or task (plus modifiers)
2. A goal of the action (plus modifiers)
3. The name of the next action (plus modifiers)
4. The goal of the next action (plus modifiers)

Each action (or task) may have a set of modifiers that include modes, constraints, and conditions. Modes specify how the action should be performed (e.g., how aggressively, how persistently, what cost is acceptable.) Constraints specify limits on the action. Constraints may include limits on velocity, acceleration, jerk, or deviation from a path or corridor through state space. Conditions specify what is required to begin or continue the action.

Each action (or task) has a goal that the action is designed to achieve or maintain. The goal is typically a state, and the action is an activity that is designed to cause a transition from one state to the next in the plan. Each goal typically has a set of modifiers that include time, priority, tolerance, and exit conditions. The time specifies how soon the goal should be achieved, or how long the goal should be maintained. Priority defines how important or valuable the goal is. Tolerance may specify how closely the goal must be approached or maintained, or how closely the time specification must be met. Exit conditions specify when the goal can be considered to be accomplished.

The set of named actions (or task names) that a BG process can perform defines a command vocabulary. Each BG process at each level of the control hierarchy has its own unique command vocabulary. We can uniquely identify the elements in a task command to a BG process by citing the level in the hierarchy and the index (or name) of the BG process within the level. For example, let $TC(i,j)$ represent a data structure for a task command to module i at level j with the following variables:

$$ac1_i^j = \text{ActionCommand for BG module } i \text{ at level } j$$

$$gc1_i^j = \text{GoalCommand state for BG module } i \text{ at level } j$$

$$gt1_i^j = \text{GoalTime for when } gc1_i^j \text{ should be achieved}$$

$$ac2_i^j = \text{NextActionCommand for BG module } i \text{ at level } j$$

$$gc2_i^j = \text{NextGoalCommand for BG module } i \text{ at level } j$$

$$gt2_i^j = \text{NextGoalTime for when } gc2_i^j \text{ should be achieved}$$

We can also uniquely identify the components of a plan generated by a BG process by citing the step in the plan, the level in the hierarchy, and the BG process index within the level for which the plan is intended. For example, let $PL(i, j-1)$ represent a data structure for a plan for module i at level $j-1$ with the following variables:

$$ap[k]_i^{j-1} = \text{action planned for step } k \text{ in the plan for}$$
$$\text{BG process } i \text{ at level } j-1$$

$$gp[k]_i^{j-1} = \text{subgoal planned for step } k \text{ in the plan for}$$
$$\text{BG process } i \text{ at level } j-1$$

$$gt[k]_i^{j-1} = \text{subgoal time planned for step } k \text{ in the plan for}$$
$$\text{BG process } i \text{ at level } j-1$$

where $k = 1, \ldots, 10$.

For example, assume that the planner in the ith BG unit at level j produces a plan for two subordinate BG units, the nth and mth BG units at level $j - 1$. Assume further that the plan for each subordinate BG has 10 steps evenly spaced in time. Let st be the time when the plan is due to start and T be the plan horizon. Then $st + 0.1kT$ is the time when subgoals $gp[k]_n^{j-1}$ and $gp[k]_m^{j-1}$ are due to be achieved.

In our simple example, a plan will be represented as a string of planned actions and planned subgoals for each subordinate BG unit. Thus the plan generated by the level i planner for the nth and mth BG units at level $j - 1$ would have the following form:

Plan for the nth BG at Level $j - 1$ Plan for the mth BG at Level $j - 1$

$$ap1_n^{j-1}, gp1_n^{j-1}, gt1_n^{j-1} = st + 0.1T \qquad ap1_m^{j-1}, gp1_m^{j-1}, gt1_m^{j-1} = st + 0.1T$$

$$ap2_n^{j-1}, gp2_n^{j-1}, gt2_n^{j-1} = st + 0.2T \qquad ap2_m^{j-1}, gp2_m^{j-1}, gt2_m^{j-1} = st + 0.2T$$

$$ap3_n^{j-1}, gp3_n^{j-1}, gt3_n^{j-1} = st + 0.3T \qquad ap3_m^{j-1}, gp3_m^{j-1}, gt3_m^{j-1} = st + 0.3T$$

$$ap4_n^{j-1}, gp4_n^{j-1}, gt4_n^{j-1} = st + 0.4T \qquad ap4_m^{j-1}, gp4_m^{j-1}, gt4_m^{j-1} = st + 0.4T$$

$$ap5_n^{j-1}, gp5_n^{j-1}, gt5_n^{j-1} = st + 0.5T \qquad ap5_m^{j-1}, gp5_m^{j-1}, gt5_m^{j-1} = st + 0.5T$$

$$ap6_n^{j-1}, gp6_n^{j-1}, gt6_n^{j-1} = st + 0.6T \qquad ap6_m^{j-1}, gp6_m^{j-1}, gt6_m^{j-1} = st + 0.6T$$

$$ap7_n^{j-1}, gp7_n^{j-1}, gt7_n^{j-1} = st + 0.7T \qquad ap7_m^{j-1}, gp7_m^{j-1}, gt7_m^{j-1} = st + 0.7T$$

$$ap8_n^{j-1}, gp8_n^{j-1}, gt8_n^{j-1} = st + 0.8T \qquad ap8_m^{j-1}, gp8_m^{j-1}, gt8_m^{j-1} = st + 0.8T$$

$$ap9_n^{j-1}, gp9_n^{j-1}, gt9_n^{j-1} = st + 0.9T \qquad ap9_m^{j-1}, gp9_m^{j-1}, gt9_m^{j-1} = st + 0.9T$$

$$ap10_n^{j-1}, gp10_n^{j-1}, gt10_n^{j-1} = st + T \qquad ap10_m^{j-1}, gp10_m^{j-1}, gt10_m^{j-1} = st + T$$

A string of planned actions, planned goals, and planned goal times is a special case of the more general form of a plan as a state graph or state table wherein each node is a planned subgoal state and each edge is a planned action in response to an observed or predicted state. This more general form admits plans that contain branching-on conditions.

The state-graph representation of plans is a canonical form. Any equivalent representation that preserves the concept of arcs and points in state space and which allows decisions based on feedback from the environment to be injected into the plans at any point so that alternative actions can be substituted during any event would be acceptable. Representations such as a series of piecewise third-order splines or clothoidre among the possibilities.

PLAN EXECUTION

Planning looks into the future and anticipates what needs to be done in the interval between a starting time and the planning horizon. Plan execution acts in the present to carry out the current plan and correct for errors [Firby89]. The planner in each BG unit accepts commands from an executor in a higher-level BG unit. The planner converts commands into plans. The executors in each BG unit accept plans from the planner in their own BG unit. Each executor cycles through its plan, converting its plan into commands, and issuing commands to its subordinate BG unit at the appropriate time. Plan execution is a point where deliberative planning merges with reactive behavior. In plan execution, feedforward planned actions are modified by feedback compensation based on an estimate of how closely the plan is being followed. For each agent in the BG process, there is an executor (EX) that issues subtasks and monitors feedback relevant to the plan for that agent. Figure 6.5 shows the internal structure of a typical executor.

The executor shown in Figure 6.5 assumes that the plan buffer contains both planned subgoals and planned actions. This allows planned actions to be used directly as feedforward actions. However, if the executor contains an inverse plant model, the plan needs to contain only planned subgoals. Planned subgoals can be input to the inverse model to obtain the feedforward actions. In Figure 6.5, the plan buffer is loaded periodically by the plan selector subprocess. The subtask sequencer cycles through the plan, issuing planned actions as feedforward actions to the control law and planned states as subgoals to the comparator. The subtask sequencer advances through the plan upon receiving a "done" signal from the compare process, and (if specified) coordination flags from peer EX modules.

The EX compare process receives three inputs:

1. A current subgoal = $gp1$ (from the plan buffer)
2. A predicted state $x0^*$ at time $t = 0$ (from the WM)
3. Predicted status $x1^*$ at the current subgoal time $gt(1)$ (from the planner at the next lower level)

When the compare process detects that the difference between the current subgoal $gp1$ and $x0^*$ is less than a goal tolerance Δg, the compare process reports to the subtask sequencer that the subgoal is done; that is,

$$\text{IF} \quad (|gp1 - x0^*| < \Delta g), \qquad \text{THEN} \quad [\text{report subgoal done}]$$

When the compare process observes that the error between the planned subgoal state and the status predicted from the lower-level planner is less than a goal tolerance, the compare process does nothing because the system is performing according to plan:

$$\text{IF} \quad (|gp1 - x1^*| < \Delta g), \qquad \text{THEN} \quad [\text{do nothing, plan is on-track}]$$

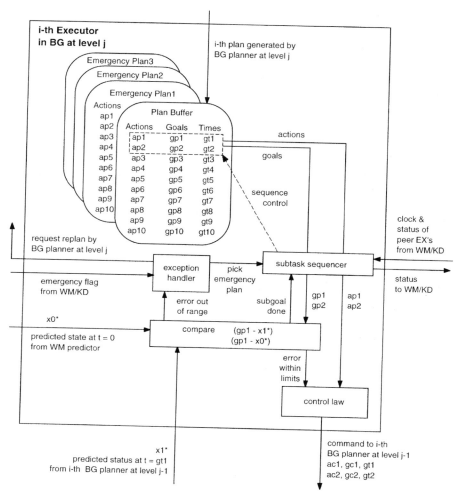

FIGURE 6.5 Inner structure of the Executor (EX) module. The plan buffer in the EX is loaded by the plan selector. Feedback to the executor arrives in the form of status reports from the subordinate BG planner, estimated (or predicted) state feedback from WM, and status from peer EX subprocesses. Output from the EX consists of commands to subordinate BG processes, status to the WM/KD, and requests for replanning to its own planner.

When the compare process observes that the error between the planned sub-goal state and the status predicted from the lower-level planner is greater than a goal tolerance but less than an emergency threshold, the compare process computes a feedback error than can be used by the control law to compensate for the error and servo the system back to the plan:

IF $(\Delta g < |gp1 - x^*1| < Te)$, THEN [compute error and
apply compensation]

where Te is an emergency threshold.

When the compare process observes that the error between the planned subgoal state and the status predicted from the lower-level planner or the WM is greater than an emergency threshold, the compare process reports that the error is out of range and branches to an emergency routine:

IF $(|gp1 - x1^*| > Te)$, THEN [activate the exception
handler and replan]

The exception handler typically will pick an emergency plan to replace the current plan. An example of an emergency plan might be to slow down and stop until the planner generates a new plan. If the planner is in the middle of a planning cycle, the exception handler may request that the planner restart a new plan based on the new information. For continuous systems, the predicted state feedback $x0^*$ and $x1^*$ may be time-dependent state variables that can be used to compute a following error. For discrete event systems, $x0^*$ and $x1^*$ feedback may report predicted status of a sensor, or busy/done from a lower-level BG process.

EMERGENCY ACTIONS

In the RCS architecture, feedback errors are detected and addressed at the lowest hierarchical level first, where the response can be quickest. At every level, EX subprocesses provide the first line of defense against failure. EX processes act immediately and reflexively, as soon as an error condition is detected, to correct it by feedback compensation, if possible, or by a preplanned emergency action, if necessary. As soon as the comparator detects an emergency or out-of-limits signal, or a problem is encountered by the subordinate planner, EX immediately requests the subtask sequencer to branch to an emergency routine and requests the planner at its level to generate a new plan. If this procedure is successful, the problem is solved and there is no need to involve higher-level BG modules.

However, if both error compensation and preplanned remedial actions are unsuccessful, and if replanning within the ith level BG unit is unsuccessful, action at a higher level is required. If an error persists or grows at the ith level, the executor at the next level $(i + 1)$ will detect it and attempt a reactive compensation strategy at that level. If this fails, the $(i + 1)$-level executor requests its planner to select or generate a new plan. If this fails, the problem will be detected at the next-higher level, $i + 2$. Thus errors are corrected at the lowest level possible. At each level, an emergency action generator buys time for its own planner to replan by executing a preplanned emergency procedure (such as stop, or take cover) until a new plan is ready. Only when this fails do higher-level BG processes become involved to change tactics or strategy.

Uncorrected errors ripple up the hierarchy, from executor to planner at each level and from lower-to higher-level BG processes, until either a solution is found or the entire system is unable to cope and a mission failure occurs. At any level, the system may request assistance from a human operator or from another system.

TIMING

Figure 6.6 illustrates the timing for plans and commands at two levels of a RCS hierarchy. Figure 6.6a illustrates how the BG planner at level j converts a command generated by an executor at level $j + 1$ into a level j plan that extends from $t = st$ to the planning horizon at $t = st + T$. st is the start time for the ActionCommand AC1. st is therefore also the start time for the level j plan. The level j plan extends from st to the planning horizon at $st + T$. The tenth subgoal, $gp10$, coincides in time with the GoalCommand GC1, which is at the level j planning horizon (i.e., $gt10 = GT1 = st + T$). Status reported to the level $j + 1$ executor from the level j planner consists of the planned subgoal $x1^* = gp10$ at the GoalTime GT1.

The jth level Planner replans more frequently than the $(j + 1)$th level. This causes the jth-level plan to advance to the right relative to the more static $(j + 1)$-level plan, as shown in Figure 6.6b. For example, in Figure 6.6a, at $t = 0$, the tenth planned subgoal $gp10$ coincides with the GoalCommand GC1. Thus the predicted status $x1^*$ at $t = 0$ is $gp10$. However, in Figure 6.6b, at $t = 0.3T$, the seventh planned subgoal $gp7$ coincides with the GoalCommand GC1. Thus the status report at $t = 0.3T$ from the level j planner to the level $j + 1$ executor consist of $x1^* = gp7$.

At each level, plans are recomputed by the planner as often as necessary for the particular application. For applications where the environment is static, replanning may need to be done only as often as necessary to replenish the plan buffer. However, for applications in a dynamic environment, the planner at each level may replan more frequently, so that the plan is always current. In Chapter 9 we examine a case where each planner replans about as fast as the executor issues commands to the next lower level, or about every $0.1T$.

Figure 6.7 illustrates some of the timing details of real-time planning and sensory feedback at the lowest level in the RCS hierarchy. In this figure we assume that the level 1 (servo level) executor behaves like a classical servo feedback system, where the world is sampled at the beginning of each compute cycle, a control function is computed, and output commands are sent to the actuators at the end of each compute cycle. At the bottom of Figure 6.7, the world state evolves along a trajectory as each compute cycle of the level 1

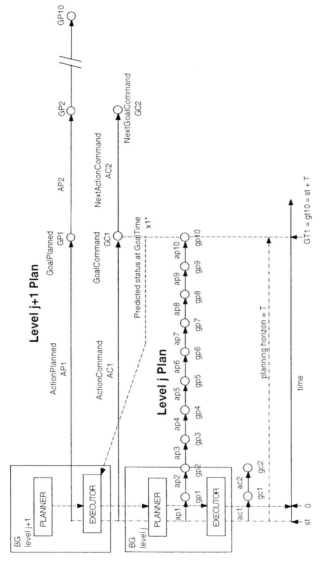

FIGURE 6.6a Planning and replanning by the level j planner. In this figure, plans and commands from level $j + 1$ are indicated by uppercase letters, and those at level j are indicated by lowercase letters. This figure shows the level j plan after the ActionCommand AC1 is first activated at $t = st$.

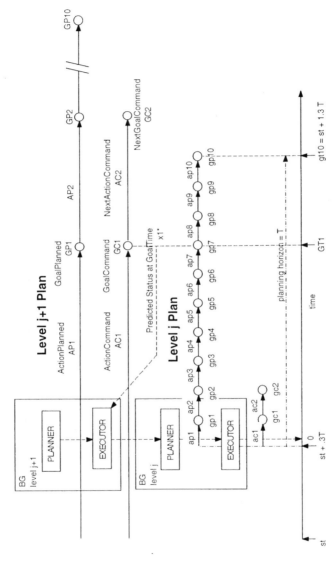

FIGURE 6.6b Planning and replanning by the level j planner. In this figure, plans and commands from level $j + 1$ are indicated by uppercase letters, and those at level j are indicated by lowercase letters. This figure shows the level j plan after replanning at $t = st + 0.3T$.

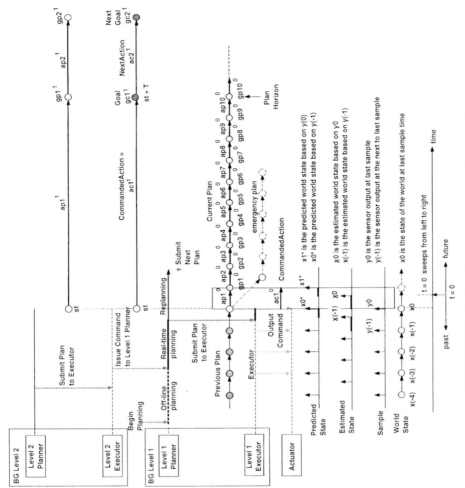

FIGURE 6.7 Timing diagram for plans and commands.

191

executor samples its input and writes its output. The world state is periodically sampled by sensors. $t = 0$ is the present. As time passes, the present sweeps from left to right. With each compute cycle, a step in a level 1 plan becomes an event in level 1 short-term memory. The index k refers to the interval during a compute cycle. $k = 0$ refers to the current cycle. $k > 0$ refers to the future. $k < 0$ refers to the past. As each compute cycle is complete, the index k is decremented so that $t = 0$ always lies within the current $k = 0$ interval.

At the top of Figure 6.7, the level 2 (primitive level) planner submits a plan to the level 2 executor. The first two steps in the level 2 plan are shown $(ap1^1, gp1^1, ap2^1, gp2^1)$. These are transformed by the level 2 executor into a command $(ac1^1, gc1^1, ac2^1, gc2^1)$ to the level 1 (servo level) planner. The level 1 planner converts this command into a level 1 plan that extends from $t = st$ to the planning horizon at the GoalTime $gt^1 = st + T$. Note that the command to the level 1 planner is issued before the commanded action is due to begin. This provides time for the level 1 planner to generate a plan. This is possible because there always exists a higher-level plan that extends far into the future compared to real-time planning delays at the lower levels. Thus an executor can always look ahead in its plan to compensate for planning delays at the lower levels. Real-time planning in the level 1 planner begins when a command is received from the level 2 executor. Off-line planning could have taken place at any previous time. Once the level 1 plan is ready, it is submitted to the level 1 executor for execution. The executor waits until the specified start time st before generating the command $ac1^0$ to the actuator. Normally, the planner attempts to find a path through state space from the starting state st to the GoalCommand state $gc1^1$ such that the tenth subgoal $gp10^0$ in the plan falls within some neighborhood of $gc1^1$. If the level 1 planner is successful in finding such a path, the level 1 executor will then use that plan to generate commands to its actuator.

As the index on each step in the plan reaches $k = 0$, the executor selects the planned action $ap1_i^0$ as a feedforward command. This is combined with feedback compensation to correct for any error between the planned subgoal $gp1^0$ and the predicted world state $x1^*$ at the subgoal time $gt1^0$. If the compare between $gp1^0$ and $x1^*$ shows an error out of range, an emergency action will be invoked and a new plan requested from the Level 1 planner. Note that at level 1, both the predicted world state $x0^*$ at $t = 0$ and the predicted state $x1^*$ at gt1 are computed by the WM and available at $t = st$.

Note also that as soon as the level 1 planner submits a plan to the executor, it begins replanning immediately. At a minimum, replanning must be completed before the current plan is completely executed. Otherwise, the executor is starved for lack of a plan. In extremely dynamic environments, replanning may be required as frequently as each step in the current plan is executed. In Figure 6.7, replanning is shown as completed during the second step in the current plan. In the case where new plans are being generated before the current plan is completed, each new plan at level j will shift to the right relative to the plan at level $j + 1$. This will continue until the planner at level $j + 1$

completes its new plan. Then the $j + 1$ plan will shift to the right relative to the level j plan. It is to accommodate for this effect that a command frame carries both the current command and the next command. This guarantees that every planner will always have a command goal (or next command goal) that lies beyond its planning horizon.

SUMMARY AND CONCLUSIONS

In this chapter we have focused on the BG hierarchy. We have outlined the functional and data flow requirements for planning and control at multiple hierarchical levels. We have indicated what must be done at each level and suggested how the interaction between planners and executors within and between BG processes at different levels can be accomplished. As can be seen from the simple examples of Figures 6.6 and 6.7, vertical interactions between planning and execution in a BG hierarchy can be subtle and complex. However, the horizontal interactions between BG, WM, VJ, and SP processes within RCS_NODEs at all levels are much more numerous and intimate. BG planning and execution processes communicate with WM and VJ processes to generate plans and make behavioral decisions within RCS_NODEs at all hierarchical levels. WM processes generate predictions for SP based on knowledge in KD, and use feedback from SP to update the KD in RCS_NODEs at all levels. Behavioral choices are enabled by values and priorities computed by VJ processes within RCS_NODEs at all levels. It is to these issues that we turn next.

7 World Modeling, Value Judgment, and Knowledge Representation

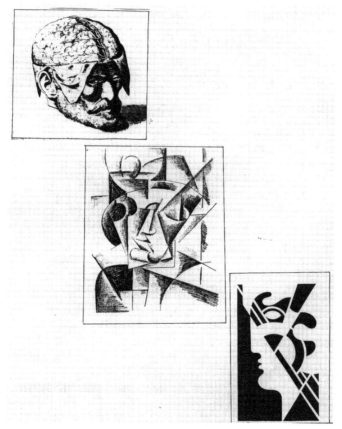

(Top) *Anatomy Becomes Respectable*, from *Tabulae Anatomicae* by Julius Casserius, published in Venice in 1627. [Reprinted by permission from *Art & Antiques*.] (Middle) *A Head* by L. S. Popova. From *Avangard Stopped in the Run*, published by Aurora, Leningrad, 1989. [Reprinted by permission of I. V. Savitsky State Museum of Art of Karakalpakstan, Pr. Doslik 127, Nukus 742000, Uzbekistan.] (Bottom) *Modern Head #5* by Roy Lichtenstein, 1970. [Reprinted by permission from the Estate of Roy Lichtenstein.]

For mind to emerge in a system, the system must have an internal representation of what it feels and experiences as it perceives entities, events, and situations in the world. It must have an internal model that captures the richness of what it knows and learns, and a mechanism for computing values and priorities that enable it to decide what it wishes to do. In this chapter we address the experimental and engineering problem of designing and building world modeling and value judgment processes and knowledge representation structures capable of supporting the perceptual, cognitive, and behavioral requirements of an intelligent system.

In the RCS reference model architecture, the sensory processing (SP), world modeling (WM), and value judgment (VJ) processes and the knowledge database (KD) are distributed throughout the set of RCS nodes. The SP and WM processes in each node maintain the KD in that node, keeping it current and consistent. The KD in each node contains the information required to support the BG and SP processes in that node. KD information is represented with the resolution and over the range in space and time required for decision making and control by VJ and BG processes in that node.

In each RCS node, the WM process performs four basic functions:

1. *Maintenance and updating of the knowledge database* (KD)

   a. The WM process updates state estimates in the KD based on variance between world model predictions and sensory observations at each node. This enables a recursive estimation process for each attribute and state variable in the KD. The recursive estimation process may operate at the data sampling rate or slower and integrate over any desired window in time and space.

   b. The WM process updates both image and frame representations and performs transformations from iconic to symbolic representations, and vice versa.

   c. The WM process enters new entities and events into the KD and deletes entities that are estimated to no longer exist in the world.

   d. The WM process maintains pointers within KD data structures that define relationships between entities, events, situations, and episodes. The WM process also maintains pointers that link images, maps, and frame representations.

2. *Prediction of sensory input*   Knowledge in the KD about object dynamics, sensor motion, behavioral actions, and projective geometry enable the WM process to predict sensory input. The WM process uses this knowledge to predict where objects in the environment will appear at the next sensory input sample. Variance between predictions and observations are used to update the KD. Correlation between predictions and observations reinforce confidence in the knowledge in the KD. Prediction of sensory input enables SP to focus attention and select sensory processing algorithms that are appropriate to the expected sensory input and useful for planning behavioral goals. Predictions of state variables

enable BG to track and anticipate target motion and maximize the performance of feedback control loops.

3. *Response to queries for information required by other processes*

   a. The WM process acts as a database server in support of BG, VJ, and SP processes in the same node and WM processes in other nodes. WM responds to "What is?" queries from the BG planning and control functions regarding the estimated state of the world or state of the self. WM responds to queries from SP regarding predicted attributes and states of entities listed as important for achieving behavioral goals.

   b. The WM transforms predicted images into the appropriate coordinate system and compensates for motion of cameras and sensor platforms that affect sensory input.

   c. The WM provides question answering and logical reasoning services required to deduce responses to queries for information that is not explicitly stored in the KD.

   d. The WM responds to queries or other inputs from the operator interface.

   e. The WM manages communications with other WM processes in other nodes within the RCS architecture.

4. *Simulation*

   a. The WM uses knowledge about the world in the KD to simulate the results of hypothesized plans generated by BG planning operations.

   b. The WM simulator responds to "What if?" queries with predictions of what would be the results if a particular string of hypothesized actions were performed. The simulator uses structural and dynamic models and rules of physics and logic to predict the results of current and hypothesized future actions.

   c. The WM simulator may use inverse models to compute the actions required to produce desired results.

   d. WM simulation results are sent to the value judgment process to be evaluated for cost, benefit, risk, and payoff.

These four functions are illustrated in Figure 7.1. The four basic WM functions are well understood in the computer science and systems engineering community [Schank and Colby73]. Recursive estimation, database management, dynamic modeling, and simulation are well practiced arts that have been applied many times in many different domains both in the laboratory and in industrial production.

The RCS architecture integrates these four processes and specifies the real-time constraints that drive the design of the computing environment necessary to implement a RCS node. The RCS design methodology makes real-time

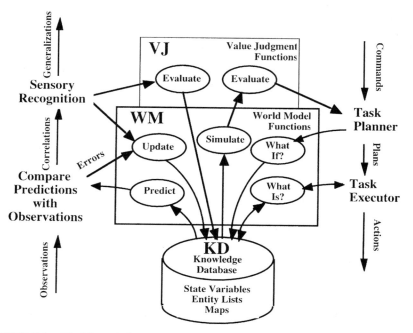

FIGURE 7.1 World modeling (WM) and value judgment (VJ) processes. WM and VJ processes typically exist in every node of the RCS architecture.

implementation feasible by distributing elements of the WM and KD over the RCS nodes such that the amount of knowledge in the KD of any node is limited to only what is needed for BG and SP processes in that node. Sufficient computing power can then be allocated to WM functions in each node to achieve the real-time performance requirements.

VALUE JUDGMENT

Value judgment (VJ) is a process that returns values for the attributes of goodness, cost, risk, benefit, or worth. These values may vary over time and may be computed in real time by VJ functions that are situation or task dependent. For example, the worth of an object such as a piece of furniture may depend on its state. A table or chair that is damaged or used may be worth less than one that is undamaged and new. The benefit of an event may depend on the situation in which it occurs. The sound of an approaching helicopter may be good or bad depending on whether it is a friend or foe.

Value judgment provides the basis for making behavioral decisions. VJ decides what is good or bad, what is attractive or repulsive, what is loved or hated, what is hoped for or feared. Value judgment is a process that computes cost, risk, benefit, and worth. It decides what is important and what is trivial. It assigns value to entities, events, and situations. It assesses the reliability of information. It calculates reward and punishment. It determines what is worth doing and what is not. It assigns priorities to tasks. For many control problems, the cost of a trajectory can be computed from an integral of cost over a path through state space. The computation of cost, benefit, and risk enables BG planners to select the best course of action. Computing the benefit or worth of a task (e.g., attacking a target or defending an object or position) enables BG executors to decide how much effort to expend or risk to accept in pursuing it.

In the biological brain, value judgment is provided by the limbic system. The limbic system occupies a sizable fraction of the volume of the brain and is tightly coupled with both the sensory and motor systems at all higher levels. In all intelligent systems, value judgment plays an important role in perception, attention, decision making, and behavior optimization.

Value judgment is crucial to planning. Computation of the expected cost, risk, and benefit of alternative plans provides the basis for selecting one plan rather than another. For each potential plan, an evaluation of the expected benefit of achieving the goal can be balanced against an evaluation of the cost and risk of pursuing the actions required for success. Various possible paths to the goal can be evaluated in the context of the expected state of the world, and the plan with the best cost/benefit ratio can be selected for execution.

Value judgment is crucial to perception. Value judgment computes what is important and where attention should be focused. Entities evaluated as important become targets of attention. Cameras can be pointed and tactile sensors directed toward objects and places in the world judged to be important.

Value judgment is crucial for assigning confidence to information in a world model. Value judgment assesses the reliability of information and judges whether current sensory experience or a priori world model information is more believable and hence a more reliable guide for behavior. Confidence and believability factors can be attached to various interpretations of entities and events. Information evaluated as reliable can be trusted. High confidence can be assigned to predictions based on that information. When confidence is high, behavior can be bold and aggressive. Information judged as unreliable or unconvincing can be treated with doubt. When confidence is low, behavior can be cautious and tentative.

Value judgment is crucial for learning. Value judgment computes what is rewarding and punishing. This enables learning. An intelligent system learns to repeat actions leading to events and situations that are evaluated as good or rewarding. It learns to avoid actions that result in punishment. Evaluation of what is important is required for deciding what to learn and what to remember. Value judgment determines what is memorable. Entities, events, and situations that are evaluated as import are stored in memory. Those that are evaluated as trivial or unimportant tend to be discarded and forgotten.

Value judgment is crucial to behavior. Value state variables assigned to entities can be used to decide how to behave toward them. Objects evaluated as attractive or valuable should be pursued or defended. Objects evaluated as dangerous should be treated with caution. Objects labeled as repulsive or situations labeled with fear should be avoided. Places labeled as safe can be used as havens of refuge. Objects labeled with hate can be attacked. Objects labeled with empathy can be assisted.

The mathematics of value judgment are well developed in several fields. In control theory, operations research, game theory, and reinforcement learning, value judgments are variously embodied in cost functions, objective functions, payoff schedules, and policies [von Neumann and Morgenstern44, Bellman57, Howard60, Sutton84, Werbos94]. Methods for computation of confidence values are particularly well developed in the fields of signal processing, digital filtering, and statistical analysis.

KNOWLEDGE DATABASE

Knowledge database (KD): a database that contains knowledge about signals, states, entities, attributes, situations, events, agents, goals, values, tasks, skills, rules, images, and maps

In Chapter 2 we discussed the kinds of knowledge that are represented in the brain and presented some of the approaches to knowledge representation that have been used in robotics and computer-integrated manufacturing. In Chapter 5 we described how the knowledge database is integrated into the RCS reference model architecture. At this point, we examine what kind of knowledge is represented in the KD and how this representation can be implemented in a computer.

Represent: to be a symbol for or an image of something in the world

Note that representation can include both symbols and images. *Symbols* may include characters, numbers, words, strings, vectors, and arrays. Symbols can be used to represent entities, events, situations, attributes, states, actions, relationships, lists, plans, and rules. *Images* are two-dimensional arrays of attributes or symbols. Images have distance and directional relationships between elements in the array that correspond to geometric and topological relationships in the world. An image may be formed by projecting the three-dimensional world onto a two-dimensional surface. Thus images can explicitly represent spatial relationships between entities as projected onto the image plane. Images can be used to represent scenes, pictures, drawings, or maps.

Temporal sequences of events can be represented symbolically by differential equations or by trajectories through state space. A temporal sequence can be represented by a sequence of states and actions, a string of characters, a series of taps on a delay line, or a trace on an oscilloscope. A temporal scenario can also be represented by a sequence of images such as are used to produce movies or television programs. A group of temporal sequences (such as the output from a bank of frequency filters or the result of a fast Fourier transform) can produce a sonogram (a two-dimensional image or array wherein frequency is assigned to the rows and time to the columns). Within images, spatial and temporal patterns can be grouped into higher-level abstractions such as entities, events or situations that can be tracked over time and recognized, classified, and named.

Data in the KD include signals, state variables, attributes, images, maps, rules, equations, schema, and recipes. Signals represent input from sensors. State variables represent measured, estimated, or predicted conditions in the world. Attributes and values describe properties. Images and maps represent geometric position and shape of entities and places. Rules and equations represent structural and dynamic models that describe how objects behave in the world. Schemata, recipes, and task frames represent how to perform tasks including what materials, tools, and resources are needed, and what instructional information is required. State variables, rules, and equations enable WM processes to predict how the world state can be expected to evolve in the future under a variety of circumstances. Images and maps enable BG processes to plan paths through the world. Task frames enable BG processes to plan tasks and control behavior.

In RCS, the KD for any particular RCS_NODE may contain any or all of the foregoing representations. A signal or attribute value may be represented in the KD by a single time-dependent scalar variable carried on a single channel or located in a single location in memory. A vector can be represented by a list of scalar variables. Images and maps can be represented by two-dimensional arrays of attributes. Objects can be represented in both symbolic and iconic form. An object can represented symbolically by a frame that contains attributes and state variables that describe the object and pointers that define the object's relationships to its parts and to the environment of which it is a part. The name of the frame is the name of the object. Objects can also be represented in images by groups of pixels, or by meshes, polygons, voxels, or other methods for representing surfaces and volumes. Dynamic relationships can be represented by four-dimensional models in space/time. Relationships that describe situations, semantics, or causality can be represented by semantic networks [Quillian68, Raphael68, Winograd72, Galambos et al.86].

ATTRIBUTE–VALUE PAIRS

Attribute: a characteristic of a person or thing

An attribute may be a physical parameter such as intensity, frequency, temperature, color, size, shape, or characteristic behavior. An attribute might be the intensity of radiation measured by a single detector, or the frequency of a tone measured by a tracking filter. An attribute might be the size of a building, the height of a tree, the weight of a person, the color of an apple, the risk of a plan, the benefit of an act, the worth of an object, or the shape of a surface. An attribute may be the range of a gun, the speed of a vehicle, or the demeanor of a person.

Attributes can have values.

Value: a quantity or quality of an attribute

Value is an overloaded term. In the context of value judgment (VJ), value indicates goodness, cost, risk, benefit, or worth. However, in the context of attribute–value pairs, value merely indicates quantity or quality of an attribute. For example, the intensity of radiation on a pixel is an attribute that may have a quantity value in units of lumens. The frequency of a tone might have a quantity value in units of hertz. The height of a tree may have a quantity value of feet or meters. The worth of an object may have a quantity value of dollars. The demeanor of a person might have a quality value of "aggressive" or "passive." The color of an apple may have a quality value of "red". Alternatively, the color of an apple may be represented by three quantity-values representing the outputs of red, green, and blue photodetectors measuring light reflected from the apple. In general, values are constants or variables that can be assigned to attributes of entities, events, situations, or plans.

An attribute–value pair consists of the name of the attribute and the value that is assigned to it. Attributes correspond to locations, names, or addresses in a data structure. Values correspond to contents of those locations. Values may be represented in a computer system in the form of data residing in data structures. Data structures include signals carried by wires, busses, or data channels, or by numbers stored in registers or locations in memory. Data types might include amplitude, frequency, or phase of signals on a wire, strings of digits on a channel, sets of bits on a bus, or by characters stored in registers or locations in memory. These are illustrated in Figure 7.2.

Attribute values can be measured directly by sensors or computed from measured attribute values. For example, the voltage on a wire may represent the temperature measured by a thermocouple. The output of a pressure sensor may be amplified and converted into a frequency on a channel that represents the pressure in a tank. A binary number in a register may represent the rotational velocity of a shaft measured by a tachometer. The contents of a particular memory location might represent the position or velocity of a tool along a coordinate axis in space. Intensity or color attribute values of pixels in an image can be represented by an array of integers in a frame grabber. The value of an attribute may vary with time.

Attribute values can be computed from other attribute values. For example, temporal derivative attributes can be computed from successive values of a

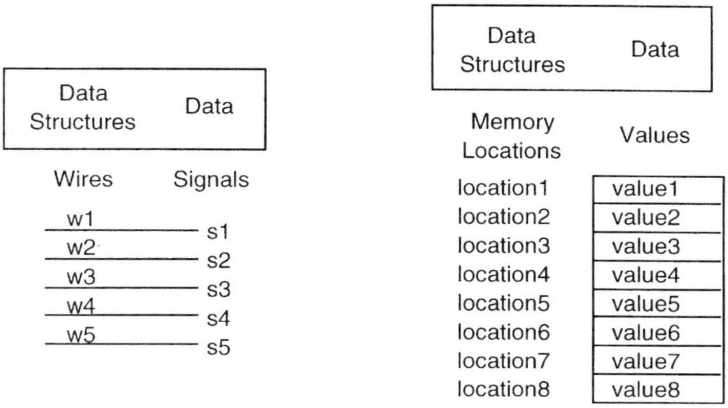

Data Structures	Data
Wires	Signals

w1 ——————— s1
w2· ——————— s2
w3 ——————— s3
w4 ——————— s4
w5 ——————— s5

Data Structures	Data
Memory Locations	Values
location1	value1
location2	value2
location3	value3
location4	value4
location5	value5
location6	value6
location7	value7
location8	value8

FIGURE 7.2 Example of primitive data structures with data contained in them. On the left is a set of five labeled wires carrying five signals or data elements. Each of these signals has a value that may vary with time. On the right is a set of eight memory locations containing eight values. These values may change at discrete points in time. The signal on each wire, or the value in each location, can represent a scalar variable. The set of values on an ordered set of wires or memory locations can represent a vector, string, character, symbol, word, attribute, state-variable, or pointer to a location.

single attribute. Attribute values may also vary over space. Spatial derivative attributes can be computed from values of neighboring attributes in an array or image. Attribute values may be filtered over time or space to remove noise or to detect frequency components. Attribute values may be integrated over regions of space and time to estimate state variables that represent position, velocity, acceleration, force, tension, pressure, or temperature of objects in the world.

Primitive data structures such as shown in Figure 7.2 may be ordered or interconnected to form higher-order data structures such as vectors, symbols, characters, strings, arrays, structures, and classes. For example, voltages on an ordered set of wires can represent a number, a character, a vector, or an array. A population of neurons in the motor cortex may represent a vector that describes the planned direction and velocity of motion for a hand at the end of an arm (see Figure 2.9). An ordered set of neurons distributed over the visual cortex may represent an attribute image. A different ordered set of neurons in a narrow column of visual cortex may represent an attribute vector for a single pixel. An array of integers in a computer memory may represent an attribute image such as a brightness, color, or range image. An array of integers may also represent an image of the spatial or temporal gradient, or difference of Gaussians, computed over an intensity, color, or range image. An array of locations in computer memory may represent a map or aerial photograph. A set of arrays can represent a set of registered attribute images or map overlays.

Higher-order data structures such as lists, frames, networks, and graphs can be configured to represent more abstract aspects of the world. Frames can be used to represent entities and events. (Entities were defined in Chapter 3. Events are defined later in this chapter.) A pointer to a frame can represent the name or address of the frame. Data in the frame can represent the state, attributes, and name of the class to which the entity or event belongs. Functions in the frame can represent behaviors of the entity. Pointers in the frame can represent relationships with other data structures. Pointers between frames can represent relationships such as *is-part-of, belongs-to, comes-before, is-caused-by, is-triggered-by, is-inside, is-outside.*

Pointers can link pixels in images or maps to entity frames, and vice versa. Pointers can link actions to subgoals in a plan. Pointers between entities can define relationships between objects and places in an image or on a map or in the world. Pointers between entities and events can define situations. Pointers can define networks of relationships that represent rules, equations, patterns, grammars, models, plans, procedures, schema, skills, and algorithms. Data structures containing attributes and pointers can be used to define geometric, mathematical, logical, syntactic, semantic, causal, or pragmatic relationships between entities, actions, words, symbols, and events [Peckham and Maryanski88].

The object-oriented computer science literature and textbooks provide many detailed examples of how to build and maintain all of these data types and structures. Entities and events are very similar to objects in the object-oriented approach [Edleman99b, Jacobson92]. Modern computer languages such as Ada, C++, and Java provide tools to realize these types of structures in software [Deitel and Deitel97,98].

IMAGES

Image: a two-dimensional surface or manifold of attribute values

Images may be generated in a number of ways. An image may be formed by a sketch or drawing, by a fine array of dots on a sheet of paper, by the projection of electrons on the face of a cathode ray tube, or by the optical projection of light from a scene in the world through a lens onto a focal plane. If the focal plane is covered with a photosensitive film, an image will be imprinted on the photosensitive film. If the focal plane is covered with an array of charge-coupled devices (CCDs) etched on the surface of a silicon chip, an image can be captured by a CCD TV camera. Images captured by the human eye are in the form of excitation voltages in an array of rods and cones in the retina.

Images may also be generated from parameters such as vectors or polygons stored in symbolic form. The commercial market for video games and virtual reality has produced a wide variety of inexpensive and high-performance graphics image hardware and software. Performance and capability per unit price is growing at an astonishing pace. Real-time shaded graphics image generation and manipulation are widely available at consumer prices in computer video games.

In a TV camera, a retina, a FLIR (forward-looking infrared camera), or a compound eye of an insect, there is an array of sensors that sample the image at an array of points. In a laser range imager (LADAR), a scanning mirror and light pulse emitter measure the range at an array of points. Each sensor integrates the energy falling on its aperture over some interval of time, producing an array of signals. The array of sensors spatially quantize, or tesselate, the image into picture elements called *pixels*. Each photodetector produces a signal that represents an intensity attribute of the pixel.

Pixel: a picture element

A pixel is the smallest distinguishable region in an image. The region within a pixel has no discernible internal structure. The signal from each sensor in a TV or FLIR image corresponds to an intensity, color, or thermal attribute value averaged over the area of the pixel. Each pixel in a LADAR image represents a range detected within the area of the pixel.

Any camera or biological eye has a field of view that is determined by the focal length of the lens and the size of the photosensitive array. (This is true of an eye with a single lens and a retina. For a compound eye, the field of view is determined by the shape of the eye and the directionality of the individual photoreceptors.) Any image has a resolution that is defined by the density of sensors in the array, the focal length, and the quality of the optics. A two-dimensional image array can be represented as a flat plane (e.g., as a typical photograph or a view through a flat windowpane). An image can also be represented as a section of a spherical surface seen from the center of the sphere: for example, by the portrayal of the sky in a planetarium, or the actual view of the sky on a clear night.

For a small field of view, there is little difference between planar and spherical representations. However, for a creature that must pay attention to entities and events in many different directions simultaneously (front, back, right, left, above, and below), the spherical representation is greatly preferred over a planar representation. A planar representation is limited to a relatively narrow field of view. At best, the planar representation can represent only half of the world as seen through a plane that goes off to infinity at 90° to the right and left (and up and down). It cannot represent the hemisphere behind the view point at all. In contrast, the egosphere is continuous and isotropic in all directions. For this reason, the celestial egosphere is the astronomer's choice for representing the positions of the stars and planets and the trajectories of the sun and moon. For centuries, astronomers have used the celestial sphere representation for mapping the heavens, and navigators have used the apparent position of heavenly bodies on the celestial sphere to compute their own position on the surface of the earth.

THE EGOSPHERE

Egosphere: a spherical coordinate system with the self (ego) at the origin

The egosphere is the most intuitive of all coordinate systems. Each of us resides at the origin of our own egosphere. Everything that we observe in the world can be described as being located at some azimuth, elevation, and range measured from the center of our self. To the observer at the center of the egosphere, the world is seen as if through a transparent sphere. Each observed point in the world appears on the egosphere at a location defined by the azimuth and elevation of that point. There is, in fact, no way for a single eye to tell whether a scene being viewed is real or is an image projected on an egosphere with resolution equal to or better than that of the eye.

The orientation of the egosphere can be defined uniquely by specifying two orthogonal directions. Typically, these are chosen to define the pole (which in turn defines an equator 90° away) and a zero azimuth on the equator. A number of different egosphere coordinate frames are useful for representing the world. These include the sensor egosphere, the head egosphere, the body egosphere, the inertial egosphere, and the velocity egosphere [Gibson50].

Sensor egosphere: an egosphere where the horizontal axis of the sensor array defines the equator of the egosphere and hence the pole. The center pixel in the sensor array defines zero azimuth on the equator.

Figure 7.3 shows a sensor egosphere for a TV camera. The optical axis defines zero azimuth and zero elevation on the egosphere. The area around the optical axis is the field of view. The equator of the egosphere is defined by the horizontal line of pixels through the center of the image. The x-axis is horizontal to the right in the image. The pole of the egosphere is defined by "up" in the image. The position of each pixel in the image is defined by its azimuth and elevation on the egosphere. Each pixel can be assigned a range attribute whose value is the estimated distance to the point from the center of the egosphere.

Each eye has its own egosphere. Attributes measured by each eye can be transformed into a head egosphere by one translation and one rotation. Head egosphere coordinates can be transformed into torso and limb coordinates and then into muscle coordinates by additional translations and rotations. All of these transformations are well defined as long as there is a good range estimate at each pixel. Dickmanns[99] describes the various coordinate frames and transformations required by a vision system used for driving an autonomous vehicle.

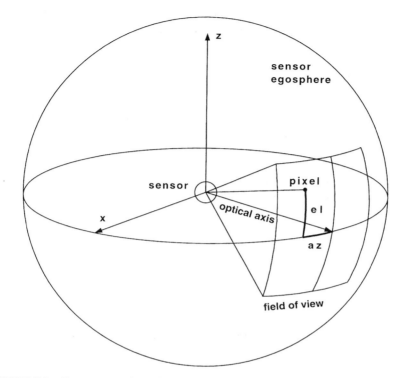

FIGURE 7.3 Sensor egosphere for a camera. The center of the imaging lens lies at the center of the sphere. The optical axis of the camera defines the zero point on the equator. The orientation of the photodetector array defines the z-axis. Each pixel has a unique azimuth and elevation. The field of view of the array covers a region on the egosphere.

Range can be estimated from a wide variety of inputs, including ocular vergence, stereo disparity, motion parallax, image flow, shading, texture, occlusion boundaries, elevation in the image, and estimates of size and speed. Range can be measured directly by laser, radar, or sonar. For an acoustic system, the direction, intensity, frequency, and other attributes of incoming sound can be represented on the head egosphere. For tactile sensors, the location and qualities of sensed surfaces can be plotted in head egosphere coordinates. The head egosphere thus provides a convenient coordinate frame for fusion of information from vision, hearing, and touch. For any resolution, there is a fixed number of pixels on the egosphere. For example, for pixel resolution of 1 square degree, there are about 41,252 pixels on the egosphere. The density of pixels on the egosphere grows as the square of the resolution.

A polar egosphere coordinate system suffers from singular points at the poles. The effect of these singularities can be minimized by placing the poles on the vertical axis so that they are far from the center of the field of view. On the other hand, the pole singularities can sometimes be used to advantage, as in the case of the velocity egosphere.

Velocity egosphere: an egosphere where the velocity vector defines the pole of the egosphere

Figure 7.4 shows a velocity egosphere. The velocity vector defines the pole of the velocity egosphere. The projected image of each point lying on a stationary surface in the world moves along a great circle arc radiating from the pole of the velocity egosphere. For constant velocity, the image of stationary points in the environment move on the velocity egosphere at an angular rate given by the simple formulae [Gibson et al.55]

$$\frac{d\alpha}{dt} = \frac{v \sin \alpha}{r} \tag{7.1}$$

$$\frac{d\beta}{dt} = 0 \tag{7.2}$$

where

α = angle between the velocity vector and the pixel on the velocity egosphere

v = velocity of the camera through the world

r = range to the point in the world

β = azimuth of the pixel relative to the intersection between the plane normal to the gravity vector and the equator of the velocity egosphere

Note the simple relationship between range, velocity, and α. Note also that this relationship is independent of β.

Additional egosphere representations are the body and inertial egospheres.

Body egosphere: an egosphere where the roll axis of the body defines zero azimuth and elevation, and the yaw axis defines the pole

Inertial egosphere: an egosphere where Earth's gravity vector defines the pole and north defines the zero azimuth

The body egosphere is useful for local path planning for locomotion or manipulation. The inertial egosphere is useful for transformations to and from world map coordinates. The inertial egosphere provides a stabilized internal representation of the world independent of rotation of the sensory array. Figure 7.5 is a top view of several egosphere representations, showing the difference

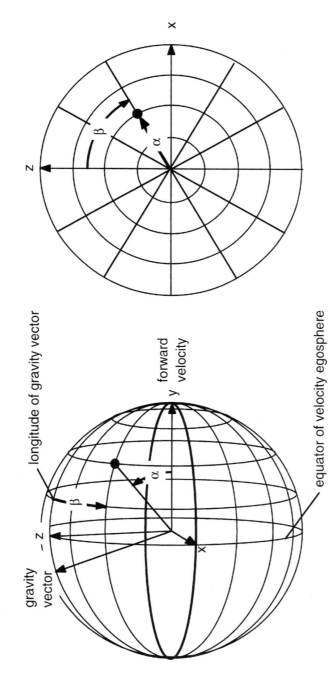

FIGURE 7.4 Two views of a velocity egosphere. On the left is a view along the velocity vector from the center. The velocity vector defines the pole of the velocity egosphere and that corresponds to the focus of expansion of flow vectors for stationary points in the image.

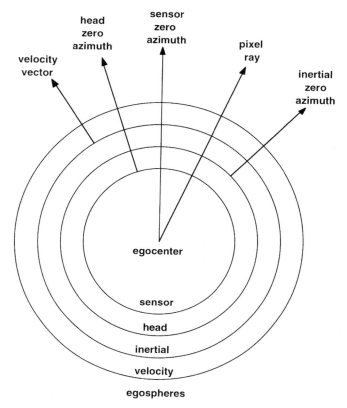

FIGURE 7.5 Two-dimensional top-down projection of four egosphere representations, illustrating angular relationships between egospheres. Pixels are represented on each egosphere such that images remain in registration. Pixel attributes detected on one egosphere may thus be inherited on others. It should be noted that pixel resolution is not typically uniform on a single egosphere, nor is it necessarily the same for different egospheres or even for different attributes on the same egosphere. [From Albus91.]

in azimuth between the head egosphere, the inertial egosphere, the sensor egosphere, a pixel ray, and the self velocity vector.

Figure 7.6 shows a Mercator projection of the human head egosphere with the field of view of both right and left eyes shown. The fovea of the two eyes are converged on the top of a rock directly ahead. Central vision for both eyes overlap and provide good stereo depth perception for objects less than about 50 m away. This means that range measurements are available for those pixels within central vision. Peripheral vision does not completely overlap, due mostly to blockage of the contralateral field of view by the nose. Near the edges of peripheral vision, resolution is low and recognition based on color and shape is poor, but perception of motion remains good. The example in

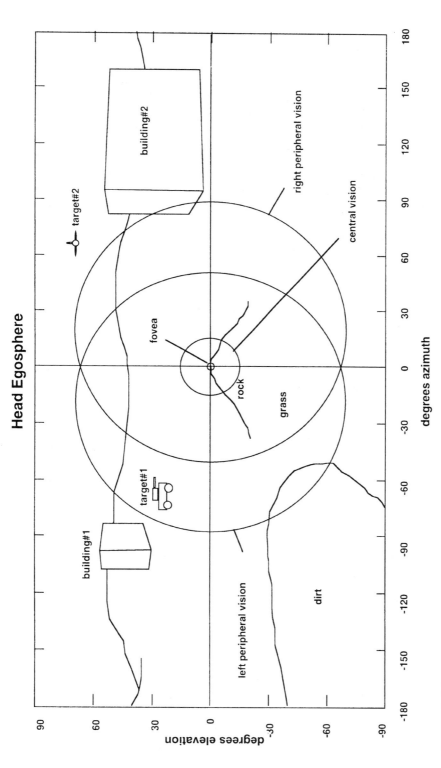

FIGURE 7.6 Mercator projection of the head egosphere showing the field of view of foveal and peripheral vision and a number of objects in the world. The two eyes are verged on the top of the rock straight ahead. [From Albus98.]

Figure 7.6 is for a vehicle driving task. The rock directly ahead is a potential obstacle that needs to be examined. Thus, the high-resolution foveal regions of both eyes are converged on the highest point of the rock. This enables high magnification and good range measurements of the top of the rock.

In Figure 7.6, the size of the human fovea relative to the peripheral field of view and to the entire egosphere is to scale. This illustrates that the fovea is essentially an optical probe that is pointed by an attention mechanism at the current most important part of the world. About one-third of the pixels in the entire visual field are concentrated in the fovea. A second third fill the region marked as central vision. The remaining third are distributed over the peripheral visual field. The relative density of pixels in human vision decreases roughly exponentially with the angle between the fovea and a pixel.

MAPS

Map: two-dimensional array of attributes and entities that are scaled to, and registered with, known locations in the world

A world map consists of a vertical projection of the surface of the earth onto the planer surface of a world map. When the position and orientation of the self is known in the world and range can be measured or estimated at each point in the visual field, it is possible to transform images from egosphere coordinates to world map coordinates. This makes it possible for entities observed in the image to be overlaid on a priori information in the world map. For example, images acquired by cameras are often transformed into world map coordinates for planning routes.

If terrain elevation data are available in the world map, it is also possible to transform world map information back into egosphere coordinates. Typical world map coordinates are latitude and longitude. World map coordinates have the advantage that stationary features on the ground do not move on the map when the self moves. Instead, the location of the self moves on the map. Thus, when stationary objects observed on the egosphere are transformed into world map coordinates, they become stationary on the map regardless of motion of the self. Thus, even though images of stationary objects move over the ego-sphere as the self moves through the world, repeated sightings of stationary objects from various view points can be integrated into a single representation on the world map. For this reason, images from cameras are typically trans-formed into world map coordinates to facilitate planning of routes for moving through the world.

Attributes of map pixels may describe the characteristics of the terrain con-tained within the pixel. Map pixel attributes may include type of ground cover, terrain elevation relative to the sea level or relative to the vehicle, and terrain roughness, slope, and traversability. Map pixels may also include names of, or pointers to, icons that represent entities such as buildings, roads, bodies of water, or landmarks that the pixel covers, or of which the pixel is a part. Maps typically have a number of overlays that are registered with the map. (See, for example, Figure 9.3.) Each overlay may represent one or more attribute or entity maps [Tomlin90]. For example, map overlays may represent roads and towns, rivers and lakes, buildings and landmarks. Map overlays may represent topographic elevation and contours of constant elevation. Map overlays may represent fields, woods, swamps, water, and sand. Map overlays can represent the surface roughness, the slope, or traversability as a function of direction at each pixel. Map overlays can represent which regions are visible from a particular view point. Battlefield map overlays can represent the deployment of friendly and enemy troops, the coverage of artillery, the location of mines, or the position of sector lines.

Military maps with multiple overlays are typically made available to soldiers as prior knowledge for military missions. However, a priori maps are too low in resolution to be used by autonomous vehicles for obstacle avoidance and too static and stale to be used to analyze and anticipate the behavior of moving targets. Most military maps have elevation postings at 30-m grid intervals. In special cases, maps with elevation postings every 3 m or even every 1 m may be available, and vector representation of features such as buildings, roads, and landmarks may be included with greater precision. But the time delay between when the terrain is sensed and a priori maps are available to the soldier for decision making and control is at best minutes to hours, and more often, days to weeks. A priori maps cannot provide the dynamic information necessary to track and predict moving objects in the world.

High-resolution dynamic information must be generated from real-time sensory data. In the vicinity of the cameras, LADAR and stereo systems can provide range information in egosphere coordinates accurate to a few centimeters. This information can be used to build local terrain maps in real time in the world model and to represent moving objects. When the position and orientation of the camera egosphere are known, local maps generated from camera data can be registered with a priori maps, such as shown in Figure 7.7 [Thrun et al.00, Dissanayke et al.00]. This enables landmark recognition and provides the information needed for path planning and task decomposition. Images in egosphere coordinates can be transformed into maps in world coordinates for route selection, path planning, and obstacle avoidance. A priori maps can also be transformed into egosphere coordinates so that terrain features and elevation can be transformed into a sensor viewpoint so that names and attributes of entities are overlaid on camera images. For example, the name of a landmark or a road can be overlaid on the image of the landmark or road. Attributes of a region or entity, such as its traversability or its value as a target or a resource to defend, can be overlaid on the image of the region or entity. This enables image-based task planning and target selection to be performed directly in egosphere coordinates.

There are many issues related to the computational power required for real-time image processing, for transformation of pixels from image coordinates

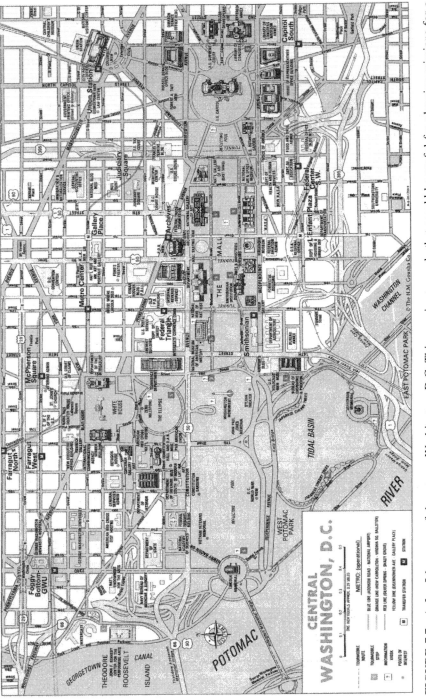

FIGURE 7.7 A 3 × 5 km map of downtown Washington, D.C. This map is at a scale that would be useful for planning missions for a small group of robot vehicles out to a planning horizon of approximately 3 minutes (assuming an average velocity of 10 m/s). This is well beyond the range of the vehicle driving sensors. It corresponds to an area that might be viewed by a camera in an air vehicle at an altitude of about 1 km. A map of this scale shows the layout of streets and the shape of large buildings, public spaces, and bodies of water. It contains symbolic references to names of streets, buildings, and agencies that occupy those buildings. A priori maps with this level of detail are rarely available except for special places such as shown above. [Reprinted by permission from Rand McNally.]

to map coordinates, and vice versa. The RCS reference model architecture addresses these issues by limiting the range and resolution of images and maps in the world model at each level of the hierarchy and by focusing attention on important regions on the egosphere. Both of these approaches reduce the number of map pixels that require updating at each level. At each hierarchical level, maps in the world model have less resolution but greater range than at the level below. At each level, maps have more resolution but less range than at the level above. See, for example, the WM maps, shown in Figure 9.3. RCS methodology uses a rule of thumb that range increases about an order of magnitude and resolution decreases by an order of magnitude at each higher level. Thus, the information density on maps in the world model remains relatively constant across levels. This produces an exponential increase in resolution and decrease in range in the space-time egosphere that can be seen in the planning maps shown in Figure 9.3. Planned waypoints at higher levels are larger (lower resolution) and farther apart (longer range) than planned waypoints at lower levels. Thus, plans for actions that are closer in space-time are more detailed and immediate than those that are more distant in space-time.

RCS also supports the use of foveal/peripheral cameras to limit the number of pixels in the image. For example, it is possible to represent a full 4π steradian egosphere field of view with resolution of $1°$ per pixel with less than 42,000 pixels. A $32° \times 32°$ field of view can be represented by a 256×256 image with resolution of $0.12°$ per pixel. A $4° \times 4°$ field of view can be represented by a 256×256 image with resolution of $0.016°$ per pixel. This is better than human 20/20 vision. A LADAR might have only about 8000 pixels with resolution of about $0.5°$ per pixel. A radar map of the ground may contain about 30,000 pixels. All of these egosphere representations put together require less than the number of pixels in a single 512×512 image.

Moreover, not all pixels need to be processed in real time. Attention mechanisms can mask out pixels that are irrelevant to behavioral goals. This may reduce the number of pixels that require real-time processing by as much as an order of magnitude. Thus, the total number of image pixels that must be processed at frame rates may be less than 100,000. To achieve real-time processing rates of 10 frames per second may require processing of less than a million pixels per second. For modern image processing technologies, this is not an prohibitive computational load.

It should be noted, however, that processing requirement for reliable full-field stereo vision can be challenging. Correlation of images from two cameras is very computationally expensive. Nevertheless, requirements for processing images can be met through parallel methods. Given the continuing increase in processing speed of computer chips and the ease of designing and fabricating special-purpose image processing hardware, the computational requirements for image processing becomes less of a problem with each passing year.

ATTRIBUTE, ENTITY, CLASS, AND VALUE IMAGES

In the RCS KD, there can be four types of images: attribute, entity, class, and value images.

Attribute image: a two-dimensional array of attribute values

In an attribute image, each pixel contains the value of the attribute that is measured or computed at that pixel. There are a number of attributes that can be computed at each pixel and hence a number of attribute images. These include intensity, color (red, blue, green), stereo disparity, range, image flow magnitude and direction, texture, surface orientation, and spatial or temporal gradients of intensity, color, or range. For each attribute, an attribute image can be constructed and maintained in registration with the other attribute images. This produces a three-dimensional array of pixel attributes as illustrated at the top of Figure 7.8.

Attribute images add information to the signals generated by the retina so as to increase, not reduce, the information content at each pixel. Attribute images can be computed in parallel and exist simultaneously in registration with the original image. Pixels with similar attributes can be grouped into entities. When a pixel is grouped into an entity, it is assigned the name of the entity. This produces an entity image.

Entity image: a two-dimensional array of entity names

An entity image is an array of names that define for each pixel the entity to which that pixel belongs. Each pixel in an entity image is either contained within the entity or contains the entity (if the image of the entity is smaller than a single pixel). An entity image is much like a "paint by numbers" drawing, where each region in the image contains the name (or identification number) of the color to be painted into that region. The name assigned to each pixel in an entity image is a pointer to an entity frame that contains the attributes of the entity. Entity attributes are characteristics and properties of the entire region occupied by the entity. Entity attributes can be computed by integrating the attributes of the pixels belonging to the entity [Adelson and Movshon82].

Entities with similar attributes can be grouped into higher-level entities. Grouping is inherently a hierarchical process that occurs at every level of the sensory processing hierarchy. Pixels can be grouped into list entities. List entities can be grouped into surface entities. Surface entities can be grouped into object entities. Object entities can be grouped into group entities. At each level of grouping, each pixel acquires a new entity pointer that points to an entity frame. Thus, each pixel acquires a set of entity names that point to the set of entity frames to which the pixel belongs. This produces a set of entity images as shown in Figure 7.8.

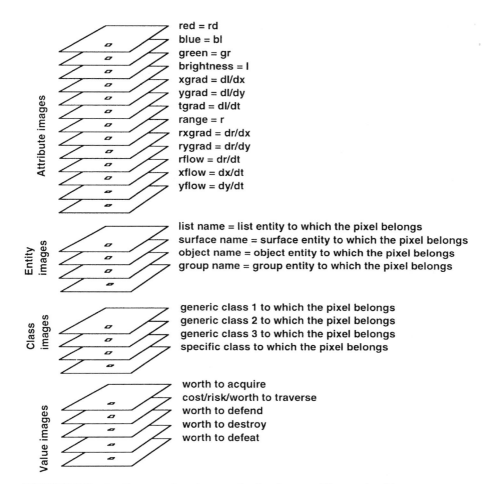

FIGURE 7.8 Attribute, entity, class, and value images. The result of image processing is to compute a set of images that are registered with the original retinal image. Value images represents information that is needed for setting priorities and making behavioral plans and decisions.

An entity image segments the input image into a set of regions, or windows, over which entity attributes can be computed. Computed entity attributes can then be compared with class attributes stored in a library of class prototypes in the KD. When computed entity attributes match the attributes of a class prototype, the entity observed can be classified or recognized as a member of the class. When an entity is classified, the pixels that are labeled with the entity name can be assigned the class name of the entity class. This produces a set of class images as shown in Figure 7.8.

Class image: a two-dimensional array of entity class names

An entity class name specifies for each region in an entity image the class of the entity that occupies that region. Class images may designate generic or specific classes. Examples of generic classes are roads, buildings, trees, or streams. Examples of specific classes are a particular road, building, tree, or stream.

Value image: a two-dimensional array of cost/risk/worth values

Entity classes can be assigned values that specify the worth of entities in that class, and the cost or risk of carrying out action on them. Cost/risk/worth values of entities can be assigned to the pixels contained in the entity class image. This generates a set of value images such as shown in Figure 7.8. These can be used for planning behaviors. For example, worth values assigned to entity classes can be used to define priorities for tasks designed to acquire them. Cost/risk values assigned to regions of terrain can be used for planning paths that traverse them. Worth assigned to assets can define how much they are worth to defend. Worth assigned to targets can specify how valuable they would be to destroy. Worth assigned to opponents can define how much it is worth to defeat them. Value images enable plans to be formulated in the image domain.

ENTITY FRAME REPRESENTATION

Entities can also be represented by entity frames.

Entity frame: a data structure that contains the entity name, a state vector, a list of attribute–value pairs, a pointer to a parent entity, pointers to subentities, a pointer to the entity image in which the entity appears, and a set of functions that define the behavior of the entity

An entity frame contains the information that the intelligent system knows about an entity. Figure 7.9 shows an example of an entity frame. The entity name is an address or index by which the entity frame can be accessed in a database or library of entities, or to which other entity frames can be linked. An uncertainty parameter can be associated with the name to indicate how certain the system is that the entity has been identified properly. The entity frame contains entity attributes that describe attributes such as color, size, and shape. Entity state variables describe dynamic properties such as position, orientation, and velocity of the entity in a particular coordinate system. These can be used as parameters in world modeling processes for prediction and simulation, and by sensory processing functions for classification, detection, and recognition. Uncertainty parameters can be associated with state variables to indicate how

NAME = entity_id (uncertainty)   *// this is the frame address in the KD*

// attributes – these are characteristics that address the question What?
color   = red, green, blue intensities
size   = length, height, width dimensions
shape   = curvature, moments, asex of symmetry, etc.

// state – these are dynamic properties that address the question Where?
position   = azimuth, elevation, range (uncertainty)
orientation   = roll, pitch, yaw (uncertainty)
velocity   =v-azimuth, v-elevation, v-range, v-roll, v-pitch, v-yaw (uncertainty)

// class – pointers to the entity image and classes to which the entity belongs
entity image   = pointer to the entity image in which the entity appears
generic class1 = pointer to the generic class1 prototype
generic class2 = pointer to the generic class2 prototype
generic class3 = pointer to the generic class3 prototype

// value or worth of the entity
worth to preserve   = value
worth to acquire   = value
worth to defend   = value
worth to defeat   = value

// pointers that define parent-child relationships
belongs to   = pointer to parent entity
has part1   = pointer to subentity1
has part2   = pointer to subentity2
has part3   = pointer to subentity3

// pointer that define situational relationships
on top of   = pointer to entity below
beside-right   =pointer to entity on right

// functions that define behavior
behavior1   = responds-to
behavior2   = acts-like

FIGURE 7.9   Structure of a typical entity frame. Each entity frame consists of a name, a list of attributes, and a set of pointers to other entities. The list of pointers includes a pointer to the entity image that contains regions labled with the entity_id.

dependable are estimates of these values. Observed state variables are typically defined in a sensor egosphere coordinate system. Estimated and predicted entity state variables may also be in egosphere coordinates or in some other, more convenient coordinate system.

The entity frame may contain pointers to the entity image in which the entity appears and to the class or classes to which the entity belongs. For example, an object entity will have a pointer to the entity image in which it appears. It may also have a generic class pointer that identifies it as a member of the class of trees, another generic class pointer that identifies it as a member of the class of pine trees, and a specific class pointer that identifies it as a particular pine tree. The entity frame may contain value attributes that define how valuable an object is as a target (if a foe), or how worth defending (if friend). Value attributes may also define how much a particular entity or situation should be feared or avoided, what its worth is as a vantage point or as a source of shelter or food, and whether it would be worth the cost or risk of acquiring it.

The entity frame may contain pointers that define inheritance relationships with other entities. Each entity frame has a pointer to the frame of a parent entity of which it is a part, and a set of pointers to frames of subentities that are its parts. For example, an object entity frame typically will have a parent_pointer to the entity frame of the group to which the object belongs. It will have a number of subentity_pointers to the surface entity frames of the surfaces and boundaries that are its parts. Similarly, each surface entity frame will have a parent_pointer to the object entity to which it belongs, and a set of subentity_pointers to the list entities that are its parts. Each list entity frame will have a parent_pointer to the surface entity frame to which it belongs, and a set of subentity_pointers to the pixels that are its parts. Inheritance pointers are established and maintained by grouping and classification operations performed by sensory processing functions discussed in Chapter 8.

The entity frame may also contain pointers that define spatial, temporal, causal, or other types of relationships that pertain to that particular entity. An entity frame may include a set of functions that describe the behavior of the entity under certain conditions or in response to certain stimuli. Simple functions may define the behavior of objects under the influence of gravity or friction. Complex functions may define how the entity might be expected to respond to an attack or a gesture of friendship. Behavioral functions may include parameters such as speed, endurance, strength, or range of weapons. Behavioral functions and parameters may be inherited from generic or specific class prototypes.

Entity frames may stand alone as data structures or be linked to form lists, strings, words, sentences, networks, and maps. Named entity frames provide the basic building blocks of language—nouns. Named entities may be used as the subjects or objects of sentences, and the agents or objects of action. In language applications, entity attributes may serve as adjectives that describe characteristics of the entities to which they are attached. Entity frames can easily be implemented as objects or classes in modern computer languages.

Entity frames can be interconnected by pointers to form causal, semantic, and situational networks. Situational networks may represent situations or geometric relationships such as *"on-top-of," "beneath," "to-the-right-of,"* and *"in-front-of."* Situational networks may also have pointers to maps or images that display spatial relationships pictorially. Causal networks represent the cause-and-effect relationships among entities, events, situations, and actions that occur in the world. Semantic networks represent the relationships among entities, attributes, situations, actions, and events that define meaning and enable reasoning, logic, and language.

IMAGES AND FRAMES

Figure 7.10 shows the relationship between entity images and entity frames. On the left is a set of attribute and entity images. A surface entity image is shown that contains three surfaces (including the background). For each pixel in the surface entity image, there is a pointer to the surface entity frame to which the pixel belongs. There also exists a pointer from each entity frame back to the entity image in which the pixels that belong to that entity are located. This two-way linkage between entity images and entity frames enables computation in both the symbolic and image domains. For example, an entity image can be used as a mask or window for focusing attention or correlating observed entity attributes with predicted entity attributes. An entity image can also be used as an integration window for integrating pixel attributes to obtain entity attributes.

If a pixel is not assigned to an entity, it can be designated as a background pixel or as unknown. Background pixels are those that require little or no additional processing. Unknown pixels may be selected for additional processing if they are judged to be important to the current task. Figure 7.11 suggests how entity frames, class images, and value images might be generated in the brain.

There are many stages in the acquisition of knowledge about entities by an intelligent system. In the visual perception systems, attributes for each pixel can be measured or computed locally and pixels can be classified based on pixel attributes. Pixels can also be grouped into entities based on gestalt hypotheses. Entity attributes can then be computed over the region occupied by each entity, and entities can be classified based on their attributes. Entities can be further grouped into higher-level entities, attributes of higher-level entities can be computed, and higher-level entities can be classified. The processes of grouping, computation, and classification may be performed at many different levels simultaneously and in parallel. To support these processes, the RCS KD provides a hierarchy of geometric entities that are generated by making and verifying gestalt hypotheses.

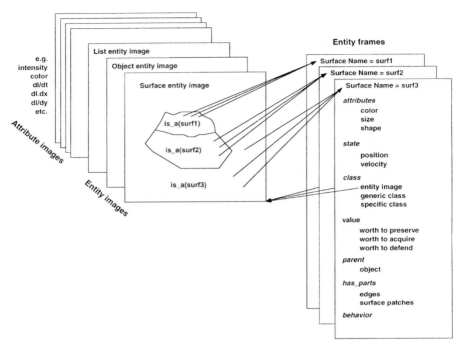

FIGURE 7.10   Relationship between entity images and entity frames. In this example, a pointer for each pixel in a surface entity image points to the frame of the surface entity to which that pixel belongs. The entity frame contains properties of the entity, including attributes, state, class, value, relationships, and behavior.

GEOMETRIC ENTITIES

Geometric entities are defined by gestalt grouping hypotheses based on geometric properties or similar attributes. Geometric entity types include pixel, list, surface, object, and group entities. Each occurrence of a geometric entity is represented in the KD by a geometric entity frame such as shown in Figure 7.9. Geometric entities are related to each other by a taxonomy of grouping relationships. List entities consist of groups of pixel entities. Surface entities consist of groups of list entities. Object entities consist of groups of surface entities. Group entities consist of groups of object entities. It should be noted that this particular taxonomy of entity classes is largely arbitrary. It was chosen for RCS because it is a useful way of organizing information about the world. Presumably, other schemes might be as good, or better.

- *Level 1: pixel entities.* Pixel entities have attributes that can be measured by a single pixel during a single sample period in time, or that can be computed at a single pixel (or over a local neighborhood centered on a

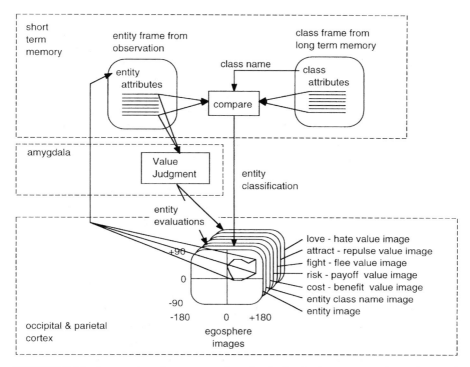

FIGURE 7.11 Interactions between regions in the brain that may produce class and value images. Entity attributes observed in entity images are held in entity frames in short-term memory. Class attributes stored in long-term memory are compared with observed entity attributes to produce class images. Entity attributes are evaluated by value judgment functions in regions of the limbic system such as the amygdala to produce values that can be assigned to pixels carrying the name of the entity class.

single pixel) for a single sample period in time. Pixel attributes describe the properties of a single pixel in an attribute image, as shown in Figure 7.8.

- *Level 2: list entities.* List entities may consist of sets of about 10 ± 7 pixel entities that satisfy some gestalt grouping hypothesis over space and/or time. List entities may include edge, vertex, and surface-patch entities. For example, an edge may consist of a set of contiguous pixels for which the first or second derivatives of intensity and/or range exceed threshold and are similar in direction. A vertex may consist of two or more edges that intersect, or a single edge with an endpoint. A surface patch may consist of a set of contiguous pixels with similar first or second derivative of intensity and/or range and similar image flow vectors.

    Entity attributes are computed over the set of pixels that comprise the entity. For example, edge entity attributes may include the orientation and the length of the edge, the sharpness of the edge, or the magnitude of the discontinuity at the edge, as well as the centroid of the group of points that make up the edge. Vertex attributes may describe the relationship between the set of edges that make up the vertex. For example, vertex attributes may define the type of vertex (e.g., V, T, Y, or endpoint), the orientation of the vertex, and perhaps the angles between lines forming the vertex. Surface-patch attributes may describe the collective properties of the set of connected points that make up the patch. For example, surface-patch attributes may define the position, and velocity of the surface patch, the texture, and the orientation of the surface patch relative to the viewing point.

- *Level 3: surface entities.* Surface entities include surface and boundary entities. Surface entities may consist of sets of about 10 ± 7 contiguous surface-patch entities that satisfy some surface gestalt grouping hypothesis. For example, a surface may consist of a set of contiguous surface patches that have similar range, orientation, texture, and color as their immediate neighbors. A boundary may consist of a set of edge entities that are contiguous along their orientation. Surface and boundary attributes are computed over the entire set of pixels that are included within the entity. Surface attributes may describe the properties of the surface, such as its area, shape, roughness, texture, color, position and velocity of the centroid, and so on. Boundary attributes may describe the properties of the boundary between two surfaces computed over the set of points that make up the boundary. For example, boundary attributes may define the shape of the boundary, its orientation, its length, its smoothness, its position and velocity, which side its corresponding surface lies on, and so on.

- *Level 4: object entities.* Object entities may consist of sets of about 10 ± 7 contiguous surface and boundary entities that satisfy some object gestalt group hypothesis. For example, an object may consist of a set of surfaces that have roughly the same range and velocity and are contiguous along their shared boundaries. Object attributes are computed over the entire set of points that are included within the object. Object attributes may describe the properties of an object, such as its volume, shape, projected size in the image, color, texture, position and velocity of centroid, and orientation and rotation about the centroid.

- *Level 5: group entities.* Group entities may consist of sets of about 10 ± 7 object entities that have similar attributes, such as proximity, color, texture, or common motion. Group attributes are computed over the entire set of objects that are included within the group. Group attributes may describe the properties of a group, such as the number of its members, size, density, position, velocity, average direction of motion, and variance from the mean.

As noted above, this particular taxonomy of classes is arbitrary. In general, all internal representations of knowledge are heuristic hypotheses that are constructed by the intelligent system to categorize the sensory input and provide information needed for successful behavior. It is not important whether the representation is "true" in some abstract logical sense. It is not necessary that the representation be optimum. What is important is that the representation is useful (i.e., efficient and effective) in predicting the future and generating successful behavior. In the biological world, natural selection rewards success and punishes failure. Natural selection does not care about truth. It favors "true" representations only to the extent that they are likely to produce success. Models that are false but lead to successful behavior will be favored over those that are true but lead to failure.

The particular hierarchy of geometric entities chosen here for the knowledge database is designed to enable close coupling between sensory processing, world modeling, value judgment, and behavior generation processes. In the RCS architecture, information can easily flow both ways, top down and bottom up. Top-down, task priorities influence the choice of grouping hypotheses that generate hypothesized entities at each level in the SP hierarchy. Behavioral goals influence the determination of what is important and how the computing resources available to sensory processing and world modeling should be focused. Bottom up, measurements of entity attributes and detection of events influence the choice of plans and the control of behavior.

The importance of top-down inputs to perception cannot be overstated. Bottom up, there are an enormous number of different possible grouping hypotheses. There are typically thousands of ways to group pixels into objects. How can the system decide which is correct? How can SP and WM decide which grouping hypothesis is most efficient and effective in generating successful behavior? For example, should the collection of entities comprising the image of a tree be designated as an object? Geometrically, the image of a tree has surfaces and boundaries, each of which consists of edges and surface patches, each of which consists of pixels. Thus, a tree could be an object. However, a tree also has parts, such as a trunk, branches, roots, and leaves. Could the trunk of a tree be an object? It also has surfaces, edges, and points. However, if the tree trunk is an object, what about a hole in the trunk? A hole also has surfaces, edges, and points. This illustrates why the definition of an object entity cannot be made bottom-up solely in terms of geometry.

The definition of what an object is depends on the task. If the task is to approach a tree, the tree is the object. If the task is to avoid hitting the trunk of a tree, the tree trunk is the object. If the task is to inspect a hole in the tree trunk, the hole is the object. If the task is to follow a tree line, the group of trees comprising the edge of the woods is an object with surfaces, boundaries, edges, and points. If a tree line is an object, individual trees are simply parts of the surface that forms the edge of the woods. At each level of the BG hierarchy, task commands are typically defined in terms of one or more objects upon which the task is to be performed. Thus, at each level, the definition of an object is determined by the nature of the task.

At each stage of perception, an intelligent system can take action based on the knowledge acquired to that point. This enables the intelligent system to act quickly and efficiently without waiting for a complete analysis and classification of every entity and event in a situation. For many behavioral tasks, segmentation of the world into geometric entities is sufficient for generating effective behavior. For obstacle avoidance, simply knowing the position, size, shape, and motion of geometrical edges, surfaces, and objects is all that is required. For example, it is sufficient for an insect flying through a complex environment simply to know where surfaces are relative to its intended flight path. To avoid crashing into books on a bookshelf, a housefly does not need to recognize that they are books. All the fly needs to know is that a region in space is occupied and therefore unavailable for flying. Simple geometry can enable an intelligent system to avoid collisions. Geometry alone can enable a vehicle to compute whether the ground is too rough or too steep to be traversed safely.

However, for many behavioral decisions, geometric calculations alone are insufficient. For example, a rock and a paper bag may have similar geometric shape. However, a large rock presents an obstacle to high-speed driving, whereas a paper bag of the same size and shape does not. To a LADAR, the top surface of a field of grass may appear to have the same texture as an expanse of rocky soil, but tall grass may conceal deep ditches or other hazards, such as logs and stumps. Therefore, it is often necessary to distinguish between geometric entities and classify them as members of generic or specific classes.

GENERIC ENTITY CLASSES

Class membership in a generic entity class is defined when attributes of a geometric entity match attributes of a generic class prototype. For any generic entity class, there exists a set of attributes that define a class template (or prototype) that exemplifies the class and provides criteria for recognition or classification. Observed geometric entities that have attributes matching those of the generic entity template may be classified as belonging to the generic entity class.

Entity class prototype: an entity image or frame that exemplifies or provides the norm for an entity class

The entity class prototype is an ideal example of the class. It is a standard by which entities can be classified. An entity class prototype may be represented in the world model knowledge database in either iconic or symbolic form, or both. For example, Figure 7.12 shows both an prototype image and an prototype frame for a generic object class—a M1 tank. An entity class may have many prototype images, taken from different perspectives under a variety

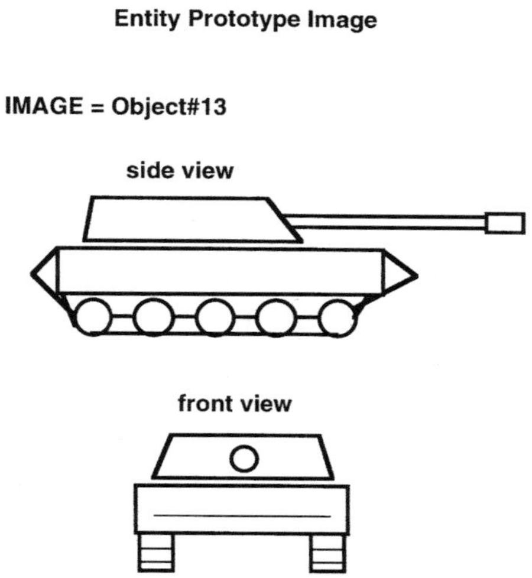

Entity Prototype Image   **Entity Prototype Frame**

IMAGE = Object#13   NAME = Object#13

side view

Type = Tank, Country of origin
Model = XYZ
Weapon = 90 mm cannon
Weapon range = 3 km
Top speed = 35 mph
Part-of = Blue force
Has-part1 = Turret
Has-part2 = Gun
Has-part3 = Body
Has-part4 = Wheels
Has-part5 = Dust cloud

front view

FIGURE 7.12   Entity prototype for image and frame representations. The entity prototype image may be a stored photograph or an image generated by a graphics engine from symbolic data such as a computer-aided design (CAD) data file. The entity prototype frame provides a symbolic description of a typical entity in the generic class. The entity frame may include information that is not available from sensory input. For example, the range of the weapon may be known from the generic class of the tank, but cannot be measured from sensory data.

of different conditions. An entity class typically has only one, or at most a few, prototype frames. This is because the attributes in the frame representation tend to be invarient with respect to perspective and viewing conditions. Thus, entity classification based on attributes in prototype frames may be quicker and more reliable than those based on prototype images, unless viewing conditions for the entity image are very similar to those of the prototype image.

Generic entity classes may include roads, buildings, grass, trees, bushes, water, dirt, sand, sky, and clouds. Generic class prototypes may exist for any geometric entity type (pixel, list, surface, object, or group entities). For example, pixels in aerial or satellite photos are often classified as belonging to a generic entity class on the basis of color or multispectral image values. A pixel may be classified as a road pixel if it has the color and texture of a generic road pixel template. Edge entities may be classified on the basis of orientation, magnitude, or type (intensity, color, or range edges). A geometric edge entity may be classified as a road edge if it has attributes similar to those of a generic road-edge class template. Surface entities may be classified on the basis of texture, orientation, shape, size, color, or motion. A geometric surface entity may be classified as the side of a building if it has attributes that match those of a generic building-wall class template. Object entities may be classified on the basis of shape, size, material, color, texture, position, orientation, or motion. Group entities may be classified on the basis of shape, size, density, or motion.

An intelligent system may have thousands of generic entity classes. The problem with a large number of classes is not the amount of memory required to store them, but the number of comparisons required to classify geometric entities as members of generic entity classes. As the number of classes grows, number of computations required to make a classification grows, as does the probability of misclassification. The number of comparisons required for classification can be addressed by parallel computational methods. The probability of misclassification can be reduced by including context and constructing a list of entities of attention. Such a list can be compiled using both top-down and bottom-up inputs to evaluate importance and likelihood of occurrence in a given task environment.

SPECIFIC ENTITY CLASSES

Class membership in a specific entity class is defined when attributes of a generic entity match attributes in a specific class prototype. A specific entity is a particular instance of a thing. A specific entity class has only one member. A specific entity is in a class all to itself. It has unique attributes that distinguish it from all others. For example, a specific object entity may be a specific person, a particular building, or a unique tree. A specific surface entity may be a particular surface of a specific object. A specific list entity may be a particular edge of a specific surface. A specific entity frame contains the attributes that uniquely define the entity as the one and only entity in that class. If the attributes of an observed geometric entity match the specific entity class attributes, the entity observed is recognized as the specific entity.

There is a taxonomy of entity classes and hence a taxonomy of entity frames. For example, a specific tree may be a member of the generic class of oak trees, which is a member of the generic class of deciduous trees, which is a member of the generic class of trees, which is a member of the generic class of plants, which is a member of the generic class of living things. This taxonomy of classes is sometimes called an *ontology* [Lenat et al.90]. An ontology can be represented in the knowledge database by a set of pointers, or links, between classes. A pointer to a more generic entity frame defines a "is_a" or "belongs_to" link. Pointers to more specific classes define "an_example_of" or "contains" links.

EVENTS

The world is a dynamic place governed by laws of energy and momentum that guarantee continuity in the position and motion of matter over time. The temporal aspect of the world is represented in the KD by events.

Event: something with temporal structure that is perceived to occur during a period of time

In computer science, switching theory, and discrete event systems, an event is defined as a change in state that occurs instantaneously in zero time. However, an instantaneous event is an artificial construct designed to simplify mathematical computation. In the real world, nothing occurs instantaneously. State changes in electronic circuits occur in nanoseconds. State changes in mechanical systems occur over periods of milliseconds or seconds. An event such as an earthquake may occur over a period of many seconds. An event such as a wedding or a funeral may occur over a period of an hour. An event such as a concert or ball game may take several hours. An event such as leaves turning color in the fall may take place over weeks. An event such as the fall of the Roman Empire may occur over a period of a century or more. An event such as the extinction of the dinosaurs may take thousands of years. The common feature among all events is that they occur over an interval in time and involve a change in state.

When we listen to the world we perceive events. We hear clicks, notes, words, and melodies. We perceive birds singing, cats meowing, dogs barking, crickets chirping, babies crying. We perceive traffic noises. We attend concerts and listen to lectures and debates. When we look at the world, we see patterns of movement. We see waves on the water. We see trees swaying in the breeze. We see patterns of motion in the ballet, in sporting events, and in human gestures. These patterns of movement can temporally be grouped and segmented into events. Events can have names and attributes and can be related to other event. Patterns of events can be grouped into higher-level events. Just as entities can be represented symbolically by entity frames, so events can be represented symbolically by event frames.

Event frame: a data structure that contains the event name, a list of attribute–value pairs, a pointer to a parent event, and pointers to subevents

An event frame contains the information that the intelligent system knows about an event. Figure 7.13 shows an example of an event frame. The event name is an address or index by which the event frame can be accessed in a database or library of events, or to which other event and entity frames can be linked. An uncertainty parameter can be associated with the name to indicate how certain the system is that the event has been identified properly. The event frame contains event attributes that describe attributes such as amplitude, shape, and spectrum. Event attributes may also characterize the event as an impulse, a string of impulses, a frequency, or a pattern of frequencies on a channel or set of channels. Event state variables describe dynamic properties such as start–end times, trigger, and length of the event. Uncertainty parameters can be associated with state parameters to indicate how dependable are estimates of these values.

```
NAME  = event_id (uncertainty)        // this is the frame address in the KD


// attributes  -  these are characteristics that address the question What?
amplitude     = magnitude of variations
shape         = time-frequency pattern, amplitude waveform, symmetry
spectrum      = Fourier transform over the event
type          = impulse, frequency, pattern of frequencies


// state – these are dynamic properties that address the question When?
boundaries   = start-end times (uncertainty)
trigger      = proximal cause of the event (uncertainty)
length       = duration of the event (uncertainty)


// class – pointer to the channel and class or classes to which the event belongs
channel              = signal pathway where the event was detected
generic_class1       = pointer to generic1 class prototype
generic_class2       = pointer to generic2 class prototype
specific_class       = pointer to specific class prototype


// value – worth of the event
benefit of event     = value
cost of event        = value


// pointers that define grouping relationships
belongs_to           = parent event
has_part             = subevent1
has_part             = subevent2
has_part             = subevent3


// pointers that define situational relationships
prior state          = conditions prior to the event
effect               = conditions resulting from the event
participants         = entities involved in or affected by the event
```

FIGURE 7.13   Structure of a typical event frame. Each event frame consists of a name, a list of attributes, and a set of pointers to other events.

The event frame may contain pointers to the signal channel on which the event was detected and to the class or classes to which the event belongs. For example, an event may have a generic class pointer that identifies it as a click, note, word, or melody. Generic events are events that have been classified but are not unique. Examples of generic event classes include shots, chirps, words, speeches, ball games, and wars. Generic classes of events include point samples of sensory signals and strings of samples that make tones. A fork event might be a point in time when two instruments begin to play two voices. A join might be a point where a duet becomes a solo. A time-frequency patch might be a sonogram of a musical chord or a speech phoneme. A boundary might be a partition between words or notes. A time-frequency surface might be a portion of an orchestral performance in which many instruments are playing many different parts simultaneously over a period of time. A specific class pointer may identify it as a particular word or melody. Examples of specific events include the shot that killed JFK, the string of words spoken by Neil Armstrong when he first set foot on the moon, and a specific concert by the Boston Pops.

The event frame may contain value attributes that define the cost or benefit of the event. The event frame may contain pointers that define inheritance relationships with other events. Each event frame has a pointer to the frame of a parent event of which it is a part and a set of pointers to frames of subevents that are its parts. For example, a word event may have a parent_pointer to the phrase or sentence to which the word belongs. It will have a number of subevent_pointers to the frames of the phoneme that are its parts. Inheritance pointers are established and maintained by grouping and classification operations performed by sensory processing functions discussed in Chapter 8.

The event frame may also contain pointers that define situational relationships such as prior conditions, effect, and entities that are involved in or affected by the event. Event effects may include changes in state produced by the event. A state change may be a change in position, orientation, velocity, color, temperature, size, or shape of entities in the world. State changes may also be changes in a signal from a sensor or the occurrence of a string or pattern of signals. It should be noted that two or more events may result in no net change in state. For example, the signal on a wire may change state from a "0" to a "1" and then change back to a "0." The net change is zero, but the historical record contains two transition events (or one "square pulse" event). An event frame may include a set of functions that describe what causes an event to occur under certain conditions or in response to certain stimuli. Behavioral functions and parameters may be inherited from generic or specific class prototypes.

Just as entities can be grouped into higher-level entities in space, so events can be grouped into longer events in time. For example, a series of acoustic sensor signal values can be grouped into a phoneme. A series of phonemes can be grouped into a word. A series of words can be grouped into a sentence. A series of sentences can be grouped into a paragraph or concept in a speech.

Event frames contain pointers that define a taxonomy of grouping relationships. Each event frame has a pointer to the frame of a parent event of which it is a part, or to which it belongs. Each event frame also contains a set of pointers to frames of subevents that are its parts or that belong to it. A level 4 event frame has a parent_pointer to the level 5 event frame to which it belongs. It has a number of subevent_pointers to the level 3 event frames that are its parts. Each level 3 event frame has a parent_pointer to the level 4 event frame to which it belongs, and a set of subevent_pointers to the level 2 event frames that are its parts. Each level 2 event frame has a parent_pointer to the level 3 event frame to which it belongs, and a set of subevent_pointers to the level 1 events that are its parts. Each level 1 event has a parent_pointer to the level 2 event frame to which it belongs. All of these pointers are established and maintained by grouping operations that are performed by sensory processing functions, discussed later.

Event frames may stand alone as data structures or be linked to form lists, strings, words, sentences, networks, and maps. Named event frames may be used to describe what happens in the world. In language applications, event attributes may serve as adjectives that describe characteristics of the events to which they are attached. Event frames can be interconnected by pointers to form causal, semantic, and situational networks. Situational networks may represent situations or temporal relationships such as *"before," "after,"* or *"simultaneous-with."* Situational networks may also have pointers to maps or images that display temporal relationships pictorially. Causal networks represent the cause-and-effect relationships between events, situations, and actions that occur in the world. Semantic networks represent the relationships between events, attributes, situations, and actions that define meaning and enable reasoning, logic, and language.

SITUATIONS AND EXPERIENCES

Much of what takes place in the world has both spatial and temporal patterns that are useful for an intelligent system to know for making behavioral decisions. To represent this aspect of the world, we define situations and experiences.

Situation: a relationship that exists between entities and events in space and time

A situation involves both a spatial and temporal configuration of entities and events. A situation consists of a number of entities interacting with each other over a period of time defined by an event or sequence of events. An intelligent system can be said to be aware of its situation when there exists a close correspondence between knowledge in the system's world model and the situation in the external world. Situation awareness enables a system to reason about the current state of the world, to analyze the historical events leading to the current state, and to plan for the future. Perception consists of the functional transformation of data from sensors, in the context of knowledge from long-term memory, into situational awareness or personal experience.

Experience: an internal representation of a situation

When we experience the world, we perceive situations. We perceive places filled with groups of objects, some of which move in complex ways. We perceive flocks of birds that swoop and turn in a flurry of beating wings. We perceive busy streets filled with vehicles and people interacting in intricate patterns of traffic. We perceive theatrical performances with scores of musicians, singers, and dancers in an auditorium filled with hundreds of people. Our perceptual systems are bombarded with visual, acoustic, and tactile experiences when we are immersed in situations in the world.

Events provide the temporal and causal structure to organize entities into situations. Event frames provide temporal pointers that link entity images and frames with semantic networks and situational graphs. The resulting experiences can be stored in short- and long-term memory and recalled when needed. For significant experiences, we can compute attributes and recognize classes that are important for making behavioral decisions. The remaining experiences may be classified as background, and largely ignored.

TEMPORAL PERSISTENCE OF REPRESENTATION

For a WM process to make efficient use of its available computing resources and memory capacity in the KD, it is useful to partition the KD into three parts. The first part supports immediate sensory experience. This provides high-speed access to data that are specific to the current focus of attention. The second part supports short-term memory. This integrates and remembers information over time intervals roughly equal to the planning horizon at that level. The third part supports long-term memory. It remembers information indefinitely and contains the entire store of knowledge accessible to the intelligent system.

Immediate Experience

Immediate experience is rich, vivid, and dynamic. Immediate experience of visual imagery is rich in detail and full of color and motion, with three-dimensional perspective. The immediate experience of hearing consists of a wide range of frequencies and intensities with a sense of direction based on amplitude and phase differences between the two ears. Immediate experience of touch can convey rich sensations of pressure, texture, vibration, and temperature. When combined with inertial sensors and proprioception, visual, auditory, and tactile experiences can provide a rich and compelling sense of position, orientation, velocity, and force.

The internal representation of immediate experience enables perception and supports the filtering and prediction mechanisms that are required for effective processing of signals and images. There are at least three distinct representations of sensory signals that contribute to immediate experience. These are:

1. *Observed signals, images, attributes, and states.* These consist of the current array of signals from sensors and the attributes that can be computed directly from sensory signals.

2. *Estimated signals, images, attributes, and states.* These are the output of filtering processes that integrate information from observed images and other sources over some interval of time and space.
3. *Predicted signals, images, attributes, and states.* These are the system's best guess of what the next sensory input will be, based on all the latest estimated state variables, plus knowledge of system dynamics and control actions.

The rich dynamic details of immediate experience are transient. They disappear as soon as the lights go out, or the sound stops, or tactile sensors lose contact with the environment. Careful psychophysical experiments designed to see what is preserved of the visual image after it no longer persists on the retina have given negative results. The visual image disappears almost immediately (except for after-images produced by sensor fatigue effects) and what was not noticed during the immediate experience cannot be recalled into view. [*Eidetic memory* (popularly known as *photographic memory*) is sometimes observed in children, but is extremely rare in adults.] Similarly, the immediate experience of sounds in an auditory sequence terminates almost immediately after the sound is gone. Once the stimulus is removed, all that remains of immediate sensory experience is a memory trace consisting of symbolic information that was specifically noticed during the experience and stored in short-term memory. All that can be recalled is what was noticed and stored or what can be reconstructed by the imagination from what was noticed and stored.

Short-Term Memory

Short-term memory differs from immediate experience in that it persists after the stimulus is removed. Short-term memory behaves much like a delay line, or dynamic random access memory (DRAM). It can store strings or sequences of symbols such as a phone number, a string of words, a musical tune, an acoustic signature, a train of thought, a plan, a procedure, or a path through state space. Short-term memory differs from immediate experience in that it contains only symbolic representations—not images, sounds, or tactile feelings. The types of information stored in short-term memory include entity attributes and state (i.e., size, shape, color, temperature, texture, position, and motion) and event attributes such as intensity, sequence, frequency, duration, and state trajectory. Short-term memory provides a space/time matrix that can support classification and prediction. The space/time matrix of entity and event attributes stored in short-term memory can be used to generate predicted attributes and state to compare with the observed attributes and state of immediate experience. Correlation between short-term memory and immediate experience provides the basis for signal detection, grouping, tracking, and spatial/temporal segmentation.

Short-term memory differs from long-term memory in that it is dynamic. It retains information by recirculation or rehearsal. If this recirculation is interrupted, or overwritten with new information, what was previously stored in short-term memory is lost. Short-term memory is limited in the number of entities it can hold. In a famous paper, "The Magical Number Seven, Plus or Minus Two," Miller[56] suggested that short-term memory can hold only five to nine entities or events as "chunks" of information. Larger numbers of things cannot be remembered individually but can be remembered by grouping them into chunks that represent higher-level entities or events.

Short-term memory provides a buffer between immediate experience and long-term memory. Entities and events detected in immediate experience can be selected by attention mechanisms to be transferred into short-term memory. While there, they can be evaluated by the VJ system. If evaluated as important, they can then be transferred into long-term memory. By this means, long-term memory can be kept up to date with entities and events that are worth remembering. Entities and events that are evaluated as irrelevant or unnoteworthy are discarded as soon as short-term memory is overwritten by subsequent input.

Short-term memory can also serve as a "cash" memory for entities and events selected by task-driven attention functions to be transferred from long-term memory into short-term memory. By this means, short-term memory can be primed with entities and events that are important to success in achieving task goals. In this case, short-term memory contains expected entities and events to be compared with entity and event observed in immediate experience. This can enable quick recognition of entities and events that are important to task goals.

Long-Term Memory

Long-term memory provides a repository of information that can accumulate and be retained over long periods of time. Long-term memory differs from short-term memory in that the information endures indefinitely. Long-term memory preserves temporal ordering through pointers and list structures such as strings or graphs, whereas short-term memory preserves temporal ordering through recirculating delay lines. In biological systems, long-term memory is the result of permanent changes in synaptic connections. These endure through sleep, unconsciousness, and electroshock therapy. In artificial systems, long-term memory is implemented by storage media such as optical or magnetic disk or tape. Information in long-term memory is retained through power outages or system shutdown.

The storage capacity of long-term memory is effectively unlimited. In biological systems, new information entered into long-term memory decays slowly with time or as the result of being overwritten but can be reinforced or embellished by rehearsal. Once firmly in place, long-term memories can endure for decades. Long-term memory may integrate many similar experiences of objects and situations into a single (or few) generic class(es). This can make it difficult to recall the details of a single undistinguished experience, especially if it is simply one of many similar experiences.

In biological systems, long-term memory apparently stores information solely in symbolic form. Of course, computer system are not subject to the same constraints as those of the biological brain. In artificial systems, mass storage technologies such as videotape and CD ROM can store maps, drawings, video records, and prototype images of generic and specific entity classes. Video recording mechanisms can record and playback sequences of images lasting many hours with no decay in fidelity over time. Video recordings can sometimes be compared and correlated with current experience to detect minute differences. This is typically not possible in biological long-term symbolic memory.

Computer systems can store images to be used as templates for image pattern matching in the recognition and classification processes. This has proven useful for recognizing images of objects taken from the same viewpoint under the same lighting conditions as those of the stored template. However, template matching has not proven very successful for recognition of objects observed under different conditions or for classification of objects that are not rigid bodies. The problem lies in the variability of the image of an object that comes from differences in position, orientation, lighting, and viewpoint. This variability makes stored images of little use except for recording what was observed for off-line viewing or processing.

KNOWLEDGE OF RULES

The long-term knowledge database contains rules of mathematics and logic that can be expressed in formulas such as equations, production rules, statements in propositional or predicate calculus, or rules of arithmetic and geometry. These are symbolic representations that describe the way the world works and how objects, events, and actions relate to each other in time, space, causality, and probability. Almost any mathematical or logical function can be expressed by a rule of the form

$$\text{IF} \quad (\text{input}) \quad \text{THEN} \quad (\text{output})$$

If the input is a proposition or predicate, the output can be the truth value of the proposition or predicate. If the input is a symbolic string, the output can be a transformed string. If the input and output are single-valued real numbers, the rule becomes a function of the form

$$\text{IF} \quad (x) \quad \text{THEN} \quad (y)$$

or

$$y = f(x)$$

If the input and output are vectors, the rule becomes

$$\text{IF} \quad (\mathbf{X}) \quad \text{THEN} \quad (\mathbf{Y})$$

or

$$\mathbf{Y} = H(\mathbf{X}^T)$$

where H is a transformation matrix.

In a digital computer, rules of mathematics and logic are typically represented in a form that can be solved by numeric or symbolic formulas, or can be computed by a function. In the brain, rules can be represented by functional mappings implemented by neural nets. In general, a function is defined as a relationship between an input and output. Thus, at least in principle, any one-to-one or many-to-one functional mapping can be represented by a table look-up where the input is the address of the location in the table where the output is stored. In other words, if the input is an address of a memory location, the output is the contents of the address. If the output is a value, the rule returns a value. If the output is a name (or pointer), the rule defines a relationship.

This suggests how rules of mathematics and logic can be implemented by neural mechanisms. The input to a neural net can be considered the address of a location in a look-up table. The neural net output is then the contents of the look-up table. For example, a CMAC (Cerebellar Model Arithmetic Computer) [Albus75a,b] or a multilayer neural net can implement any relatively smooth nonlinear function of a small number of input variables. Procedural methods can also be implemented in neural nets. If the output of the neural net is an input to another neural net, the output is a pointer to another location. Pointers can be used to define relationships between data structures. Pointers can also implement a string of function calls, finite-state automata, or a procedural program. Of course, procedural programs are simple to implement in a digital computer.

There are practical limitations to the representation of mathematical functions by table look-up. These limitations are typically related to the dimensionality and resolution of the input space and to the precision required of the output. One method the brain uses to overcome this limitation is through population encoding. In this case, the value of a vector or the class of an entity is encoded by the activity of an ordered set of neurons. Precise values can be achieved through population encoding by integrating over a large population of neurons that are broadly tuned with overlapping receptive fields [Hinton84, Snippe and Koenderink92, Georgopoulos95]. Barlow[94] shows that the dimensionality of the input can be quite large if the code is sparsely distributed over the population.

A second method the brain uses to overcome the limitations of dimensionality and resolution is to invoke numerical procedures or symbolic algorithms. Although it is possible to memorize simple functions such as the multiplication tables, most humans solve complex problems in mathematics or logic by learning how to carry out numerical or symbolic procedures such as solving equations or writing computer programs. This is the method typically used for computer algorithms.

Rules of Physics

Rules of physics are often expressed in compact mathematical formulas such as differential equations that describe the relationships between force, mass, and acceleration, velocity, and position. These may be embedded in control laws, prediction generators, and simulators used for control, recursive estimation, and planning. Rules of physics can also be formulated in terms of

$$IF \quad (input) \quad THEN \quad (output)$$

Hence, they can be represented either by equations in digital computers or by table look-up in neural nets. For everyday experience such as running, jumping, throwing and catching a ball, or driving a car, humans learn what to expect from the environment by trial and error. This is the type of knowledge that can be stored and retrieved by neural nets in table look-up form. For complex problems, procedural methods must be invoked to solve equations and execute programs.

Rules of physics may be embedded in structural or dynamic models.

Structural models: rules and equations that describe how physical structures are kinematically connected and how forces and stresses are distributed

Dynamic models: rules and equations that describe how forces and inertia interact with each other in time and space

Structural and dynamic models enable WM to simulate the results of hypothesized actions for planning. They also enable WM to predict the evolution of state variables for recursive estimation and predictive filtering.

SIMULATION AND IMAGINATION

The WM process uses the knowledge stored in the KD to predict what will happen in the future. Knowledge of entities, events, situations, and rules of mathematics and physics enable the world model to simulate the world and predict the results of hypothesized actions before they are executed. This capability is necessary for planning, recursive estimation, and predictive filtering. The benefit of simulating the future is that problems can be anticipated and preemptive action can be taken to avoid dangerous or unpleasant situations.

The brain has evolved the process of imagination, which enables simulation and visualization of the future as well as recall and analysis of the past. In the imagination, intelligent creatures can predict the results of hypothesized actions and evaluate the probable cost and benefit of future plans. Using the power of imagination, the brain can recall and replay past experiences from symbolic representations in long-term memory. These can be put into short-term memory, where they can be used to generate spatial/temporal images.

Entity attributes stored in memory include shape, size, color, position, and velocity. Event attributes include sequencing, duration, and temporal characteristics. This information enables WM simulation processes to construct sequences of graphic images in the mind.

Imagination is a critical factor in intelligence, in both species and individuals. The ability of higher species to envision the future and make long-range plans gives them a significant survival benefit. The ability of humans to generate sequences of images in their minds is obvious. People are able to imagine themselves in many types of situations and to visualize various scenarios. At an early age, children develop the ability for make-believe. They create imaginary friends and visualize themselves in fairytale situations. As humans mature, they acquire more sophisticated models of the world and develop the ability to make increasingly longer-range plans.

Imagination also provides a mechanism for dreaming. People commonly report dreams that involve images. It has been observed that during dreaming eye movements are consistent with viewing sequences of images. How the brain does this is a subject of considerable debate. Kosslyn[90] argues convincingly that the mind generates images during the process of recall and that these images can be searched for information in response to questions regarding what is remembered. Others [Lea75, Pylyshyn81] argue that the experimental data are also consistent with theories of memory that involve symbolic information stored in linked lists or graph structures. These viewpoints would not be in conflict if imagination were understood to be a process that uses symbolic knowledge to construct images in the brain.

It is clear, however, that whatever can be reconstructed from long-term memory is a pale shadow of immediate experience. Except for pathological cases of hallucinations, there is no mistaking what is imagined or dreamed for what is directly experienced. The imagination may be able to recall some details about an earlier experience, but nothing like what can be directly observed via immediate experience. For example, subjects are typically able to remember the number (up to about seven) of windows in a familiar wall by imagining standing in the room and counting windows. But one can never duplicate from memory exquisite details, or vivid colors, or the complexity of dynamic movement that occurs during immediate sensory experience of the natural world.

SUMMARY AND CONCLUSIONS

Figure 7.14 summarizes the images, maps, entity, and event data structures that make up the knowledge database. The total amount of memory required by the entire KD can be estimated from the assumptions in Figure 7.14. For a typical application, it is assumed that the library of entity and event classes stored in the long-term memory might contain on the order of 100,000 entries, where each entry represents a generic or specific class of object, event, or situation.

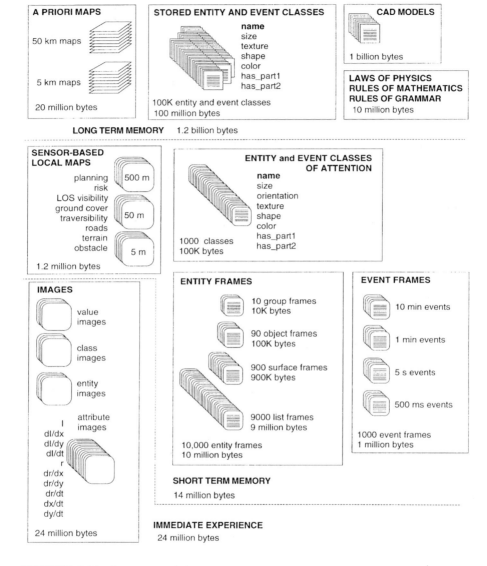

FIGURE 7.14 Summary of data structures in the KD. The KD is partitioned into immediate experience, short-term memory, and long-term memory. Images reside in immediate experience. Entity and event frames reside in short-term memory along with expected entity classes and sensor-based local maps. A library of stored entity and event classes reside in long-term memory along with a priori maps and CAD models.

Assuming that each entity has on the order of 100 attributes or pointers, each of which uses 10 bytes of memory, this requires 100 million bytes of storage. In a manufacturing environment, some objects may be defined using CAD solid models with up to 10 million bytes per object. One hundred such objects would require 1 billion bytes of memory. For some vehicle applications, 5- and 50-km maps may be required with a resolution of 1000×1000 pixels and perhaps 10 overlays. The memory required for these types of maps is 20 million bytes. The amount of memory required to store laws of physics and rules of mathematics and grammar is almost surely less than 10 million bytes. Thus, the total long-term memory requirements might be on the order of 1.13 billion bytes of storage. This requirement is easily met with current hard drive or optical disk technologies. Larger maps or more complex parts may require larger amounts of data, but 10- to 100-billion-byte storage media are readily available.

The amount of memory required for short-term memory is significantly less than for long-term memory. The 500-, 50-, and 5-m maps can all have resolution of about 200×200 pixels. Assuming 10 overlays for each map with one byte per overlay produces a requirement of only about 1.2 million bytes. If we assume that short-term memory also contains about 10,000 entity frames and 1000 event frames, each with 1000 bytes per frame, this produces a requirement for about 11 million bytes. The three entity images each contain 40,000 bytes. Thus, the total short-term memory requirements is less than 14 million bytes.

The amount of memory required for immediate experience is determined by the number of sensors. Typically, the majority of memory is required by imaging cameras. Assuming 10 cameras, each with 200×200 pixels, produces a requirement for 400,000 pixels. If each camera image is represented in terms of observed, estimated, and predicted images, the number of pixels triples to 1.2 million pixels. If there are 20 one-byte attributes or pointers for each pixel, the total requirements for immediate experience might be 24 million bytes. The total amount of RAM memory required for both short-term memory and immediate experience is less than 40 million bytes. Ten-billion-byte hard drives and 40 million bytes of RAM are readily available in off-the-shelf PC technology. Thus, the total amount of memory required to build the KD proposed in this chapter appears to be well within the capacity of a current desktop computer.

The amount of computation required to maintain the KD for a real-time intelligent system is considerably more challenging. The 24 million pixels in immediate experience need to be updated about 20 times per second. Assuming 100 operations per pixel, this yields a computational requirement of about 48 billion operations per second. Most of the approximately 14 million bytes of short-term memory need updating less frequently, on average perhaps only once per second. Assuming 1000 operations per byte yields a computational requirement of about 13 billion operations per second. The long-term memory portion of the KD can be serviced much less frequently, and can easily be maintained by less than 1 billion operations per second.

Thus, the total computing capacity required to maintain the KD may be on the order of 60 billion operations per second. This is well within the capacity of a typical modern supercomputer and can be approached with special-purpose hardware. State-of-the-art supercomputers can perform 100 billion operations per second [Moravec98]. Small special-purpose hardware systems can do more than 25 billion operations per second [Burt and Adelson83, Hansen98]. Current desktop computers can achieve only about 1 billion operations per second. However, the effective use of attention might reduce the amount of computing required by as much as a factor of 10. If attention can focus the available computing resources on the 10% of the KD that is most important, the remaining 90% might need updating 10 times less frequently, or with three times less spatial resolution. In this case the computational requirements would drop by an order of magnitude. Thus, attention could make it possible to maintain a real-time KD with only 6 billion operations per second. If this is true, the necessary computing power is currently available in a network of less than 10 desktop PCs.

Today, the biggest challenge appears to be in the limited bandwidth that is available for communication of information between computers. Current bus backplanes and local area networks present a major bottleneck. However, this is an engineering problem that will be solved. High-speed communications are being developed. Within a decade, single desktop computers will be capable of 100 billion operations per second. Then a network of 10 PCs will be able to generate one trillion operations per second. Within two decades, one trillion operations per second will be available on a single desktop computer.

8 Sensory Processing

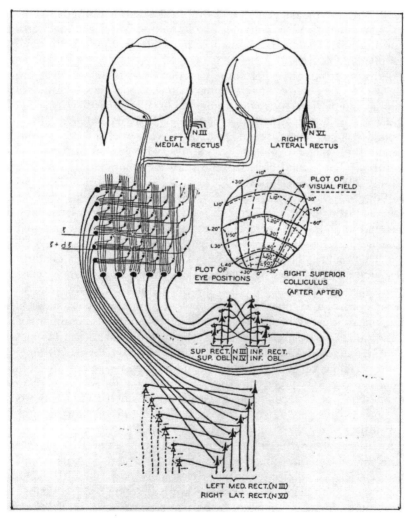

From W. Pitts and W. McCulloch, How to know universals: the perception of auditory and visual forms, *Bulletin of Mathematical Biophysics*, 9.

Sensory processing (SP) is the set of computational processes within each RCS node that keeps the world model knowledge database up to date. SP extracts from sensory signals the state variables, images, entities, attributes, events, symbols, maps, strings, arrays, classes, pointers, and other data structures necessary to generate and maintain an internal representation of the world that is useful for generating successful behavior. In the visual subsystems, SP performs the operations required to compute attributes such as brightness, color, range, spatial and temporal gradients, and image flow at each pixel. This results in a set of attribute images such as shown in Figure 7.8. SP performs spatial and temporal filtering of image data. For example, SP may compare observed attribute images generated from sensor data with predicted attribute images generated by WM. Variance is returned to WM to update estimated attribute images. SP also compares pixel attributes with class attributes, and assigns pixels to classes.

SP performs grouping operations to generate entity images such as shown in Figure 7.8. SP computes entity attributes such as shape, size, texture, position, orientation, and motion. Attributes computed from estimated entity images can then be compared with stored class attributes. This enables classification of entities into classes such as roads, ditches, fences, rocks, trees, bushes, dirt, sand, mud, fire, smoke, explosions, buildings, wire, bodies of water, automobiles, trucks, people, and animals. A survey of computational methods for performing all of these types of operations on images can be found in Ulman[96].

In the acoustic subsystem, sensory processing may detect frequency components and temporal patterns. Time and frequency patterns computed from the acoustic input are compared with patterns represented as stored class attributes. This enables the classification of sounds and acoustic signatures from sources such as helicopters, airplanes, trucks, birds, insects, dogs, cats, footsteps, or voices. At various levels in the SP hierarchy, acoustic sensory processing may recognize tones, phonemes, words, tunes, phrases, and sentences in a language. Sensory processing may also perform phase comparisons and correlation that can be used to compute the direction of arrival for incoming acoustic signals. A variety of computational methods for performing these types of operations on acoustic signals can be found in Green[76]. In the tactile subsystem, sensory processing may detect the position, orientation, texture, and temperature of surfaces being touched. In the smell and taste subsystem, sensory processing may detect the presence of pheromones or recognize the scent of food.

SP focuses attention on entities and events of importance to task goals and priorities and masks out sensory data that are irrelevant. Signals from sensor platform encoders and from inertial sensors may be processed and used to stabilize images and determine sensor pointing direction and tracking velocity in inertial coordinates. This enables sensory processing to infer the position and motion of objects in the external world and to transform objects observed in sensor coordinates into world map coordinates. It also enables entities in a world map to be projected back into sensor coordinates, to be overlaid on, and compared with, signals from sensors. This enables SP to track targets and analyze dynamic situations. In all these ways, SP provides WM with the information necessary to maintain the KD as a current, accurate, and relevant model of the world.

FIVE PROCESSING FUNCTIONS

Within the SP process in each RCS node, there are five basic processing functions: (1) focusing attention, (2) grouping, (3) computing group attributes, (4) filtering group attributes and confirming grouping hypotheses, and (5) classifying, recognizing, and identifying grouped entities and events [Ullman96, Riseman and Hanson90].

Focusing Attention

Attention selects (or windows) the regions of space and the intervals of time over which SP processes will operate on sensory inputs. The remainder of the input can be masked out or ignored. Windowing allocates the available computational resources to the entities and events that are most important for success in achieving behavioral goals. Regions within windows can be assigned priorities and be allocated computational resources in proportion to their relative importance.

At the servo level, focusing attention is accomplished by pointing sensors in a particular direction or positioning them at a particular location to optimize the observation of an object or region of interest. Visual images window the world from the viewpoint of the eye. The size, shape, and position of the visual window depend on the size of the sensor array, the focal length of the imaging optics, where the sensor array is located in position, and how it is oriented in roll, pitch, and yaw. Tactile images window the world from the viewpoint of the skin. Orienting the ears causes the acoustic environment to be windowed onto the cochlea. At various levels in the SP hierarchy, windowing can restrict processing to signals that emanate from regions of interest.

Windowing is controlled by an attention function that determines what regions and entities in the world are important and therefore worthy of attention. The attention function defines the shape, position, and duration of spatial and temporal windows and masks. It decides where the window of attention should be positioned and oriented, how tightly it should be focused, how it should move, and how dwell time should be allocated to various parts of the egosphere.

Figure 8.1 illustrates windows of attention in a visual scene. The solid rectangular windows positioned over the edges of the road represent selected regions of the image to be processed by an edge finder. The dashed windows positioned over the car on the road in front represent the fields of view

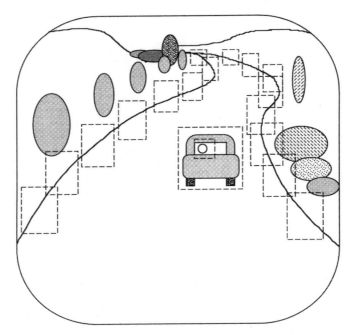

FIGURE 8.1 Typical image seen by a set of cameras mounted in an autonomous highway vehicle. Windows of attention are shown as a set of dashed boxes that cover the road edges and the car on the road in front. Note that all the windows are not of the same size. Windows for distant objects are smaller than those for near objects. There is a large window of attention around the car in front and a smaller window of attention focused on the driver of the car. Sampling frequency in space and time may differ from window to window.

of two additional cameras, a high-resolution camera focused on the car, and an ultrahigh-resolution camera focused on the driver of that car. Spatial and temporal sampling are also attention mechanisms that allocate computing resources to regions of space-time that are important. Regions of space or periods of time that are sampled at low frequency (or resolution) are effectively allocated a relatively small amount of sensory processing resources. Regions of space or time that are sampled at high resolution are allocated a relatively large amount of processing resources. For example, pixels in the fovea are densely packed so as to sample the visual image at high resolution, yielding high magnification. Tactile sensors in the lips, tongue, and fingertips are also densely packed so as to sample the tactile space at high resolution. Within the brain, approximately the same amount of computing resources is dedicated to processing each sensor signal. Thus, regions such as the fovea that are densely sampled are allocated a larger percentage of computing resources than those that are less densely sampled. The fovea is sequentially scanned over points of interest in the world. The dwell time and visit frequency for each point also determine the amount of computing resources allocated to that point.

Top Down versus Bottom Up. Attention processes that control windowing are driven by both top-down and bottom-up inputs. Top down, attention is driven by behavioral goals and priorities established by value judgment processes that assign value to entities and events and evaluate alternative courses of behavior. Entities and events may be placed on an attention priority list in order of their importance for achieving task goals and priorities [Ullman96, p. 147]. At each node in the organizational hierarchy, an attention function may specify the range (i.e., field of view) and scale (i.e., resolution) of windows of attention for SP processes in that node. For example, upper-level task goals may define a camera viewpoint or sequence of saccades and dwell points that is optimal for planning a manipulation or locomotion task, or for monitoring task execution. Task goals may also guide the alignment of internal models with sensory images [Ullman96, p. 57]. At all levels, task goals may specify selected regions within a visual image, or within an acoustic sonogram, for further processing. Low-level task goals may specify the particular points or trajectories on the egosphere where attention should be focused.

Bottom-up attention mechanisms are driven by entities and events that fall outside an expected norm or are designated as dangerous or otherwise note-worthy. For example, a region or entity in a visual image may be worthy of attention if it moves unexpectedly relative to a stationary background, or if it exhibits looming suggestive of an attack or eminent collision. A region in an image may also be worthy of attention if it exhibits shrinkage suggestive of flight. A region of the skin becomes worthy of attention if it reports pain or itches. A region of the egosphere may become notable if a loud or threatening noise emanates from that direction.

Grouping

Grouping aggregates or clusters lower-level entities and events into higher-level entities and events that can be assigned labels or names. At the lowest level in the vision system, each picture element (or pixel) aggregates all the incoming energy from the region of space imaged upon it over a time period determined by the sampling interval of the photodetector. This yields an intensity image such as shown in Figure 8.2. At higher levels in the vision system, the process of grouping partitions (or segments) images into spatial regions that can be assigned entity labels, or identifiers. For example, pixels may be grouped into list entities such as edges, lines, vertices, or surface patches. Figure 8.3 shows list entity images generated by grouping edge pixels computed from Figure 8.2*a* and *c*.

At the next level of SP, list entities can be grouped into surface entities such as surfaces and boundaries. For example, list edge entities that are contiguous and have similar slope can be grouped to form surface boundary entity

(a)   (b)

(c)

FIGURE 8.2   Three intensity images of suburban scenes. The gray level of each pixel is equal to the intensity at that pixel. [From Riseman and Hanson90.]

images. The two images on the right in Figure 8.3 are examples of surface boundary entity images. Figure 8.4 is an example of a surface entity image formed by grouping surface patch list entities. Whenever a grouping operation creates an entity, each pixel in the group is assigned a pointer to (i.e., name of, or identifier of) the entity to which it belongs. This creates an entity image consisting of pixels labeled with entity identifiers. In Figure 8.4, each surface entity is labeled with a surface identifier.

For each region in an entity image, an entity frame can be defined and labeled with the name of the region. The name is a pointer that links the pixels in the image to the entity frame in symbolic memory. The entity frame is a data structure that can store a list of the attributes computed over the region occupied by the entity. The grouping operation sets pointers in the entity frame that point to subentities and to the parent entity of which it is a

FIGURE 8.3 Edge pixels are grouped into list entities. The two images on the left show all list edges whose gradient magnitude in Figure 8.2*a* and *c* exceeds 10 gray levels per pixel. The images on the right are filtered on the basis of edge length. [From Riseman and Hanson90.]

part. The entity frame also contains a pointer back to the entity image in which the entity appears. This is illustrated in Figure 8.5. Each pixel in the surface entity image has a pointer that identifies the location in memory of the surface entity frame to which it belongs and each surface entity frame has a pointer back to the surface entity image in which it appears. Similarly, each pixel in the object entity image has a pointer to the object entity frame to which it belongs, and each object entity frame has a pointer back to the object entity image in which it appears. In addition, each surface entity frame has a pointer to the object entity frame to which it belongs, and the object frame has a set of pointers to the surface entity frames that are part of it. All of these pointers are established by the SP grouping operation that groups pixels into entities.

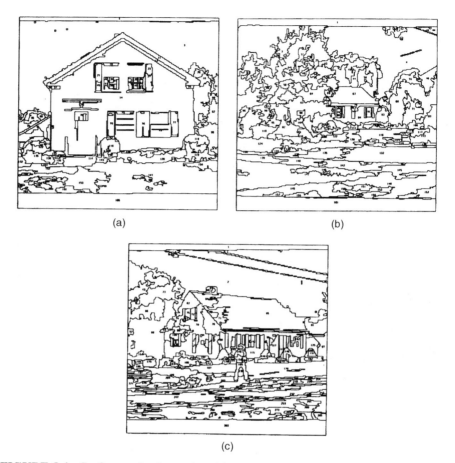

(a)   (b)

(c)

FIGURE 8.4   Surface entity image in which regions are labeled with geometric surface entity identifiers. The regions are not yet identified as members of a generic or
specific entity class. Thus, they are not yet labeled with class names. [From Riseman
and Hanson90.]

Temporal grouping can partition sequences of subevents into events (and
events into strings) that can be assigned labels or identifiers. For example,
a series of notes can be grouped into a melody, a series of drum beats can
be grouped into a rhythm, and a series of phonemes can be grouped into a
word. The grouping of events establishes a hierarchy of events. For example,
notes can be grouped into melodies, melodies into verses or refrains, verses
into songs, and a group of songs can be assembled into a concert. At the
lowest level in the speech-understanding SP system, each acoustic transducer
produces a voltage that depends on the amplitude and phase of the incoming
signal. At the next higher level, a filter aggregates strings of voltages into a

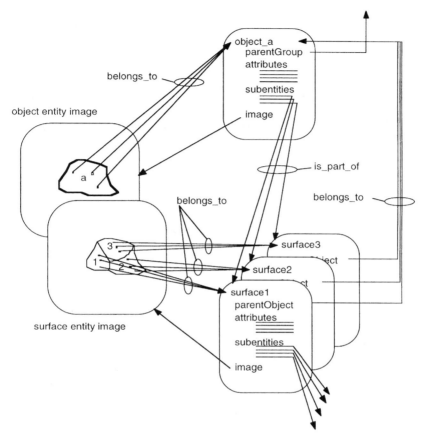

FIGURE 8.5   Grouping of pixels into surface and object entities. In the surface entity image, pixels are grouped into surface entities. In the object entity image, pixels are grouped into an object entity. Pointers in the entity frames describe the relationship between entities and provide links back to entity images.

spectrum of frequencies. At the next higher level, strings of frequencies can be grouped into phonemes. At still higher levels, strings of phonemes are grouped into words, words are grouped into phrases, phrases into sentences, and sentences into concepts. Figure 8.6 illustrates temporal grouping of subevents into an event.

Events are arrayed along the time line. Thus, a string of events can be represented by the contents of a delay line, a string of characters in a queue, a string of states output from a finite-state automata, a recording of sensory signals, or a historical record complied over a long period of time. Once a string of events is grouped into a higher-level event, characteristics of the string can be summarized and represented as attributes of the higher-level

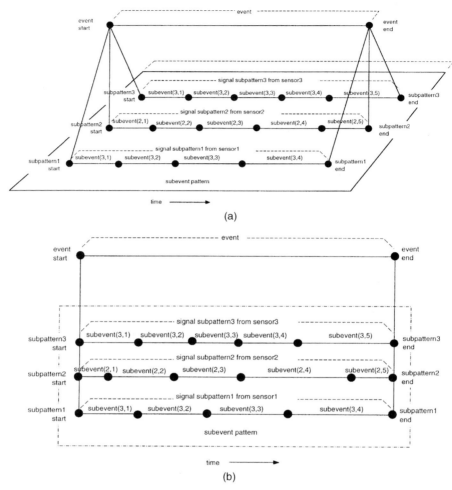

FIGURE 8.6 Temporal grouping of subevents into an event. An event consists of a set of subevents that have been grouped together. Each event has some duration and is bounded by points on the time line. (*a*) Event from an oblique perspective; (*b*) the same event projected onto the time line such that coincident subevents are aligned in a vertical column. Note the similarity between the event representation in part (*b*) and notes and bars in musical notation.

event. Grouping relationships establish the pointers that link subevents with events, and vice versa.

In Figure 8.6, three channels each carry a series of subevents that are grouped together into a single higher-level event. Each event has a start time and an end time. The event occurs during the interval between the start time and end time. In signal processing, the establishing of event boundaries cor-

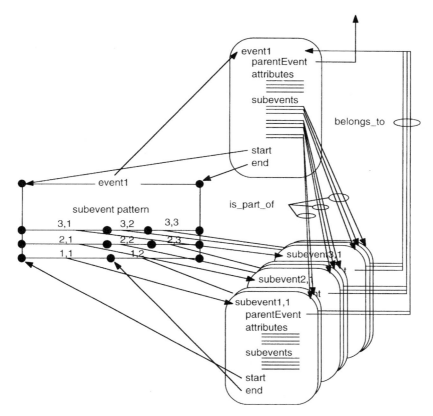

FIGURE 8.7 Events and event frames. Each observed event has a pointer to a frame that contains a list of its attributes, a set of pointers to subevents that belong to it, and a pointer to the higher-level event to which it belongs. These pointers are established by a grouping process.

responds to the establishing of a synch signal for bit, word, and frame boundaries. Synchronization is typically a first step in the detection of signal events.

Each event has an identifier that points to an event frame. Each event frame contains attributes of the event such as its start and end points, duration, pattern, tempo, intensity, spectral components, and meaning. Each event frame also contains pointers to subevents that are parts of it and the parent event of which it is a part. This is illustrated in Figure 8.7. Note that there are many similarities between events in the temporal domain and entities in the spatial domain.

Typically, there are many possible groupings of subevents into events, just as there are many possible groupings of subentities into entities. Any particular grouping is a hypothesis based on a set of assumptions about the world. For example, the spatial grouping of pixels into edges, or edges into surface boundaries, is typically based on a hypothesis that edge pixels in the image correspond to edges or surface boundaries in the external world. Both spatial and temporal grouping hypotheses are typically generated by gestalt heuristics [Biederman90, Kohler29, Koffka35]. These include:

1. *Proximity.* Subentities are close together in the image. Subevents are clustered along the time axis.
2. *Similarity.* Subentities have similar attributes, such as color, texture, range, or motion. Subevents have similar spectral properties or sequential relationships.
3. *Continuity.* Subentities have directional attributes that line up, or lie, on a straight line or smooth curve. Subevents have patterns with temporal continuity.
4. *Symmetry.* Subentities are evenly spaced or are symmetrical about a point, line, or surface. Subevents have regular rhythm or endings that rhyme.
5. *Simplicity.* Subentities are grouped in a way to produce the simplest figure. Subevents are grouped so as to produce the simplest composition.

There is no guarantee that any grouping hypothesis produces an internal entity that corresponds to an external entity in the real world. For example, two pixels that are in close proximity in an image may lie on completely different objects in the world. Two pixels of the same color and intensity in an image may lie on different objects, and two pixels with different colors and intensities may lie on the same object. Edges that line up in an image may or may not lie on the same edge in the world. Similarly, there is no guarantee that any temporal grouping hypothesis produces an internal event that corresponds to an external event in the real world. Grouping is often ambiguous, and many different grouping hypotheses may be plausible.

Grouping hypotheses need to be tested, and confirmed or rejected, by observing how well predictions based on each grouping hypothesis match subsequent observations of sensory data over time under a variety of circumstances. Hypothesis testing may be performed by comparing observations with predictions. This may be accomplished locally through filtering, recursive estimation, and correlation between observations and predictions, or globally through reasoning about consistency between what is perceived and what is consistent with physical, logical, or mathematical rules.

The fact that many different groupings are possible means that many different hypotheses may need to be tested before a good match is discovered. It remains a research issue how this can best be accomplished in real time with a finite amount of computational resources. However, there are many possible approaches, including parallel computation, multiresolutional search, continuity constraints, estimation theory, situation-dependent expectation, and task-driven focus of attention. See Ullman[96] for an extended review of this active field of research.

Computing Group Attributes

For each entity or event created by a grouping hypothesis, a set of attributes can be computed. For example, entity attributes such as position, velocity, orientation, area, shape, color, texture, and pattern of subentities can be computed by integrating subentity attributes over the region in an image covered by the hypothesized entity. Event attributes such as waveform, duration, motion, frequency components, and pattern of subevents can be computed by integrating subevent attributes over the duration of the hypothesized event.

For example, a brightness attribute can be computed for each pixel entity by integrating the photons within a spectral energy band falling on a photodetector in the image plane during a sample interval of time. A color attribute for a pixel may be computed from the ratio of brightness in two or more different spectral bands. Pixel attributes of spatial or temporal gradient of brightness or color can be computed from spatial or temporal differences between adjacent pixels in space or time. Pixel attributes of second, third, or fourth derivatives of brightness or color can be computed by convolution with spatial or temporal filters over larger neighborhoods.

A length attribute for an edge entity in an image can be computed by counting the number of pixels along the edge. The area of a surface entity in an image can be computed by counting the number of pixels contained in it. The cross-sectional area of an object entity in the world can be computed by multiplying the area of its projection in the image by the square of the ratio of its range to the focal length of the camera. The lateral velocity of an object entity can be computed by multiplying its estimated range by the angular velocity of its center of gravity in the image. The radial velocity can be computed from range-rate measurements. The orientation of an edge can be computed by regression of edge pixels onto a tangent line segment. The curvature of an edge can be computed by regression of edge pixels on a curved line segment. For surface entities, attributes such as area, texture, color, shape, and orientation can be computed. For surface boundary entities, attributes such as length, shape, and type (i.e., intensity, color, or range discontinuity) may be computed. For object entities, attributes such as size, shape, color, texture, and state including position, orientation, and velocity can be computed by integrating pixel attributes over the region occupied by the entity. Entity attributes and states can be represented in the list of attribute–value pairs contained in entity frames, as illustrated in Figure 7.9.

Filtering

Filtering is a process that reduces noise, enhances signal quality, and eliminates ambiguity. The computed values of entity attributes can be filtered over intervals of space by averaging or by convolution with spatial filters. Entity attributes and states can be filtered over intervals of time by phase-locked loops, by correlation with Fourier components, or by recursive estimation techniques such as Kalman filtering. Recursive estimation operates on each new sensory observation as it occurs to compute a new "best estimate" (over a window of space and time) of entity attributes and states. Each sensory measurement adds new information to what was known previously about entities and events in the world. Recursive estimation operates by comparing a prediction based on the current best estimate with an observation based on sensory input. Variance between observations and predictions are used to update the best estimate and to compute a confidence in the best estimate. Variance can also be used to compute confidence in sensory input data.

When variance between observed and predicted attributes is small, confidence in the grouping hypothesis is increased. When the variance between the observed and predicted attributes is large, confidence is reduced. When the confidence factor for entity attributes rises above a confirmation threshold, the grouping hypothesis that generated the entity is confirmed. When the confidence falls below a denial threshold, the grouping hypothesis is rejected and a new grouping hypothesis must be selected. Thus, confidence in estimated entity attributes can be used to confirm or deny the grouping hypothesis that created the entity. An entirely similar process can be used to compute best estimates for events and to confirm or deny event hypotheses. Hypothesis testing can be carried out in parallel for a number of hypotheses at multiple levels of resolution.

Classifying, Recognizing, and Identifying

Classification, recognition, and identification are all processes by which entities or events are classified into classes with names. The terms *classification, recognition*, and *identification* are often used interchangeably. In our discussion we use the term *classification* for the assignment of entities or events to classes (generic or specific) and reserve the terms *recognition* and *identification* for the assignment of entities or events to specific classes. The term *detection* often refers to classification of temporal events.

Any intelligent system has a library of classes stored in a knowledge database against which observed entities and events can be compared. For each class stored in the KD, there exists a set of attributes (i.e., an attribute vector) that is characteristic of that class. This characteristic class attribute vector is called the *prototype* of that class. Confirmed geometric entities and events can be classified into generic or specific classes on the basis of the degree of similarity between their measured attribute vectors and prototypes of the classes in memory. Classification occurs when the degree of similarity between an observed attribute vector and a class prototype satisfies a recognition criteria or exceeds a classification threshold. At this point, the confirmed entity or event is recognized as belonging to the entity class whose attribute vector it matches. The frame of an observed entity or event then can be labeled with the class name.

One possible method for computing similarity between an observed entity and a class is to calculate the dot product between the observed entity attribute

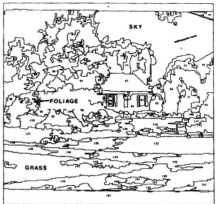

FIGURE 8.8 Three images in which surface regions are labeled with the class name of the entity occupying that region. [From Riseman and Hanson90.]

vector and prototypes of entity classes in the KD. Correlation is typically used to compute similarity between events and prototypes in the KD. Many other methods have been proposed. Ullman[96] provides an excellent survey of research results in classification. Figure 8.8 illustrates a surface image in which each of the regions is classified as belonging to a generic class such as sky, foliage, and grass.

Once an entity or event frame is recognized as a member of the class, it can inherit characteristics of the class. Furthermore, attributes of the class can be backprojected onto the image of the entity. Because the grouping process assigns to each pixel a pointer to the frame to which it belongs, all the pixels labeled as belonging to a frame can inherit the characteristics of that frame. This means that each pixel in an entity image (and in a corresponding region on a map) can be labeled with a class name and can inherit the attributes of the class. Thus, regions in images and maps can be labeled as roads, buildings, bridges, trees, and people, and inherit attributes that are characteristic of these classes. Thus, classification and recognition enable the mind to project class attributes, including preconceptions and characterizations onto entities and events in the world.

Many class attributes may not be observable by sensors or detectable by sensory processing. For example, the classification of a region in an image as a road carries with it a high probability that a moving vehicle may be encountered even though there are none currently within sight. The recognition of a particular road may carry with it the name of the road and attributes such where the road leads and what other roads connect to it, even though the end of the road is not observable and no intersections can be seen.

Labeling regions and objects in the world enables the intelligent system to select goals and plan behavior relative to objects in the world. Image and map overlays can be generated that contain information such as names and probable contents of buildings, types and capabilities of vehicles, types of vegetation and ground cover, and value of objects to be attacked, defended, approached, or avoided. For example, when an object on the egosphere is labeled as being prey, its value as food can be assigned to the region in space occupied by that object. Behavior can then be generated based on the benefit of the food, the cost and risk of the pursuit, and the likelihood of success. If an object is labeled as being an enemy, the risk of an attack can be weighed against the cost of running away given the current situation in the environment. Once regions in an image or map are labeled with names and class characteristics, BG processes can select appropriate behavior and plan paths through the world.

FUNCTIONAL MODULES AND DATA FLOW

Figure 8.9 is a functional module and data flow diagram of the SP and WM processes within each RCS node. At the upper right side, a task goal, priorities, and other parameters are specified by a command to a BG process. This information enables the WM to select a list of entity classes that are important for the task from a library of entity class frames that resides in KD. These are arranged in priority order in a list of entities of attention. This list can be used by another process within WM to compute *what* the entities on the list are expected (or known) to look like and *where* in the image (or on the egosphere) they are expected (or known) to appear. *What* entities look like can be computed from the attributes in the entity class frames in the WM. *What* information provides guidance to the heuristic selection of gestalt hypotheses that will be used to control the grouping of subentities into entities. *Where* important entities can be expected to appear in the image can be computed from the state variables in the entity class frames in the WM. *Where* infor-

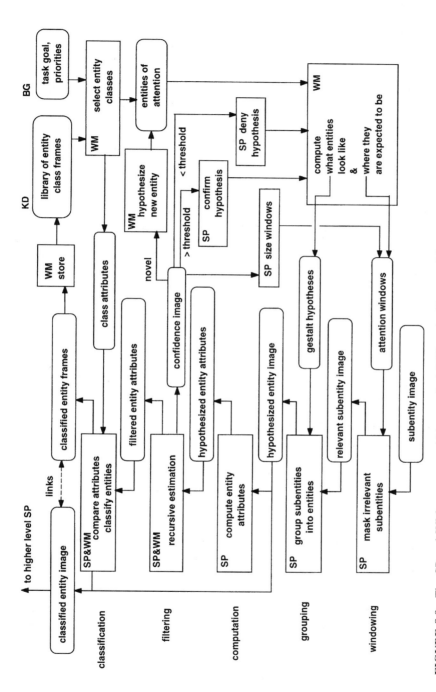

FIGURE 8.9 The SP and WM processes within a typical RCS node. The set of five basic processing functions that make up SP functionality for image processing are shown on the left. The boxes with rounded corners represent information. Boxes with square corners represent functional processes.

mation provides guidance to the heuristic processes that define windows of attention to be used to control pointing, tracking, and masking operations. At the bottom left, subentity images enter SP to be processed. At the lowest level, the subentity image is simply an array of pixels from the camera. At higher levels, the subentity image might be a list entity image, a surface entity image, or an object entity image. Subentity images are windowed to obtain image regions that are relevant to entities of attention. Subentities that reside within relevant regions are then grouped into hypothesized entities and assigned a group label. The labeled hypothesized entity image then goes two places: (1) it is output to the next-higher level, and (2) it is forwarded to the SP computation process where attributes of hypothesized entities are computed. Computed attributes of hypothesized entities are forwarded to a recursive estimation process that filters attribute values and generates a level of confidence in each entity hypothesis. (Note that a confidence factor for each entity produces a confidence image that is registered with the rest of the attribute and entity images.) The confidence image is used by three functions: (1) a SP process that confirms or denies the entity hypothesis, (2) a SP process that broadens or narrows the window size, and (3) a WM process that puts novel regions onto the list of entities of attention so that the novel regions can be tracked and identified. The confidence image can also be used to compute confidence values for each of the hypothesized entities.

Finally, at the top left of Figure 8.9, filtered attributes of confirmed entities are forwarded to a SP/WM classification process where they are compared with class prototype attributes. When the attributes of an entity match the attributes of a class prototype, the entity is classified as a member of the class and the class pointer in the entity frame is set to the name (or address) of the class frame. Classified entities can be stored in the KD, and each pixel in the entity image can inherit class attributes through its link to the entity frame.

The goals and priorities of the current task being planned and executed in the BG hierarchy affects the processing of sensory information in at least three ways:

1. Task knowledge influences the selection of attention functions. Those regions that are likely to contain information important to the task are windowed for processing. Those regions that are unlikely to contain relevant information are masked out and ignored.

2. Task knowledge influences the selection of gestalt hypotheses that perform grouping functions. Those grouping hypotheses that support planned behavior are favored.

3. Task knowledge defines a set of expected entities, events, and situations that are relevant to the task. This generates a set of expected entities and events and narrows the search for a match between observed entities and stored entity classes.

A summary of the relationships and interactions among the BG, WM, KD, SP, and VJ processes in a typical node of the RCS architecture is shown in Figure 8.10. The behavior generator (BG) process contains the job assignor (JA), schedulers (SC), plan selector (PS), and executor (EX) processes. The planner (PL) process includes the JA, SC, PS processes, the WM simulator, and VJ plan evaluator. The PS process selects the best plan for execution by the EX processes. The world modeling (WM) process supports the knowledge database (KD) that contains both long- and short-term symbolic representations as well as images that support immediate experience.

The WM contains the plan simulator, where alternative plans generated by JA and SC are tested and evaluated. The WM has a predictor mechanism for generating predicted states to be used by EX for anticipatory feedback error compensation. The WM can also generate predicted images that can be compared with observed images. The SP contains windowing, grouping, and filtering algorithms for comparing predictions generated by the WM process with observations from sensors. SP also has algorithms for recognizing entities and labeling entity images. The value judgment (VJ) process evaluates plans, assigns value to recognized entities, and computes confidence factors based on the variance between observed and predicted entity attributes.

Figure 8.10 illustrates how task commands can influence WM selection of SP attention and grouping functions. The task command to BG specifies the object on which the task is to be performed. This causes the WM to select from long-term memory a frame containing class attributes of the task object. This entity frame is then moved into short-term memory, where it can be used to generate expectations about where the task object is and what it looks like. This information, plus task goals and priorities, can be used to establish windows of attention and select grouping algorithms.

Each node of the RCS hierarchy closes a control loop. Input from sensors is processed through SP and used by the WM to update the KD. This provides a current best estimate $\underline{X}$ and a predicted state X^* of the world. The predicted state X^* is used in three ways:

1. The EX process uses X^* to compute the compensation required to minimize the error between the desired state Xd and the predicted state X^*.
2. The JA, SC, and WM plan simulator functions use X^* to perform their respective planning computations.
3. The WM predictor uses X^* to predict the next sensory input.

Within each node SP provides WM the information needed to update the KD. Lower-level entities and events are windowed and grouped into higher-level entities and events. Attributes are computed and filtered, and entities and events are classified and recognized. Labeled entity images may be used to generate labeled maps (not shown in Figure 8.10) that can be used by BG for path planning. Within each node, BG planning and execution processes use

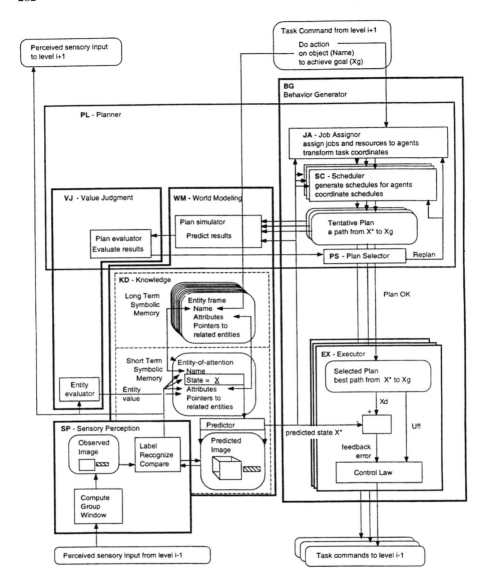

FIGURE 8.10 Relationships within a typical node at level i of the RCS architecture. A task command from level $i + 1$ specifies the goal and object of the task. The object specification selects an entity from long-term memory and moves it into short-term memory. The planner generates a plan that leads from the starting state to the goal state. The EX processes execute the plan using predicted state information in the KD short-term memory. SP operates on sensory input from level $i - 1$ to update the KD within the node and sends processed sensory information to level $i + 1$.

the knowledge maintained in the KD to plan and control behavior that achieves commanded goals.

RECURSIVE ESTIMATION

The interaction between the SP and WM is dynamic and complex. The WM generates a prediction based on a hypothesized best estimate of the state of the world. This WM prediction is compared with information derived from sensors or lower-level SP processes. The variance between what is predicted and what is observed is returned to the WM to update the hypothesized best estimate of the state of the world. This process is called *recursive estimation*. Kalman filtering is a particular type of recursive estimation.

Hypothesize and test by recursive estimation is the fundamental paradigm of the four-dimensional approach to dynamic vision pioneered by Dickmanns et al. over the past 15 years [Dickmanns92,95, Dickmanns et al.94]. Dickmanns [92] explains Figure 8.11 as follows:

At the upper left, the real world is shown by a block. Control inputs to the self vehicle may lead to changes in the visual appearance of the world either by changing the viewing direction or through egomotion. The continuous changes of objects and their relative position in the world over time are sensed by CCD-sensor arrays (shown as converging lines to the lower center, symbolizing the 3D to 2D data reduction). They record the incoming light intensity from a certain field of view at a fixed sampling rate. By this imaging process the information flow is discretized in two ways: There is a limited spatial resolution in the image plane determined by the pixel spacing in the camera, and a temporal discretization determined by the scan rate of the camera.

Instead of trying to invert this image sequence for 3D scene understanding, a different approach of analysis through synthesis has been selected, taking advantage of the available recursive estimation scheme after Kalman. From previous experience, generic models of objects in the 3D-world are known in the interpretation process. This comprises both 3D shape, recognizable by certain feature aggregations given the aspect conditions, and motion behavior over time. In an initialization phase, starting from a collection of features extracted by low level image processing (lower center left in Figure 8.11), object hypotheses including the aspect conditions and the motion behavior (transition matrices) in space have to be generated (upper center left in Figure 8.11). They are installed in an internal 'mental' world representation intended to duplicate the outside real world. This is sometimes called 'world 2,' as opposed to the real 'world 1.'

Once an aggregation of objects has been instantiated in the world 2, exploiting the dynamical models for those objects allows the prediction of object states for that point in time when the next measurements are going to be taken. By applying the *forward* perspective projection to those features which will be well visible, using the same mapping conditions as in the CCD sensor, a model image can be generated which should duplicate the measured image if the situation

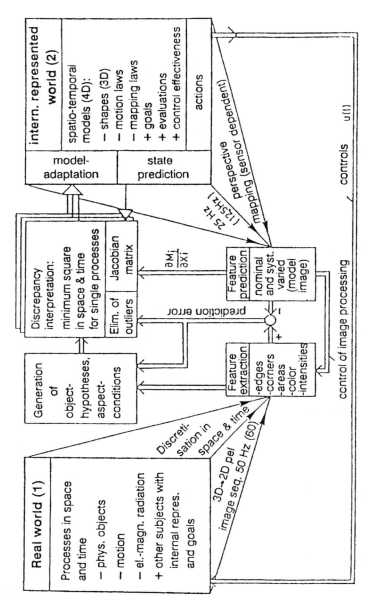

FIGURE 8.11 Basic scheme for four-dimensional image sequence understanding by prediction error minimization. [From Dickmanns92.]

has been understood properly. The situation is thus 'imagined' (right and lower center right in Figure 8.11). The big advantage of this approach is that due to the internal 4D-model not only the actual situation at the present time but also the Jacobian matrix of the feature positions and orientations with respect to all state component changes can be determined (upper block in center right, lower right corner). This need not necessarily be done by analytical means but maybe achieved by numerical differentiation exploiting the mapping subroutines already implemented for the nominal case.

This rich information is used for bypassing the perspective inversion via recursive least squares filtering through feedback of the prediction errors of the features. More details are available in Dickmans and Graefe[88].

A precise mathematical statement of the recursive estimation process in a sampled data system is

$$\underline{\mathbf{x}}(k \mid k) = \mathbf{x}^*(k \mid k-1) + \mathbf{K}(k)(\mathbf{y}(k) - \mathbf{y}^*(k \mid k-1)) \tag{8.1}$$

where

$$\underline{\mathbf{x}}(k \mid k) = \text{estimated state of the world at time } k$$
$$\text{after a measurement at time } k$$

$$\mathbf{x}^*(k \mid k-1) = \text{predicted state of the world at time } k$$
$$\text{after a measurement at time } k-1$$

$$\mathbf{y}(k) = \text{observed sensory input at time } k$$

$$\mathbf{y}^*(k \mid k-1) = \text{predicted sensory input at time } k$$
$$\text{based on estimated state at time } k-1$$
$$\text{plus control output at time } k-1$$

$$\mathbf{K}(k) = \text{inverse measurement model and}$$
$$\text{confidence factor}$$

SP processes receive observed sensory input $\mathbf{y}(k)$ directly from sensors or from lower-level SP processes. WM processes simultaneously generate predicted sensory input $\mathbf{y}^*(k \mid k-1)$ based on the previous best estimate of the state of the world $\underline{\mathbf{x}}(k-1 \mid k-1)$ stored in the knowledge database. SP processes compare observed attributes with predicted attributes and compute variance. WM processes then use the variance $\mathbf{y}(k) - \mathbf{y}^*(k \mid k-1)$ to update the knowledge database, producing a new best estimate $\underline{\mathbf{x}}(k \mid k)$. Statistics are kept on the variance to assess the confidence in the model.

A block diagram of the recursive estimation loop between the SP and WM modules is shown in Figure 8.12. The entity attributes in the entity frame format are shown at the right. This is the estimate at time $k-1$ after the measurement at $k-1$. This estimate is projected forward in time by the matrix $\mathbf{A}(k-1)$ and combined with the effect of the control input $\mathbf{u}(k-1)$ to pro-

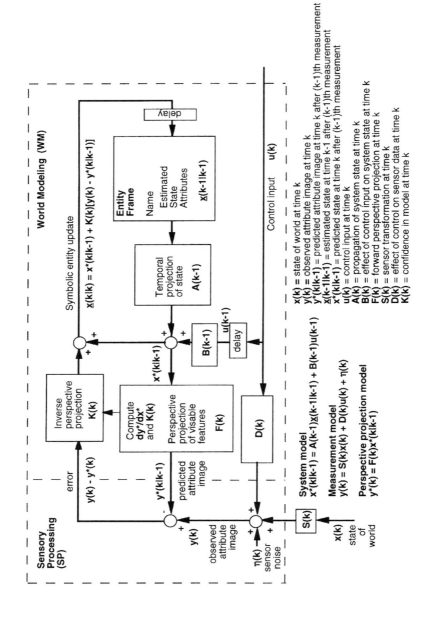

FIGURE 8.12 Recursive estimation loop. This loop updates entity attributes in the WM knowledge database (KD) based on differences between predicted images and observed images computed in the SP modules.

vide a prediction of the entity attributes at time k, based on information from measurements up to and including time $k - 1$. The entity attributes are then transformed into image coordinates through the perspective projection $\mathbf{F}(k)$, resulting in a predicted attribute image $\mathbf{y}^*(k)$. This is compared with the observed attribute image $\mathbf{y}(k)$. This comparison may consist of correlation or difference operations. The result is an error signal $\mathbf{y}(k) - \mathbf{y}^*(k)$. This error is then transformed back into attribute coordinates through the inverse Jacobian $\mathbf{K}(k)$. The forward Jacobian can be computed numerically from the forward perspective projection $\mathbf{F}(k)$ as described by Dickmanns[92]. The transformed error is then added to the predicted entity attribute vector $\mathbf{x}^*(k \mid k - 1)$ to produce the new estimated entity attribute vector $\mathbf{x}^*(k \mid k)$. The text included in Figure 8.12 contains mathematical expressions for the system model, the measurement model, and the forward and inverse perspective projection models.

INITIALIZATION

A recursive estimation loop requires initialization. There must be an initial estimate at time $k = 0$ in order to make an initial prediction. In simple cases the initial estimate can be zero. This will produce an initial prediction of zero and an initial error equal to the negative of the initial observation. Thus, the observation at $k = 0$ becomes the estimate at time $k = 1$. The recursive estimation process then proceeds to refine the $k = 1$ estimate with each new measurement of the state of the world.

In more complex cases, part of the initial estimate can be provided top down by a priori information. Additional information can be derived through a series of hypothesis and test procedures designed to build up a reliable model of the situation that exists in the world. For example, when an intelligent system first wakes up from an unconscious state, it must first initialize itself to a known starting state and then initiate procedures that answer such basic questions as "Where am I?", "What is the situation in my immediate vicinity?", and "What am I supposed to do?"

An intelligent system must always have answers to these basic questions. If it does not, it must invoke procedures designed to discover the answers. For example, to discover where it is in the world, the system typically must measure its orientation with respect to gravity. It must use its visual, tactile, and acoustic sensors to explore its immediate environment. Is it light or dark? Where is the ground? What is the ground made of? What is its shape? Are there objects nearby? If so, where? Are any objects moving? What are they? How many are there? It may be necessary to look around and move about. The system may need to probe the environment—to touch things and see how they feel. It may need to push on objects and see if they move, or look at objects from many sides. Turn them over. Pick them up. Smell them. Taste them. Squeeze them. Thump them.

To answer the question "What should I do next?" the system must look to its command input. "What is the commanded goal for the next second, minute, hour, day, week, month, year?" Intelligent systems cannot function without a goal. If an intelligent system has no goal, it must request a goal from an external source, pick a goal from a library of goals, or use a default goal. Once the initialization process is complete, the system knows where it is, what is going on in the world around it, what it is doing, and what it is supposed to do for the foreseeable future. It then can begin to decompose its high level goals into plans for action. As part of this process, it must decide what is important and plan how to behave in order to accomplish its goals in the safest, most efficient, and cost-effective manner.

When a RCS control system is powered up initially, the main program creates the control modules included in a header file and establishes the NML communications network specified in a configuration file. The NML communications system then initializes all of the RCS modules using shell script. This establishes the topography of the interconnections between modules and initialized each of the RCS modules. Sensors are turned on and sensory data processed to discover the state of the world and initialize the world model. Once the world model is initialized, the system is ready to accept commands and generate plans.

Analysis of the current situation and initialization of the world model are bottom-up processes. Where is the gravity vector? Where is north? What is the date and time? Where is the system located on the map? What entities can be observed in the immediate environment? What is the current situation in the world?

Accepting commands and generating plans are top-down processes. The top-level control module generates a plan out to its highest-level commanded goal. This provides a planning umbrella that extends well beyond (typically, 10 times further than) the planning horizon of the next-lower level. The first two steps in the highest-level plan are then passed to the next-lower level for planning purposes. This process is repeated at each level. As planning ripples down the hierarchy, each level begins planning within the context of an overarching plan from above that stretches 10 times further into the future. Once all levels down to the lowest level have generated plans, the system is ready to begin work.

As plans are executed, reaction to sensory feedback is initiated at the lowest possible level, where reaction latencies are the shortest. When errors are detected or unexpected conditions are recognized, error compensation, replanning, and emergency actions are triggered immediately. Only if reactive behavior at the lower levels is unable to cope with errors is action required by the higher levels. Thus errors ripple up the hierarchy, with more global reactive behavior evoked at each level. In all cases, reactive behavior takes place within the context of a hierarchy of plans that extend to the mission goal of the highest level. Failure occurs only when a problem cannot be corrected by compensation, replanning, or emergency action at any level.

HIERARCHY OF PERCEPTION

We turn now to the vertical interconnection between SP processes in RCS nodes at different hierarchical levels. The spatial grouping operations in the sensory processing hierarchy define levels of abstraction in the WM and KD hierarchy. The windowing, grouping, filtering, computation, and classification operations performed in the SP and WM hierarchy are largely defined by the requirements of planning and control in the BG hierarchy. The behavior generation hierarchy requires many different levels of abstraction and of range and resolution in the knowledge database to plan and control behavior at many different levels of control. Thus, a hierarchy of SP processes is required to extract information at many different levels of abstraction and at many different resolutions in space and time. The knowledge database (KD) is organized as a hierarchy of entities, events, classes, images, and maps that provide the BG processes with the knowledge of the world at the level of abstraction and resolution required to accomplish task goals at each level. Thus, at all levels, SP, BG, and WM processes work together to construct and maintain a distributed multiresolutional world model.

As illustrated in Figure 5.10, SP processes are in many ways a mirror image of the planning and control processes in the BG hierarchy. BG processes generate behavior by decomposing tasks into subtasks for a number of agents over an extended period of time in the future, while SP processes generate perception by integrating information from a number of sensors over an extended period of time in the past. The BG hierarchy generates a hierarchy of tasks and plans while the SP hierarchy generates a hierarchy of entities and events. The detection of events and the tracking of entities in the SP hierarchy provides the timing for sequencing subtasks in BG. The definition of tasks and goals in the BG hierarchy provides expectations for entities and events in SP.

Computations in the BG hierarchy are primarily top down, decomposing high-level conceptual representations of goals and tasks for a unified self into strings of low-level actions by a large number of subsystems and actuators. Top-down inputs from BG to SP enable focusing of attention, masking of unimportant signals, and selective application of sensory processing algorithms. Computations in the SP hierarchy are primarily bottom up, integrating low-level signals from a multitude of sensors into high-level gestalt perceptions of situations and concepts. Bottom-up inputs from SP to BG enable reactive or reflexive behavior.

At each level, interactions between SP and WM processes can generate a variety of windowing, masking, grouping, filtering, prediction, detection, and model-matching processes that enable classification, segmentation, scene understanding, and learning. For the SP processes to compare observations with predictions, it is necessary for observed attributes be comparable with predicted attributes. This means that hypothesized entities in the SP hierarchy should be comparable with the classes of entities in the WM and KD hierarchy.

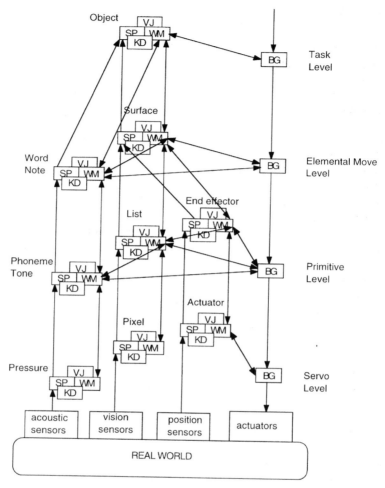

FIGURE 8.13 Three SP/WM/VJ/KD hierarchies that provide information about the world to a BG hierarchy for an intelligent vehicle. At each successively higher level, information from different sensory inputs is integrated and grouped into higher-level, more global entities and situations. Note that the acoustic and vision systems do not interact with each other or the BG hierarchy at the servo level. The entity class hierarchy on the left can be orthogonal to the BG process hierarchy on the right. This is enabled by vertical connections between WM modules. See, for example, Figure 9.4.

It also means that the observed entities in SP shuld be expressed in the same coordinate system as the predicted entities from the WM.

Each of the nodes in the RCS hierarchy is built around a BG process, with the SP, WM, KD, and VJ functions necessary to support that BG process. This is illustrated in Figure 8.13. At each hierarchical level in the BG/WM hierarchy, task goals and task objects are of different kind and scale and there are different types of planning and control requirements. Similarly at different levels of the SP/WM hierarchy, there are different kinds of computational algorithms and data representations. For example at the servo level, task objects are state variables, each of which can be computed from one or more sensor signals at a single point in space and time. Task goals are desired values of these state variables. State variables are expressed in sensor and actuator coordinates. At the primitive level, task objects may be edges, vertices, or curves that represent features in space and time such as road edges, object edges, and so on. Task goals are expressed in terms of desired states in configuration space (i.e., desired positions, velocities, and orientations of task objects) expressed in a vehicle (or manipulator) coordinate frame. At the subsystem level, task objects may be surfaces of objects or terrain features. Task goals are desired relationships between the vehicle (or manipulator) and task objects. Goals may be expressed in coordinates that are defined by the surfaces of task objects. At the vehicle (or task) level, task objects may be real-world objects such as vehicles, buildings, trees, bushes, and people. Task goals may be desired relationships between the self object (vehicle or manipulator) and task objects expressed in task coordinates. Thus, tasks and world model knowledge are expressed in terms of different types of entities and event at different levels of range and resolution in different coordinate systems at different hierarchical levels.

EXAMPLE OF A SP HIERARCHY

Figure 8.14 illustrates how RCS generates an image processing hierarchy by grouping lower-level entities into higher-level entities at several levels. Pixel entities are grouped into list entities, list entities are grouped into surface entities, and surface entities are grouped into object entities. As a result of each grouping process, pointers are established between entity images and entity frames at each level. In addition to the grouping processes at each level, SP processes include windowing, computation, filtering, and classification. The following is an example of windowing, grouping, computation, filtering, and classification processes at five levels of a RCS sensory processing hierarchy designed for an autonomous vehicle. At each level, entities represent segmented regions in an image and events represent bounded intervals in time.

Level 1 Regions: Individual Sensor Signals

Inputs to level 1 of the SP hierarchy are signals from individual sensors integrated over a single sample period. A single sensor signal may represent the position, velocity, or force of a single actuator. A sensor signal may represent the brightness measured by a single photodetector in a CCD or FLIR camera, or a single range measurement from a LADAR camera or radar antenna.

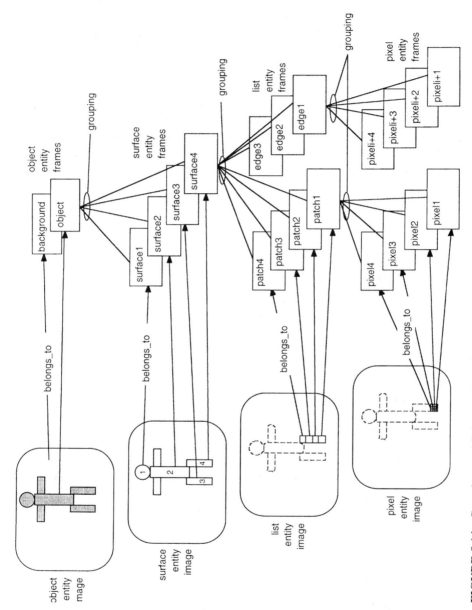

FIGURE 8.14 Grouping establishes a hierarchy of entity images and frames. At each level, pixels are identified in entity images by pointers to the entity frames to which they belong. Pointers between entity frames express grouping relationships.

272

A signal may also represent a single measurement by an acoustic sensor, an accelerometer, a gyro, speedometer, or GPS receiver.

Level 1 Windowing. At level 1, a window of attention defines the region of space and time that is sampled by each sensor during each sample interval. An attention window may move about over the egosphere and an attention schedule may define how frequently each sensor is sampled. For acoustic or microwave sensors, windowing may be performed by pointing directional microphones or antennae at high-priority targets. For position, orientation, velocity, acceleration, and force sensors, windowing may consist of selecting those signals that are relevant to the current servo task command for each actuator. For vision sensors, windowing includes pointing and focusing a camera photodector array on the region of the egosphere that is most worthy of attention.

There may be several cameras of different types with different resolutions, fields of view, and frame rates. The attention system will focus the highest-resolution cameras on the most important region in the image while using lower-resolution cameras to observe surrounding and less important regions. The attention window may track the highest priority entity of attention, or saccade quickly from one high-priority entity to the next as the priority list is updated.

Level 1 Grouping. At level 1, grouping simply means that each sensor integrates, or groups, all the energy impinging on it during a sample interval. For example, each photodetector in a CCD camera integrates the incoming visible radiation over the spatial region occupied by that photodetector and over the temporal interval of an exposure time period.

Level 1 Computation. A number of attributes can be directly measured for each sensor signal. For an acoustic sensor, the amplitude of the acoustic signal can be computed. For a black-and-white TV camera, the intensity (I) of radiation at each pixel is measured. In a color TV camera, the intensity of red, blue, and green light at each pixel is measured. In a laser range imager (LADAR), or in an imaging radar, a value of range (r) is measured at each pixel. For a forward-looking infrared (FLIR) camera, the temperature at each pixel is measured.

Additional attributes for each pixel in an image can be computed from measured attributes in the local neighborhood. For example, the x- and y-intensity gradients ($dI/dx, dI/dy$) may be computed at each pixel by calculating the difference in intensity between adjacent pixels in the x- and y-directions, or by applying a gradient operator such as a Sobel operator on a neighborhood about a pixel. The temporal intensity gradient (dI/dt) can be computed by subtracting the intensity of a pixel at time $t - 1$ from the intensity at time t or by applying a temporal gradient operator. Second-, third-, or fourth-order derivatives can be computed by convolution with spatial or temporal filters over larger neighborhoods. Differences in range at adjacent pixels can be used to compute spatial range gradients $(dr/dx, dr/dy)$ and surface roughness, or texture (tx). Spatial range gradients can also be computed from shading $(dI/dx, dI/dy)$ when there is knowledge of the incidence of illumination. Temporal differences in range at a point can yield range rate (dr/dt).

For a pair of stereo images, disparity can be computed at each pixel. When combined with camera vergence and knowledge of camera spacing, disparity can be used to compute range for each pixel at every point in time. If the camera image is stabilized (or camera rotation is precisely known), spatial and temporal gradients can be combined with knowledge of camera motion derived from inertial and speedometer measurements to compute image flow vectors $(dx/dt, dy/dt)$ at each pixel (except where the spatial gradient is zero; even in this case, the flow rate sometimes can be inferred from the flow of surrounding pixels). Under certain conditions, range at each pixel and time to contact (TTC) can also be computed from image flow [Albus and Hong90, Raviv and Herman94]. Each attribute that can be measured or computed at each pixel forms an attribute image. The set of attributes that can be measured or computed at each pixel forms a pixel frame, or attribute vector, for each pixel.

Level 1 Filtering. Each sensor signal can be filtered by recursive estimation to reduce noise and to improve signal-to-noise ratio. Figure 8.15 is an example of how recursive estimation might be used to filter pixel attribute images. In this example, there are two recursive estimation loops: an outer loop where image flow rate at each pixel is estimated and an inner loop where pixel attributes at each pixel are estimated. In the inner loop, each observed (i.e., measured or computed) pixel attribute image is compared with a predicted attribute image. The difference between a pixel attribute in an observed attribute image $\underline{A}(i,j,t)$ and a predicted image $A^*(i,j,t)$ can be used to update the estimated attribute image $\underline{A}(i,j,t)$.

The prediction process uses the predicted pixel motion at each pixel to predict where an attribute estimated at time $t-1$ will occur in the predicted attribute image at time t. For example, in Figure 8.15, a predicted attribute image $A^*(i,j,t)$ is generated from an estimated attribute image $\underline{A}(i,j,t-1)$ using predicted image motion at each pixel $dx^*(i,j,t)$, $dy^*(i,j,t)$, and $dr^*(i,j,t)$. Predicted image motion is computed by combining information from two sources:

1. Estimated image flow rates $(d\underline{x}/dt, d\underline{y}/dt)$ and range rate $(d\underline{r}/dt)$, which describe the current motion of entities in the image
2. Commanded camera motions and other actions that affect the motion of entities in the image

Estimated image flow rates are computed in the outer loop of Figure 8.15. Predicted attribute images can be perturbed in space systematically to generate a correlation function between observed and predicted attribute images. The

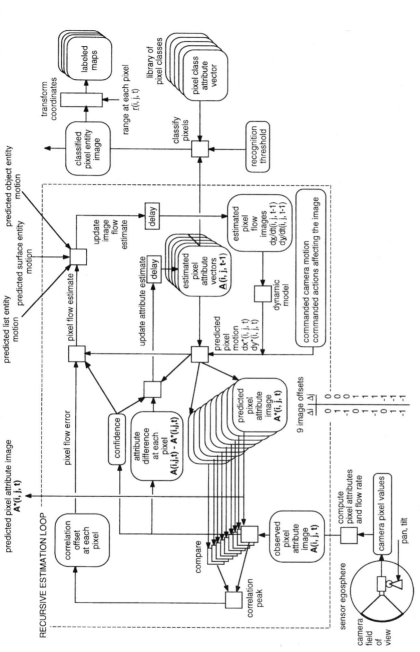

FIGURE 8.15 Image processing at level 1. Pixel signals from cameras are used to compute observed pixel attribute images. These are compared with predicted pixel attribute images and the difference is used to update estimated pixel attribute images. Perturbed attribute images are used to compute a correlation function over a set of image offsets. The peak of the correlation function is used to compute errors in the estimated pixel flow. This is combined with higher-level estimates of entity motions to estimate pixel flow at each pixel. Predicted pixel attribute vectors are compared with pixel class attribute vectors to classify pixels and generate a classified pixel entity image.

275

offset in the peak of the correlation function at each pixel is a measure of the error in predicted position caused by errors in estimated image flow rates. Thus, the correlation offset at each pixel can be used to update the flow rate estimates for each pixel.

Of course, the computation of image flow at a single pixel at a single point in time is notoriously unreliable and noisy [Warren98]. However, filtering over an interval in time has been shown to provide considerable improvement in flow rate estimation [Camus97, Camus et al.99]. In addition, integration over a group of points comprising an entity can further improve the flow rate estimation. In Figure 8.15, the computation of image flow includes information from several different levels, each of which performs recursive estimate on entity flow rates. Thus pixel flow rate at each pixel is computed from a combination of estimates, including:

1. Correlation offset between a pixel in predicted attribute images and the corresponding pixel in observed attribute images
2. Level 2 entity flow estimate for the list entity to which the pixel belongs
3. Level 3 entity flow estimate for the surface entity to which the pixel belongs
4. Level 4 entity flow estimate for the object entity to which the pixel belongs

Once a reliable estimate of image flow $dx/dt, dy/dt$ is known, the value of an attribute of a pixel at position (i, j) at time t can be predicted from the estimated value of the attribute at a pixel at position $(i - dx, j - dy)$ at time $t - 1$. This can be expressed as

$$\mathbf{A}^*(i,j,t) = \underline{\mathbf{A}}(i - dx, j - dy, t - 1)$$

Thus in Figure 8.15, a predicted attribute image $\mathbf{A}^*(i,j,t)$ is generated from an estimated attribute image $\underline{\mathbf{A}}(i - d\underline{x}, j - d\underline{y}, t - 1)$ by computing the expected image motion for each pixel based on image flow estimates $dx/dt(i,j,t)$, $dy/dt(i,j,t)$, and $dr/dt(i,j,t)$ at each pixel, plus the effect of control signals sent to the camera pan/tilt unit, plus other control actions affecting the image.

Level 1 Classification. The final SP function performed at each level is the recognition of classes. This is illustrated on the right side of Figure 8.15 and in more detail in Figure 8.16. The set of attributes for each pixel defines a pixel attribute vector for each pixel. At level 1, classification is accomplished by comparing the predicted pixel attribute vector with pixel class attribute vectors from a library of pixel classes. When the inner product between the predicted pixel attribute vector $\mathbf{A}^*(i,j,t)$ and the class attribute vector $\mathbf{A}(Cn)$ exceeds threshold, the pixel (i,j) is recognized as belonging to the class Cn.

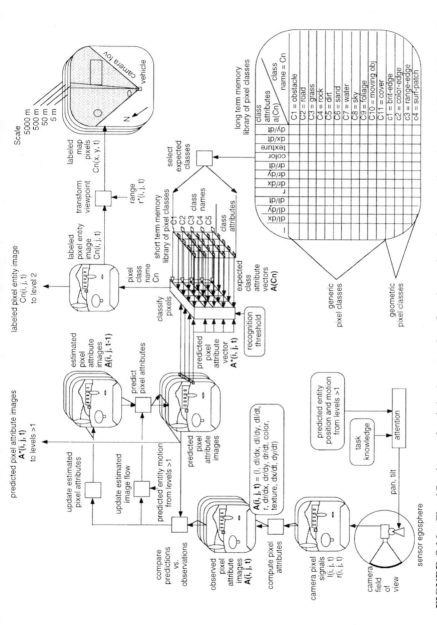

FIGURE 8.16 Classification at level 1. An attention function points the camera based on task knowledge and predicted entity position and motion from higher levels. Pixel attributes are computed in a recursive estimation loop as described in Figure 8.15. Predicted pixel attribute vectors are compared with class attribute vectors from a library of pixel classes. When a match is achieved, the name of the class is entered into a pixel entity image. If the range is known or can be estimated, labeled pixels can be transformed into a map at one or more levels of resolution.

277

The pixel class library contains a set of geometric classes. These may include brightness-edge, color-edge, range-edge, range-gradient edge, and surface-patch. Edge class attributes may include orientation, size, position, and velocity of the edge. Edge classes are important because pixels belonging to these classes tend to lie on the edges of objects or on the boundaries of surfaces. Surface-patch class attributes may include surface orientation, color, texture, temperature, position, and velocity. Surface-patch classes are important because pixels belonging to these classes have attributes that describe the surfaces and objects to which they belong.

The pixel class library may also contain a set of generic classes. These may include obstacle, road, grass, rock, dirt, sand, water, sky, foliage, moving object, or cover. Cover is defined as a clear region beneath overhanging branches that provides concealment from observation from above. Generic class attributes include surface orientation, color, texture, temperature, position, and velocity. Thus, pixels classified into surface-continuity classes may be further classified into generic classes. Upon classification, each pixel acquires one (or more pointers) that is(are) the name(s) of the pixel class(es) to which it belongs. The classification process thus generates one or more new pixel entity images where each pixel in the entity image(s) carries the name(s) of the pixel class(es) to which the pixel belongs.

Output from level 1 to level 2 is thus a set of pixel attribute images plus one or more pixel entity images wherein each pixel contains the name or names of the geometric and generic class(es) to which it belongs. The set of predicted pixel attribute images may also be available to all higher levels of image processing. When there exists a range attribute for each pixel, attribute images may be transformed into a map of the ground in the region that can be seen by the camera. This map may then be transformed into one or more maps at different levels in the RCS hierarchy, with range and resolution at each level appropriate to the planning and control requirements at that level.

Summary of Level 1

1. A portion of the egosphere is windowed onto each sensor.
2. The energy falling on each sensor is integrated over a sample interval in space and time.
3. Observed values for a set of attributes are computed for each pixel, producing a pixel attribute vector for each pixel.
4. Estimated and predicted values are computed for the flow rate and attributes of each pixel by a recursive estimation filter process.
5. Pixels are classified as belonging to one or more classes, and one or more pixel entity images are formed in which each pixel has a pointer (or pointers) to the class (or classes) to which it belongs.
6. Maps are generated containing classified pixels.

Where and What

Note that in Figures 8.15 and 8.16 there are two recursive estimation loops: one for estimating state (image flow) at each pixel and another for estimating attributes of each pixel. The two recursive estimation loops are analogous to the WHERE and WHAT channels that are known to exist in the visual processing hierarchy in primate brains. In the RCS architecture:

1. The WHERE channel computes estimated attributes of position and motion (i.e., state) of entities in the image. Estimated state is used to predict where to expect entities to appear next in the image.
2. The WHAT channel computes attributes that describe distinguishing characteristics of entities in the image. These include brightness, color, size, shape, texture, sound, taste, feel, or smell.

Both WHERE and WHAT attributes can be used to identify, recognize, classify, or name entities in the image, and by projection, in the external world. Both WHERE and WHAT channels also provide information that is useful in windowing, grouping, and segmentation. WHERE information may be used to group subentities into entities (or regions) based on how subentities are connected or how they move relative to each other in the image. WHAT information may be used to group subentities based on similarity of characteristics or attributes.

Level 2 Regions: List Entities and Events

At level 2 of the vision SP hierarchy, groups of a few (on average about 10) classified level 1 regions (pixels or events) can be grouped, processed, and analyzed as level 2 regions (list entities or events). For example, a group of pixel entities with contiguous brightness or color gradients might be grouped, processed, and analyzed as a list entity such as an edge, vertex, or surface patch. A temporal string of filtered attributes from a single sensor can be grouped, processed, and analyzed as a level 2 event. For example, a temporal signal or string of intensity attributes from an acoustic sensor might be grouped as a tone, a frequency, or a phoneme.

Level 2 Windowing. Windowing consists of placing windows around regions in the image that are designated as worthy of attention. At level 2 the placement of windows may depend on the goal and priorities of the current level 2 task. The placement of windows may also depend on the previous detection of higher-level entities. For example, the placement of windows on the edge of the road or on the vehicle shown in Figure 8.1 depends on previous detection and classification of the road edge and vehicle, respectively. The size and shape of each window is determined from the size and shape of recognized entities. The size of each window is also determined by the confidence factor computed by the level 2 recursive filtering process. If the confidence value for the entity in a window is high, the window will be narrowed to only slightly larger than the set of pixels in the entity. If the confidence value is low, the window will be significantly larger than the hypothesized entity. Until a set of recognized list entities exists, level 2 windows are set to their maximum size.

Level 2 Grouping. Grouping causes an image to be segmented, or partitioned, into regions (or sets of pixels) that correspond to higher-order entities. At level 2, grouping is a process by which neighboring pixels of the same class with similar attributes can be grouped to form higher-level entities. As illustrated at the lower left of Figure 8.17, labeled pixels are windowed and grouped into level 2 regions that form a hypothesized list entity image. Grouping is performed by a heuristic algorithm based on gestalt principles such as contiguity, similarity, proximity, pattern continuity, or symmetry. For example, contiguous pixels in the same geometric or generic class with similar attributes of range, orientation, and flow rate might be grouped into level 2 regions.

A grouping is a hypothesis that the pixels in the group all derive from the same entity in the world. The choice of which gestalt heuristic to use for grouping may depend on an attention function that is determined by the goal of the current task. It may also depend on the confidence factor developed by the recursive estimation process as well as on attributes of previously recognized list entities. A grouping hypothesis can be tested by a recursive estimation filtering algorithm applied to each of the hypothesized entities. If the recursive filtering process is successful in predicting the observed behavior of a hypothesized group in the image, the grouping hypothesis for that entity is confirmed. On the other hand, if the recursive filtering process is not successful in predicting the behavior of the hypothesized entity, the grouping hypothesis will be rejected and another grouping hypothesis must be selected. If the computational resources are available, a number of grouping hypotheses may be tested in parallel and the grouping hypothesis with the greatest success in predicting entity behavior will be selected. The result of level 2 grouping is a hypothesized list entity image in which each list entity is assigned a label and a frame in which a set of entity attributes can be stored.

Level 2 Computation. Each of the regions defined by the grouping hypotheses has attributes that can be computed. Entity attributes differ from pixel attributes in that entity attributes are computed over the entire group of pixels comprising the entity. For example, edge entities have attributes such as edge-length, edge-curvature, edge-orientation, and position and motion of the edge center of gravity. Surface-patch entities have attributes such as area, texture, average range, average surface gradient, average color, and position and motion of the surface center of gravity. For each hypothesized level 2 grouping, computed entity attributes fill slots in an observed list entity frame.

Level 2 Filtering. At level 2, a recursive estimation process compares the observed list entity image with a predicted list entity image as illustrated in Figure

8.17. Predicted entity images are generated from estimated entity images. Of particular importance for prediction are state attributes of position, orientation, and motion. Entity motion in an image can arise from three sources:

1. Estimated dynamic image flow rates $(d\underline{x}/dt, d\underline{y}/dt)$ and range rate $(d\underline{r}/dt)$ that describe the motion of entities in the image
2. Estimated motion of the camera platform
3. Commanded actions that affect motion of entities in the image

Comparison of observed entity images with predicted entity images produces correlation and difference values. Difference values are used to update the estimated entity position and motion. Estimates of image flow rates of level 2 entities are generated from a combination of correlation offsets generated at level 2, plus estimates of flow attributes from higher levels. A confidence level for each estimated entity attribute value is computed as a function of the correlation and difference values. If predictions based on estimated entity images track the behavior of observed entity images successfully, the confidence level rises. When the confidence level of the recursive estimation filter rises above threshold, the hypothesized list entity grouping is confirmed. If the confidence level of the recursive estimation filter falls below threshold, the hypothesized level 2 grouping is rejected and another grouping hypothesis must be selected.

Level 2 Classification. The set of attributes for each list entity define a list entity attribute vector that is represented in a list entity frame. At level 2, classification is accomplished by comparing the estimated list entity attribute vector with list entity class attributes from a library of list entity class frames. When the inner product between the estimated list entity attribute vector and a list entity class attribute vector exceeds threshold, the list entity is recognized as belonging to the class. This is illustrated on the right side of Figure 8.17 and is shown in more detail in Figure 8.18.

There are two libraries of list entity class frames: one resides in long-term memory (LTM), where frames for all the list entities known to the system are stored; the second resides in short-term memory (STM), where a subset of expected list entity frames is stored. Entity frames are transferred from long-term to short-term memory based on expectations from task knowledge. Entity frames are transferred from short-term to long-term memory based on evaluations of importance from value judgment processes.

The list entity class library contains several types of entity classes. One is a set of geometric classes. These include brightness-edge, color-edge, range-edge, range-gradient-edge, brightness-vertex, color-vertex, range-vertex, range-gradient-vertex, and surface-patch. Edge and vertex class attributes may include orientation, size, position, and velocity of the edge or vertex. These are important because pixels belonging to these classes tend to lie on the edges or corners of objects or on the boundaries of surfaces. Surface-patch

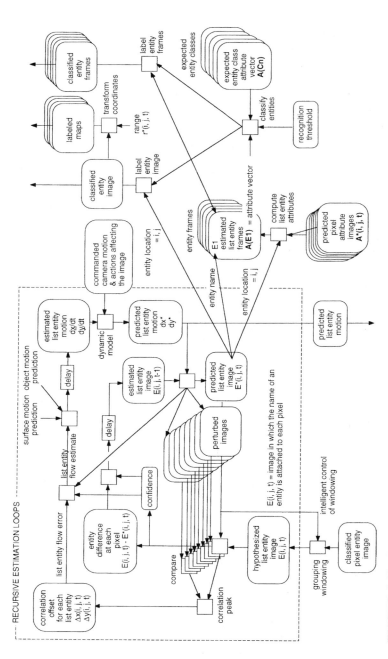

FIGURE 8.17 Recursive estimation in image processing at level 2. Classified pixels from level 1 are windowed and grouped into level 2 regions to form a hypothesized list entity image. This is compared with a predicted list entity image generated from an estimated list entity image. Perturbed list entity images are used to compute a correlation function over a set of image offsets. Offset in the peak of the correlation function is used to compute errors in estimated flow for the list entities in the image. This is combined with higher-level estimates of entity motions to generate estimated list entity motion. The predicted list entity image is used to compute list entity attributes that are represented in list entity frames. These attributes are compared with expected list entity classes to classify entities. The result is a set of classified entity frames and a classified list entity image.

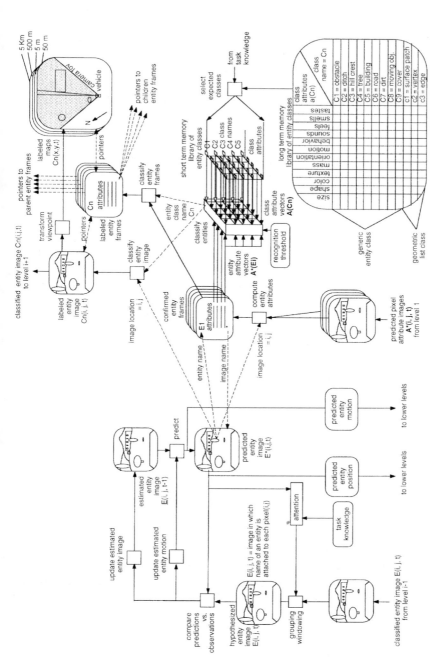

FIGURE 8.18 Classification at level 2. An attention function windows and masks the incoming entity image from level 1 based on task knowledge and predicted entity position and motion. Pixel attributes are computed in a recursive estimation loop as described in Figure 8.15. Predicted entity attribute vectors are compared with class attributes from a library of entity classes. When a match is achieved, the name of the class is entered into a labeled list entity image. Labeled pixels can be transformed into a map at one or more levels of resolution.

class attributes may include surface orientation, color, texture, temperature, position, and velocity. These are important because pixels belonging to these classes have attributes that describe the surfaces of objects to which they belong.

The list entity class library also contains generic classes. A list entity classified into a geometric class may be further classified into a generic class. Generic classes may include the top, bottom, or sides of an obstacle. They may also include the near edge of a ditch, the far edge of a ditch, the crest of a hill, the edge of a tree trunk, the side of a building, the edge of surface of a road, a grassy area, a rock, a patch of dirt, sand, or water, the sky, foliage of a tree or bush, or a moving object. Generic class attributes include surface orientation, color, texture, temperature, position, and velocity.

Upon classification, each pixel acquires one or more pointers that contain the name (or names) of the list entity class (or classes) to which the pixel belongs. Output from level 2 includes a list entity image wherein each pixel points to a list entity frame that contains the attributes of the list entity to which the pixel belongs. Each list entity frame also has a pointer back to the list entity image that contains the pixels that belong to it.

When there exists a range attribute for each pixel, attribute images may be transformed into a map of the ground that can be seen by the camera. This map may then be transformed into one or more maps at different levels in the RCS hierarchy with range and resolution at each level appropriate to the planning and control requirements at that level.

Summary of Level 2

1. Regions of the image containing list entities of attention are windowed.
2. Pixels with similar attributes are tentatively grouped into level 2 regions that comprise hypothesized list entities.
3. The attributes of each hypothesized list entity are computed.
4. The attributes of hypothesized list entities are filtered by recursive estimation, and each grouping hypothesis is either confirmed or rejected.
5. The attributes of each confirmed list entity are compared with the attributes of a set of list entity classes. Those that match are labeled with list entity class names. Pixels in the classified regions are given pointers to the list entity frames to which they belong. Each frame also points back to the list entity image.

Level 3 Regions: Surface Entities and Events

At level 3 of the SP hierarchy, groups of list entities with geometric attributes that correspond to a surface entity class can be grouped, processed, and analyzed as level 3 regions called *surface entities*. For example, a set of contiguous surface patch list entities with similar range and motion might be grouped, processed, and analyzed as a surface entity. A set of edge and vertex entities that are contiguous along their orientation might be grouped, processed, and analyzed as a surface boundary entity. Temporal strings of level 2 events with attributes that correspond to level 3 events can be grouped into level 3 events such as sonograms or two-dimensional surfaces in frequency and time. For example, a temporal string of phonemes might be grouped as a word, or a temporal string of outputs from a set of frequency filters as a note or phrase in an acoustic signature.

Level 3 Windowing. At level 3, windowing consists of placing windows around regions in the labeled list entity image that are designated as worthy of attention. The selection of which regions to window is made by an attention function that depends on the goal and priorities of the current level 3 task, or on the detection of noteworthy attributes of estimated surface entities, or on inclusion in a higher-level entity of attention. The shapes of the windows are determined by the recognized regions in the surface entity image. The sizes of the windows are determined by the confidence factor associated with the degree of match between the predicted surface entities and the observed surface entities.

Level 3 Grouping. At level 3, recognized list entities in the same class are grouped into level 3 regions called *surface entities* based on gestalt properties such as contiguity, similarity, proximity, pattern continuity, and symmetry. Surface patches in the same class that are contiguous at their edges and have similar range, velocity, average surface orientation, and average color may be grouped into surface entities. Edges and vertices in the same class that are contiguous along their orientation with similar range and velocity may be grouped into boundary entities. Grouping is a hypothesis that all the pixels in the group have the common property of imaging the same physical surface or boundary in the real world. This grouping hypothesis will be confirmed or rejected by the level 3 filtering function.

Level 3 Computation. Observed level 3 regions have attributes such as area, shape, position and motion of the center of gravity, range, orientation, surface texture, and color. Surface boundary entities have attributes such as boundary type, length, shape, orientation, position and motion of the center of gravity, and rotation. Boundary types may include intensity, range, texture, color, and slope boundaries. For each level 3 region, computed entity attributes fill slots in an observed surface entity frame.

Level 3 Filtering. A recursive estimation process compares observed surface entity attributes with predicted surface entity attributes generated from estimated surface entity attributes, planned results, and predicted attributes of parent entities. Comparison produces correlation and difference values. Difference values are used to update the estimated surface entity attributes. Correlation values are used to update estimates of entity motion. A confidence level associated with each estimated entity confirms or rejects the surface entity grouping hypothesis that created it.

Level 3 Classification. The classification process establishes a match between estimated surface entities and a generic or specific class of surface entities in the world model knowledge database. Typical generic surface entity classes are the ground, tree foliage, tree trunk, side of building, roof of building, road lane, fence, or lake surface. As a result of classification, a labeled surface entity image is formed in which each pixel has a pointer to (or the name of) the surface entity to which it belongs. Attribute images are transformed into maps that are registered with WTM world map coordinates.

Summary of Level 3

1. Portions of the image containing surface entities of attention are windowed.
2. List entities with similar attributes are tentatively grouped into level 3 regions, or surface entities.
3. The attribute values of each topological surface and boundary entity are computed.
4. Attributes of each hypothesized surface entity are estimated by recursive estimation and each grouping hypotheses is either confirmed or rejected.
5. The attributes of each confirmed surface entity are compared with the attributes of a set of surface entity classes, and those that match are assigned to surface entity classes. Each pixel in the classified regions is given pointers to the surface entity frames to which they belong, and each frame has a pointer to the surface entity image containing the pixels that belong to it.
6. Maps are generated wherein each pixel carries the name of the generic surface class to which it belongs.

Level 4 Regions: Object Entities and Events

At level 4 of the SP hierarchy, groups of level 3 entities with attributes such as range, motion, orientation, color, and texture that correspond to an object entity class can be grouped, processed, and analyzed as level 4 regions, or object entities. For example, a group of surfaces with coincident boundaries and similar or smoothly varying range and velocity attributes might be grouped into an object such as a building, a vehicle, or a tree. Attributes of surface entities comprising each object entity are combined into object entity attributes. Also at level 4, temporal strings of level 3 events such as words might be grouped into a sentence, or acoustic signatures might be grouped into a level 4 event.

Level 4 Windowing. At level 4, windowing consists of placing windows around regions in the image that are classified as worthy of attention. The selection of which regions to so classify depends on the goal and priorities of the current level 4 task, or on the detection of noteworthy attributes of groups of level 3 regions. The shape of the windows is determined by the set of pixels in the recognized object entity image. The size of the windows is determined by the confidence factor generated by the level 4 recursive estimation filter.

Level 4 Grouping. At level 4, recognized surface entities in the same class are grouped into level 4 regions, or object entities, based on gestalt properties such as contiguity, similarity, proximity, pattern continuity, and symmetry. Surfaces in the same class that are contiguous along their boundaries and have similar range and velocity may be grouped into object entities. Boundaries that separate surfaces with different range, velocity, average surface orientation, and average color are used to distinguish between different objects. Level 4 grouping is a hypothesis that all the pixels in the group lie on the same physical object in the real world. This grouping hypothesis will be confirmed or rejected by the level 4 filtering function.

Level 4 Computation. For each of the hypothesized object entities, entity attributes such as position, range, and motion of the center of gravity, average surface texture, average color, solid-model shape, projected area, and estimated volume can be computed. For each hypothesized object entity, computed attributes fill slots in an observed object entity frame.

Level 4 Filtering. A recursive estimation process compares observed object entity attributes with predictions based on estimated object entity attributes. Included in the prediction process are estimated motion of object entities and expected results of commanded actions. Comparison of observed entity attributes with predicted entity attributes produces correlation and difference values. Difference values are used to update the attribute values in the estimated object entity frames. A confidence level is computed as a function of the correlation and difference values. If the confidence level of the recursive estimation filter rises above threshold, the object entity grouping hypothesis is confirmed.

Level 4 Classification. The classification process compares the attributes of each confirmed object entity with attributes of object entity classes stored in the knowledge database. A match causes a confirmed geometrical object to be classified as a generic or a specific object entity in the world model knowledge database. For example, in the case of a ground vehicle performing a typical task, a list of generic object classes may include the ground, the sky, the horizon, dirt, grass, sand, water, bush, tree, rock, road, mud, brush, woods, log, ditch, hole, pole, fence, building, truck, tank, and so on. There may be several tens, hundreds, or even thousands of generic object classes in the KD. A list of specific object classes may include a specific tree, bush, building, vehicle, road, or bridge. As a result of classification, a new object entity image is formed in which each pixel has a pointer to (or the name of) the generic or specific object entity to which it belongs. Attribute images are transformed into maps that are registered with WTM world map coordinates.

Summary of Level 4

1. Portions of the image containing object entities of attention are windowed.
2. Surface entities with similar attributes are tentatively grouped into level 4 regions, or object entities.
3. The attribute values of each hypothesized object entity are computed.
4. Attributes of each hypothesized object entity are estimated by recursive estimation and each grouping hypotheses is either confirmed or rejected.
5. The attributes of each confirmed object entity are compared with the attributes of a set of object entity classes, and those that match are assigned to object entity classes. Pixels in the classified regions are given pointers to the object entity frames to which they belong, and each frame has a pointer to the object entity image containing the pixels that belong to it.
6. Maps are generated containing classified objects and regions.

Level 5 Regions: Group Entities and Events

At level 5 of the SP hierarchy, groups of object entities with similar range, motion, and other attributes are grouped, processed, and analyzed as a group, collection, or assemblage of objects. Attributes of object entities comprising each group entity are combined into group entity attributes. Also at level 5, temporal strings of level 4 events are integrated into level 5 events.

Level 5 Windowing. At level 5, windowing consists of placing a window around regions in the image that are classified as group entities of attention. The selection of which regions to so classify depends on the task and on detection of entity attributes that are worthy of attention. The shape of the windows is determined by the collection of pixels in the recognized group entity image. The size of the windows is determined by the confidence factor associated with the level 5 recursive estimation process.

Level 5 Grouping. Object entities in the same class are grouped into level 5 regions, or group entities, based on gestalt properties such as contiguity, similarity, proximity, pattern continuity, and symmetry. Objects in the same class that are near each other and have similar range and velocity may be grouped into group entities. These grouping hypotheses will be confirmed or rejected by the level 5 filtering function.

Level 5 Computation. For each of the hypothesized group entities, attributes such as position, range, and motion of the center of gravity, density, and shape can be computed. These computed attributes become observed entity attributes in an observed group entity attribute frame.

Level 5 Filtering. A recursive estimation process compares observed group entity attributes with predictions based on estimated group entity attributes and expected results of commanded actions. Comparison of observed group entity attributes with predicted entity attributes produces correlation and difference values. Difference values are used to update the attribute values in the estimated squad entity frames. A confidence level is computed as a function of the correlation and difference values. If the confidence level of the recursive estimation filter rises above threshold, the group entity grouping hypothesis is confirmed.

Level 5 Classification. The classification process compares the attributes of each confirmed group entity with attributes of group entity classes stored in the knowledge database. A match causes a recognized geometric group entity to be classified as a generic or a specific group entity in the world model knowledge database. For example, in the case of a ground vehicle, a list of generic group classes may include woods, fields, groups of vehicles, groups of people, and clusters of buildings. There may be several tens, hundreds, or even thousands of generic group classes in the KD. A list of specific group classes may include a specific group of humans, trees, bushes, buildings, vehicles, or the intersection of two or more roads. As a result of classification, a new entity image is formed in which each pixel has a pointer to (or the name of) the generic or specific group entity to which it belongs. Attribute images are transformed into maps that are registered with a priori maps provided in WTM world map coordinates.

Summary of Level 5

1. Portions of the image containing group entities of attention are windowed.
2. Object entities with similar attributes are tentatively grouped into level 5 regions, or group entities.
3. The attribute values of each observed group entity are computed.
4. Attributes of each hypothesized group entity are estimated by recursive estimation and each grouping hypotheses is either confirmed or rejected.
5. The attributes of each confirmed group entity are compared with the attributes of a set of group entity classes, and those that match are assigned to group entity classes. Pixels in the classified regions are given pointers

to the object entity frames to which they belong, and each frame has a pointer to the object entity image containing the pixels that belong to it.

6. Maps are generated containing classified objects and regions.

To summarize, SP establishes correspondence between internal entities and events in the world model and external entities and events in the world. SP grounds the KD in the real world. SP uses attention to focus sensors and computing resources on what is important to behavioral goals. SP integrates information from many sources and makes hypotheses that disambiguate the many possible interpretations of the sensory signals. SP then uses recursive estimation to confirm or deny its hypotheses. Finally, SP matches observed attributes against stored class attributes to recognize correspondence between what is observed and what is known. A review of the psychological literature supporting this basic concept of sensory processing was recently published by Richardson[99].

THE COMPLEXITY OF PERCEPTION

The apparent ease with which the human mind perceives the world is deceiving. Most of the sensory processing in the brain goes on below the level of consciousness. People typically perceive entities and events immediately without apparent difficulty. Usually, objects and relationships are easily recognized and classified. It was not until researchers began attempting to program computers to perceive the world that it became clear how complex and computationally intensive the process of perception really is. Only recently have researchers begun to appreciate the complexities of establishing and maintaining an internal world model that corresponds to entities and events in the external world.

One of the biggest problems is that sensory input is almost always ambiguous. Typically, there are many (often an infinite number of) possible configurations of the world that might produce any single piece of sensory data. For example, a brightness image on a retina contains no direct measurement of range at each pixel. The range to a surface in the world must be inferred from stereo disparity, image flow, or other information which may also be ambiguous. Many attributes of objects in the world cannot be measured directly and must be inferred. For example, the size, shape, mass, temperature, position, orientation, and velocity of objects typically cannot be measured directly but must be inferred from a multiplicity of measurements and from prior knowledge. In most cases, hypothesis and test is the only way to verify that the internal world model is an adequate representation of the external world.

But hypothesis and test can be extremely complex. The number of possible hypotheses can be enormous. Until a suitable grouping hypothesis is found, attributes of entities and events cannot be compared with stored classes. But there are many different ways that sensory data can be grouped into entities and events. A typical image may contain tens of objects, hundreds of surfaces, and thousands of list entities. Objects in the world can appear in an infinite variety of different positions, orientations, and scales under many different lighting conditions and in different relationships to each other. Many objects move and change shape over time. For example, consider the many different shapes that a cat may present to a viewer under different lighting conditions, in various locations, while performing different activities such as running, sitting, fighting, and sleeping.

The number of classes stored in memory is also very large. For humans, the number of stored classes must be at least as large as the number of words in the speaking vocabulary. Human vocabularies typically contain several thousand words, and each word can have many modifiers [Fischler and Firschein87]. There are several hundred thousand entries in the unabridged dictionary. Thus, the classification problem of comparing objects in the image with classes stored in memory has a potential for combinatorial explosion. It is impossible to formulate and test all possible hypotheses. For hypothesis and test to be successful, it is necessary to limit the number of hypotheses to be considered. For classification to occur within a reasonable period of time using a feasibly amount of computational resources, the search space of possible matches must be limited.

This combinatorial problem can be addressed through a variety of approaches:

1. Information from many different sensors and sensor modalities can be integrated so as to reduce ambiguities and limit the number of possible interpretations of sensory input. For example, two stereo cameras can eliminate the range ambiguity at each pixel. Alternatively, range data from radar or LADAR can be registered and overlaid on visual brightness images to eliminate range ambiguities. Knowledge of camera translation and rotation derived from inertial, GPS, and dead reckoning systems can be used to remove ambiguities in the computation of range from image flow. Range can be combined with angular measurements from the image plane to eliminate ambiguities in object size. Range and angular rate measurements can be combined to eliminate ambiguities regarding object velocity.

2. The number of attributes required in comparisons between observed entities and stored entity classes can be limited by selecting attributes for comparison that are invariant under translation and rotation. Invariant attributes include size, shape, color, mass, and surface texture. It should be noted, however, that invariant properties of objects in the world do not necessarily produce invariant attributes in the image. For example, the apparent color of an object depends on the color of incident light. The apparent texture may depend on range in space and resolution in the image. Apparent size and shape depend on the position and orientation of the objects in the world with respect to the camera.

3. The number of entity classes that need to be searched during comparison can be limited by using task command goals and objects to generate prioritized lists of entities of attention. This reduces the number of entity classes to be considered and assures that the most important classes are addressed first.

4. Parallel computation can be used to perform many comparisons simultaneously.

5. Ambiguities in grouping, feature extraction, and image segmentation can be limited by tracking entities over time. Tracking geometric entities before they are recognized also eases the requirements for rapid recognition and allows the classification process to take place over an extended time interval. Tracking objects after they have been recognized means that classification may be required only once for each object.

6. Content addressable memory can be used to speed the recall of stored classes. Content addressable memory enables the attribute vector of a geometric object to be used as an address into a memory where the name of the class is stored. The class name can then address the location where the class attribute vector is stored. The stored class attribute vector can then be compared with the observed attribute vector to verify the classification. A number of content addressable memory methods have been developed. For example, Hopfield[82] nets use a simulated annealing process to recall an image from memory given only a piece of the image as input. Grossberg[87] and Carpenter and Grossberg[88,92] developed an adaptive resonance process to establish a link between a pattern in memory and a suggestion of that pattern in the input. Kohonen[77] has also developed content addressable memory mechanisms.

7. The classification process can be partitioned into a hierarchy of partial solutions such that the number of hypotheses and the size of the search space can be limited at each level.

Thus, there exist a number of heuristic methods and computing mechanisms whereby ambiguities can be eliminated and classification can be accomplished in real time for a useful number of classes using a feasible amount of computational resources. Ullman[96] provides an excellent review of these mechanisms. It should be noted, however, that the problem is by no means solved. Machine vision is nowhere near as robust as biological vision. Perception remains an active field of research.

SUMMARY AND CONCLUSIONS

Perception is arguably the most complex of all operations performed by the brain. Almost half of the neurons in the human brain are dedicated to perception. Sensory processing may turn out to be the most difficult of all brain functions to duplicate in computers. What has been described in this chapter is only a brief outline of what will be required to implement the full range of sensory processing capabilities that are common in biological systems. We have addressed only the rudiments of image processing. Processing of acoustic information, and in particular understanding of spoken language, is at least as complex and demanding as the understanding of images. It will be many years before artificial systems can approach the capabilities of biological systems in building and maintaining a rich and dynamic internal model of the external world.

Nevertheless, there is much that is known and much more that could be done than has been done with current technology. We believe that the proper approach was proposed by Alan Turing in 1950 when he suggested that one could "provide the machine with the best sense organs that money can buy, and then teach it to understand" [Turing50]. This, of course, would be an expensive and complex undertaking, and partly for that reason has not been the course of investigation pursued by the artificial intelligence community over the past half century. However, we predict that in the end it will be the only way to achieve the goal of engineering of mind. In the next chapter we describe an ongoing effort to pursue this approach through the design of an unmanned ground vehicle for military scouting missions.

9 Engineering Unmanned Ground Vehicles

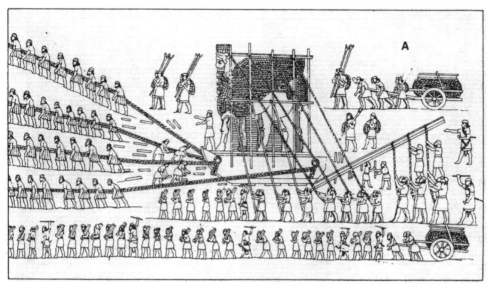

Stone image being moved on a sled and rollers pulled by a large number of people, an early mobility project that inspires appreciation for how far vehicle technology has advanced in 29 centuries. [An Assyrian relief, ninth century B.C.; from *The History of the Machine* by S. Strandh, published by Dorset Press, NY, reprinted by permission of AB Nordbook, Gothenburg, Sweden.]

We now illustrate how the RCS architecture can be applied to a practical application using the Army Research Lab Demo III Experimental Unmanned Ground Vehicle (XUV) program as an example. The Demo III program is designed to culminate in the year 2001 with four experimental unmanned scout testbed vehicles participating in a Battle Lab Warfighting Experiment [Shoemaker et al.99]. A photograph of an early prototype of the Demo III XUV is shown in Figure 9.1.

The primary focus of the Demo III program is the development of autonomous mobility technology to enable a small, survivable unmanned ground vehicle (UGV) to maneuver tactically over rugged terrain as an integral part of

FIGURE 9.1 The Demo III XUV has four-wheel drive and four-wheel steering. It is powered by a 78-hp turbocharged diesel engine. The dome in the center of the vehicle is a stabilized reconnaissance, surveillance, and target acquisition package. Over the left front wheel is a pan/tilt head with four cameras: a stereo pair of FLIRs and a stereo pair of color cameras. The fifth camera port at the center-top is currently unused. Over the right front wheel is a LADAR laser range imaging camera. The white panel in the front center is planned for an imaging radar. The front bumper is instrumented to detect obstacles hidden in tall weeds. [Photo courtesy of General Dynamics Robotic Systems.]

a mixed military force containing both manned and unmanned vehicles (e.g., M1A2 Abrams tanks, M2/M3 Bradley Fighting Vehicles, and HMMWVs). Implicit in this goal is the requirement for the UGV to maneuver at speeds comparable to manned vehicles, employ appropriate driving skills and tactical behaviors, function reliably without requiring additional specialized personnel, and use intuitive operator-friendly human–machine interfaces that can readily be coupled into existing military command-and-control systems.

The technical objectives of the Demo III program are to develop a semiautonomous XUV system with intelligent, modular, open software architecture [Albus98,99,00]. It is intended to be operated by soldiers in the field to accomplish useful military missions such as scouting, reconnaissance, surveillance, target acquisition, laying smoke, positioning sensors, sensing for nuclear or biological agents, and carrying supplies or weapons. Each Demo III XUV is designed to weigh less than 1500 kg, be transportable by helicopter, and operate in daylight or darkness, in clear or adverse weather. It should be capable of driving without human supervision on roads at speeds up to 64 km per hour (km/h) and off-road at speeds appropriate to terrain and vehicle dynamics up to 32 km/h in day and 16 km/h at night while avoiding obstacles. The vehicle should be capable of executing useful military missions up to 24 hours in duration. It must be safe to operate with manned or unmanned units among mounted or dismounted troops. The Demo III program will deliver four vehicles with operator control units and communications such that all four vehicles can be commanded by a single human operator riding in a HMMWV (Highly Mobile Multipurpose Wheeled Vehicle).

The program is being funded by the Office of Secretary of Defense through the DOD Joint Robotics Program and managed by Army Research Lab with participation by the Tank and Automotive Command, Training and Doctrine Command, Mounted Maneuver Battle Lab, and the Joint Program Office for Unmanned Ground Vehicles. Technology development support is provided by National Institute of Standards and Technology, Jet Propulsion Lab, Army Research Lab, David Sarnoff Labs, SAIC Center for Information System, and Universität der Bundeswehr München. It is being coordinated with similar unmanned vehicle programs in Germany through a memorandum of understanding with the German Ministry of Defense. Systems integration and project management is provided by General Dynamics Robotic Systems.

4-D/RCS

4-D/RCS is the reference model architecture currently being developed for the Demo III XUV program [Albus98]. 4-D/RCS integrates the NIST (National Institute of Standards and Technology) RCS (Real-Time Control System) with the German (Universität der Bundeswehr München) VaMoRs 4-D approach to dynamic machine vision [Dickmanns et al.94]. The 4-D/RCS architecture consists of a hierarchy of computational nodes each of which contains behavior generation (BG), world modeling (WM), sensory processing (SP), and value judgment (VJ) processes [Albus95, Albus and Meystel96]. Each node also contains a knowledge database (KD) and an operator interface. These computational nodes are arranged such that the BG processes represent organizational units within a command and control hierarchy such as that shown in Figure 9.2.

Each BG process includes a planner module that accepts task command inputs from its supervisor and generates coordinated plans for subordinate BG processes. The BG planner hypothesizes tentative plans, WM predicts the probable results, and VJ evaluates the results of each tentative plan. The BG planner then selects the tentative plan with the best evaluation to be placed in the plan buffers in the BG executors. There is an executor that services each subordinate BG unit, issuing subtask commands, monitoring progress,

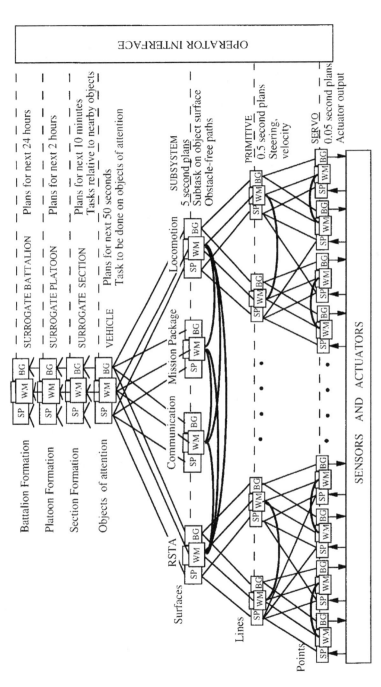

FIGURE 9.2 4-D/RCS reference model architecture for an individual vehicle. Processing nodes are organized such that the behavior generation (BG) processes form a command tree. Information in the knowledge database (KD) is shared between world modeling (WM) processes in nodes within the same subtree. (KD structures are incorporated within WM processes and not shown in this figure.) On the right are examples of the functional characteristics of the behavior generation (BG) processes at each level. On the left are examples of the type of entities recognized by the sensory processing (SP) processes and stored by the WM in the KD at each level. Sensory data paths flowing up the hierarchy typically form a graph, not a tree. Value judgment (VJ) processes are partly visible behind WM processes. A control loop is typically closed through every node. An operator interface may provide input to, and output from, processes in every node. Note: The entity class hierarchy on the left can be orthogonal to the BG process hierarchy on the right. This is enabled by vertical connections between WM modules. See, for example, Figure 9.4.

compensating for errors and differences between planned and observed situations in the world, and reacting quickly to emergency conditions with appropriate actions. Feedback from a real-time knowledge database KD enables the planners and executors to generate reactive behavior. SP and WM processes update the KD with images, maps, entities, events, attributes, and states necessary for both deliberative and reactive behavior. Coordination between subordinate BG processes is achieved by cross-coupling among plans and sharing of information among executors through the KD. BG processes are described in detail in Chapter 6. WM, KD, and VJ processes are described in Chapter 7, and SP processes are described in Chapter 8.

The reference model architecture shown in Figure 9.2 has four subsystems: locomotion, mission package, communication, and RSTA (reconnaissance, surveillance, and target acquisition). Each subsystem consists of one or more primitive units, each of which has one or more servo units that sequence and coordinate one or more actuator controllers. For example, the locomotion subsystem consists of a primitive-level driver unit and a gaze control unit. The primitive-level gaze control unit has a servo-level control unit for the stereo sensor platform and another for the LADAR platform. The stereo platform control unit synchronizes actuator controllers for pan, tilt, zoom, focus, iris, and vergence actuators. The LADAR platform control unit controls only a tilt actuator controller. Each actuator controller receives feedback from one or more sensors. The communication subsystem consists of a message encoding subsystem, a protocol syntax and encryption generator, and communications bus interface, plus antenna-pointing and band selection actuators. The mission package might be a smoke generator, a dispenser of surveillance sensors, or a weapons system consisting of loading, aiming, and firing subsystems each with a number of sensors and actuators. The RSTA subsystem consists of an acoustic array and a RSTA package containing a variety of sensors and sensor-pointing actuators, all of which need to be coordinated to achieve behavioral goals successfully.

The operator interface (OI) provides the capability for the operator to interact with the system at any time at a number of different levels: to insert commands, change missions, halt the system, adjust parameters, alter priorities, change speeds, select or verify targets, perform identification friend-or-foe (IFF), authorize the use of weapons, or monitor any of the system functions. The OI can send commands or requests to any BG process, or display information from any SP, WM, or VJ process. It can display any of the state variables in the KD at a rate and latency dictated by the communications bandwidth. Using the OI, a human operator can view situational maps with topographic features and both friendly and enemy forces indicated with overlays. The operator may use the OI to generate graphics images of motion paths, or display control programs (plans) in advance or while they are being executed. The OI may also provide a mechanism to run diagnostic programs in the case of system malfunctions [Morganthaler et al.2000].

In Figure 9.2, three levels of control are shown above the node representing the individual vehicle. These three additional levels represent a surrogate chain of command that exists above the individual vehicle. Because each vehicle is semiautonomous, it carries a copy of the control nodes that otherwise would exist in its superiors if those superiors were tightly coupled in an integrated control structure. On the battlefield, individual vehicles are physically separate and occasionally may be out of contact with each other and their superiors. Often, communications are restricted to low-bandwidth or unreliable communication channels. Therefore, each vehicle carries a surrogate chain of command, so that when it is out of radio contact, it can perform for itself the higher-level command and control functions normally performed for it by its superiors in the command chain.

The surrogate chain of command serves four purposes. First, it provides each vehicle with an estimate of what its superiors would command it to do if they were in direct communication. Second, it enables any vehicle to assume the duties of any of its superiors should that become necessary. Third, it provides a natural interface for human commanders at the section, platoon, or battalion level to interface with the vehicle at a level relevant to the task being addressed. Fourth, it enables each vehicle to dedicate a separate node to handle each layer of higher-level tasks. In this example, the surrogate chain of command consists of three levels with three different planning horizons (10 minutes, 2 hours, and 24 hours). These three levels deal with external objects and maps at three different scales and ranges. There may, of course, be more than three levels above the vehicle.

The functionality of each level in the 4-D/RCS reference model hierarchy is defined by the characteristic timing, bandwidth, task vocabulary, and algorithms available at each level for decomposing tasks and goals. Typically, these are design choices that depend on the nature and dynamics of the processes being controlled. The numerical values shown in Figure 9.2 represent planning horizons appropriate for a ground scout vehicle. For other types of systems, different numerical values would be derived from different design specifications.

The horizontal curved lines between WM processes indicate the main avenues for sharing of state information in the world model between nodes within subtrees in order to synchronize related tasks. However, the absence of lines in the diagram should not be interpreted as a prohibition on communication between modules. Any 4-D/RCS module can access any public data variable or data structure in any other module or in the KD by establishing an NML channel. Of course, data that must be routed through servers and across buses and local area networks are subject to the timing and bandwidth restrictions of the physical medium. It is the responsibility of the system designer to assure that processes that require high-bandwidth communications between them be located on the same high-speed data bus, or within the same computer if necessary, or even within the same software task or thread.

SCOUT PLATOON EXAMPLE

To illustrate the types of issues that will be addressed by the 4-D/RCS reference model architecture, an example is given below of a seven-level hierarchy for a scout platoon attached to an armor battalion. The specific numbers and functions described in this example are illustrative only. They are given simply to illustrate how the generic structure and function of the 4-D/RCS reference model architecture might be instantiated.

4-D/RCS Level 7: Battalion

An armor battalion is a unit that consists of a group of M1 or Bradley companies and a scout platoon. A 4-D/RCS node at level 7 corresponds to a battalion headquarters unit, consisting of a battalion commander, several company commanders, a scout platoon leader, and support staff. (In principle, any or all of these could be humans or intelligent agent software processes. In current practice, they are all humans.) Intelligent agents within the battalion headquarters unit plan activities and allocate resources for the armored companies and the scout platoon in the company. Incoming orders to the battalion are decomposed by the battalion commander into assignments for the armor companies and the scout platoon. Resources and assets are allocated to each subordinate unit, and a schedule is generated for each unit to maneuver and carry out its respective assigned operations. Together, these assignments, allocations, and schedules comprise a plan. The plan may be devised by the battalion commander alone or in consultation with his subordinate unit leaders. The battalion-level planning process may consider the exposure of each unit's movements to enemy observation and the traversability of roads and cross-country routes. The battalion commander typically defines the rules of engagement for the units under his command and works with his unit leaders to develop a schedule that meets the objectives of the mission orders given to the battalion. In the 4-D/RCS battalion node, plans are typically computed for a period of about 24 hours, recomputed at least once every 2 hours, and modified more often when necessary. Plans include both desired activities and desired positions for each of the subordinate units at about 2-hour intervals.

When the vehicle is not in direct communication with its chain of command, its surrogate battalion node performs the functions of the battalion headquarters unit. The surrogate battalion node plans activities and waypoints for the surrogate platoon mode at 2-hour intervals and estimates what platoon-level operations should be executed to follow that plan. The surrogate battalion node considers the exposure of scout platoon operations to enemy observations and the traversability of roads and cross-country routes. At the battalion level, the 4-D/RCS world model maintains a knowledge database that contains names and attributes of friendly and enemy forces and of the force levels required to engage them. Maps have a range of 1000 km (i.e., more than the distance that the battalion is likely to travel in a 24-hour day) with a resolution of about

400 m. Maps describe the terrain and location of friendly and enemy forces (to the extent that they are known), and roads, bridges, towns, and obstacles such as mountains, rivers, and woods. Objects such as bridges, targets, and road intersections on the maps typically are linked to symbolic data structures that describe their attributes and define position in the world to much better than 400-m resolution. Battalion-level maps may be updated at irregular intervals from intelligence reports.

4-D/RCS sensory processing at the surrogate battalion level integrates information about the movement of forces, the level of supplies, and the operational status of all the units in the battalion, plus intelligence about enemy units in the area of concern to the battalion. This information is used to update maps and symbolic data structures in the knowledge database so as to keep it accurate and current. The surrogate battalion headquarters node also contains value judgment functions (e.g., for calculating the risk of casualties) that enable the battalion commander to evaluate the cost and benefit of various tactical options. To the extent that the information and algorithms in the surrogate battalion node is identical with that in the real battalion headquarters, the surrogate battalion node will make the same decisions as the real battalion headquarters.

An operator interface to the surrogate battalion node allows human operators (either on-site or remotely) to visualize information such as the deployment and movement of forces, the availability of ammunition, and the overall situation within the scope of attention of the surrogate battalion node. The operator can intervene to change priorities, alter tactics, or redirect the allocation of resources. Output from the battalion node through the company commanders and scout platoon leader comprise input commands to the company/platoon nodes. Armor company commanders and the scout platoon leader are expected to assign job responsibility within their respective units, monitor how well their units are following the battalion plan, and make adjustments in job assignments as necessary to keep on plan. New output commands from the battalion node to the scout platoon node may be issued at any time and typically consist of tasks expected to require about 2 hours to complete.

Level 6: Platoon

A scout platoon is a unit that typically consists of 10 HMMWVs or Bradley vehicles organized into one or more sections. For the Demo III project, a scout platoon may consist of six manned HMMWVs and four unmanned XUVs. A 4-D/RCS node at the platoon level corresponds to a scout platoon headquarters unit. It consists of a platoon commander plus his or her assistant commander and section leaders. (Any of these could be humans or intelligent agent software processes, in any combination.) The platoon commander and section leaders plan activities and allocate resources for the sections in the platoon. Platoon orders are decomposed into job assignments for each section. Resources are allocated, and a schedule of activities is generated for each section. Movements for each section are planned relative to major terrain features and are coordinated with the activities of other sections within the platoon. Inter-section formations and tactical behaviors are selected on the basis of tactical goals, stealth requirements, and other priorities in the battalion-level plan. At the platoon level, each step in the battalion-level plan is decomposed into a platoon-level plan computed for a period of about 2 hours into the future. Replanning is done about every 10 minutes, or more often if necessary. Section activities are planned for intervals of about 10 minutes' duration and waypoints about 10 minutes apart on the map are computed.

The surrogate platoon node in each vehicle performs the functions of the platoon headquarters unit when the vehicle is not in direct communication with the chain of command. It plans activities for the surrogate section node on a platoon-level time scale and estimates what section-level maneuvers should be executed in order to follow that plan. Movements are planned relative to major terrain features and other vehicles within the platoon. At the platoon level, the 4-D/RCS world model symbolic database contains names and attributes of targets, and the weapons and ammunition necessary to attack them. Maps with a range of about 100 km (i.e., more than the distance a platoon is likely to travel in 2 hours) and resolution of about 40 m describe the location of objectives, and routing between them. Sensory processing integrates intelligence about the location and status of friendly and enemy forces. Value judgment evaluates tactical options for achieving section objectives. An operator interface allows human operators to visualize the status of operations and the movement of vehicles within the section formation. Operators can intervene to change priorities and reorder the plan of operations. Section leaders are expected to sequence commands to their respective sections, monitor how well their sections are following the platoon plan, and make adjustments as necessary to keep on schedule. The output from the platoon level to the section leaders are commands issued to sections to maneuver and perform reconnaissance, surveillance, and target acquisition operations over particular sectors of the battlefield. Output commands may be issued at any time, but typically are planned to change only about once every 5 minutes.

Level 5: Section

A scout section is a unit that consists of a group of individual scout vehicles such as HMMWVs and UGVs. A 4-D/RCS node at the section level corresponds to a section leader and vehicle commanders (humans or intelligent software agents). The section leader assigns duties to the vehicles in his section and coordinates the vehicle commanders in scheduling cooperative activities of the vehicles within a section. Orders are decomposed into assignments for each vehicle, and a schedule is developed for each vehicle to maneuver in formation within assigned corridors taking advantage of local terrain features and avoiding obstacles. Plans are developed to conduct coordinated maneuvers and to perform reconnaissance, surveillance, or target acquisition functions. At the section level, plans are computed for about 10 minutes into the future, and replanning is done about every 1 minute, more often if necessary. Vehicle waypoints about 1 minute apart are computed.

The surrogate section node in each XUV performs the functions of the section command unit when the XUV is not in direct communication with the section commander. The surrogate node plans activities for the XUV on a section-level time scale and estimates what vehicle-level maneuvers should be executed in order to follow that plan. At the section level, the 4-D/RCS world model symbolic database contains names, coordinates, and other attributes of other vehicles within the section, other sections, and potential enemy targets. Maps with a range of about 10 km and a resolution of about 40 m are typical. Maps at the section level describe the location of vehicles, targets, landmarks, and local terrain features, such as buildings, roads, woods, fields, fences, ponds, and so on. Sensory processing determines the position of observed landmarks and terrain features and tracks the motion of groups of observed vehicles and targets. Value judgment evaluates plans and computes cost, risk, and payoff of various alternatives. An operator interface allows human operators to visualize the status of the battlefield within the scope of the section or to intervene to change priorities and reorder the sequence of operations or selection of targets. Vehicle commanders issue commands to their respective vehicles, monitor how well plans are being followed, and make adjustments as necessary to keep on plan. Output commands to individual vehicles to engage targets or maneuver relative to landmarks or other vehicles may be issued at any time, but on average are planned for tasks that last about 1 minute.

Level 4: Individual Vehicle

The vehicle is a unit that consists of a group of subsystems, such as locomotion, attention, communication, and mission package. A manned scout vehicle may have a driver, vehicle commander, and a lookout. Thus, a 4-D/RCS node at the vehicle level corresponds to a vehicle commander plus subsystem planners and executors. The vehicle commander assigns jobs to subsystems and (possibly in collaboration with subsystem controllers) schedules the activities of all the subsystems within the vehicle. A schedule of waypoints is developed by the locomotion subsystem to avoid obstacles, maintain position relative to nearby vehicles, and achieve desired vehicle heading and speed along the desired path on roads or cross-country. A schedule of tracking activities is generated for the RSTA subsystem to track obstacles, other vehicles, and targets. A schedule of activities is generated for the mission package and the communication subsystems. Task activities about 5 seconds (s) apart out to a planning horizon of 50 s are replanned every 5 s, more often if necessary. At 10 m/s, autonomous mobility subsystem waypoints are computed about 50 m apart out to a planning horizon of 500 m.

At the vehicle level, the world model symbolic database contains names (identifiers) and attributes of objects—for example, the size, shape, location, orientation, and surface characteristics of roads, other vehicles, intersections, fences, ground cover, or objects such as rocks, trees, bushes, mud, and water. Maps are generated from on-board sensors with a range of about 500 m and resolution of 4 m. These maps are registered and overlaid with 40-m resolution data from a priori section-level maps. Maps represent object positions (relative to the vehicle) and dimensions of road surfaces, buildings, trees, craters, and ditches. Sensory processing measures object dimensions and distances and computes relative motion. Value judgment evaluates trajectory planning and sensor dwell-time sequences. An operator interface allows a human operator to visualize the operational status of the vehicle and to intervene to change priorities or steer the vehicle through difficult situations. Subsystem controller executors sequence commands to subsystems, monitor how well plans are being followed, and modify parameters as necessary to keep on plan. Output commands to subsystems may be issued at any time, but typically are planned to change only about once every 5 s.

Level 3: Subsystem Level

Each subsystem node is a unit consisting of a controller for a group of related primitive-level nodes. A 4-D/RCS node at the subsystem level assigns jobs to each of the subordinate primitive-level nodes and coordinates planned activities between them. A schedule of primitive-level mobility waypoints and actions is developed to avoid obstacles and achieve the smoothest possible ride. A schedule of primitive-level pointing commands is generated for aiming cameras and sensors. A schedule of primitive-level messages is generated for communications, and a schedule of primitive-level actions is developed for operating controllers within the mission package. Planned primitive actions are about 500 ms apart on the time line out to a planning horizon of about 5 s in the future. At 10 m/s, planned primitive-level mobility waypoints fall about 5 m apart on a local planning map out to a planning horizon of about 50 m. A new plan is generated about every 500 ms.

At the subsystem level, the world model symbolic database contains names and attributes of environmental features such as road edges, potholes, lane markings, obstacles, ditches, targets, road signs, other vehicles, and pedestrians. Vehicle-centered maps with a range of 50 m and resolution of 40 cm are generated using data from range sensors. These maps represent the shape and location of terrain features and obstacle boundaries. The Demo III LADAR and stereo cameras measure position and range of surfaces in the environment out to about 50 m. Sensory processing computes surface properties such as dimensions, area, orientation, texture, and motion. Value judgment computes goodness-of-path plans and provides priority for pointing cameras and aiming guns. Value judgment also evaluates sensor data quality and model data reliability. An operator interface allows a human operator to visualize the state of the vehicle or to intervene to change mode or interrupt the sequence of operations. Subsystem executors sequence commands to primitive-level nodes, monitor how well commands are being followed, and modify parameters as necessary to keep on plan. Output commands to primitive-level nodes may be issued at any 200-ms interval, but typically are planned to change about once every 500 ms.

Level 2: Primitive Level

Each node at the primitive level is a unit consisting of a group of controllers that plan and execute velocities and accelerations to optimize dynamic performance of servo-level nodes, taking into consideration dynamical interaction between mass, stiffness, force, and time. At the primitive level, communication messages for servo-level nodes are encoded into words, numbers, and strings of symbols. Velocity and acceleration set points for the servo-level nodes are planned every 50 ms out to a planning horizon of 500 ms. At 10 m/s, planned servo-level mobility waypoints are computed about 50 cm apart out to a planning horizon of 5 m. The world model symbolic database contains names and attributes of state variables and features such as target trajectories and objects that are within 5 m. Maps are generated from camera data. Five-meter maps have a resolution of about 4 cm. Driving plans can be represented by predicted tire tracks on the map, and visual attention plans by predicted fixation points in the visual field.

Sensory processing computes image features such as occluding edges, boundaries, and vertices and detects strings of events. Value judgment cost functions support dynamic trajectory optimization. An operator interface allows a human operator to visualize the state of each controller and to intervene to change mode or override velocities. Primitive-level executors keep track of how well plans are being followed and modify parameters as necessary to keep within tolerance. Primitive executors issue commands to servo-level nodes to generate dynamically optimized trajectories for vehicle and sensor platforms. Output commands are issued to servo-level nodes every 50 ms.

Level 1: Servo Level

Servo-level nodes transform commands into actuator coordinates and compute motion or torque commands for each actuator. Each node at the servo level is a unit consisting of a group of single-axis controllers that plan and execute coordinated actuator motions for vehicle steering, velocity, and acceleration, or for pointing sensors, or aiming weapons platforms. Desired forces, velocities, and discrete outputs are computed at 5-ms intervals out to a planning horizon of 50 ms.

The world model symbolic database contains values of state variables such as actuator positions, velocities, and forces, pressure sensor readings, position of switches, and gear-shift settings. Sensory processing detects events, and scales and filters data from individual sensors that measure position, velocity, force, torque, and pressure. Sensory processing also computes pixel attributes in images such as spatial and temporal gradients, stereo disparity, range, color, and image flow. An operator interface allows a human operator to view state variables or to change mode, set switches, or jog individual actuators. Executors servo actuators and motors to follow planned trajectories. Position, velocity, or force servoing may be implemented and in various combinations. Servo executors compute at a 200-Hz clock rate. Motion output commands to power amplifiers specify desired actuator torque or power every 5 ms. Discrete output commands produce switch closures and activate relays and solenoids.

The example above illustrates how the 4-D/RCS multilevel hierarchical architecture assigns different responsibilities and duties to various levels of the hierarchy with different range and resolution in time and space at each level. At each level, sensory data are processed, entities are classified, world model representations are maintained, and tasks are decomposed into parallel and sequential subtasks, to be performed by cooperating sets of agents. At each level, feedback from sensors reactively closes a control loop allowing each agent to respond and react to unexpected events.

At each hierarchical level, there is a characteristic range and resolution in space and time, a characteristic bandwidth and response time, a characteristic planning horizon, and a characteristic scale of detail in plans. The 4-D/RCS architecture thus organizes the planning of behavior, the control of action, and the focusing of computational resources such that functional processes at each level have a limited amount of responsibility and a manageable level of complexity [Meystel et al.98].

COMMAND VOCABULARIES

A command vocabulary is the set of named actions or tasks that a 4D/RCS behavior generation (BG) process can perform. Each BG process at each level of the control hierarchy has its own unique command vocabulary. Examples of the command vocabularies at various levels of the Demo III hierarchy are:

Section-Level Commands

Init

E-stop

Suspend/Resume(task T)

RetroTraverse(to point P)

CooperativeSearch(of area A)

PerformRouteReconnaissance(along route R)

PerformAreaReconnaissance(of area A)

ConductScreen(for unit U)

PerformObstacleRestrictedRecon(over area A)

ReconnoiterBuiltUpArea(area A)

ConductTacticalMovement(to point P)

ConductTacticalRoadMarch(along route R)

EstablishObservationPost(at point P)

Section-level commands are expressed in UTM WGS84 world coordinates. Parameters may include goal positions to be occupied, desired paths to be traversed, required regions to be observed. Parameters may also include specifications for performance, such as speed, time of completion, required precision, and choice of formation (e.g., line, wedge, vee, column, staggered.) Mode parameters may include level of aggressiveness, priority, probability of enemy contact, and acceptable risk or cost. Constraint parameters may specify corridor boundaries and speed limit. Condition parameters may specify what is required to begin or continue. Typical intervals between section-level commands are 10 minutes.

Vehicle-Level Commands

Init

E-stop

Suspend/Resume(task T)

RetroTraverse(to x, y by t)

SendImage(between $x1, y1$ and $x2, y2$)

ReportStatus

NavigateToGoalPoint(at x, y by t)

PerformRoadMarch(to x, y by t)

OccupyOverwatchPosition(at x, y by t)

OccupyObservation/ListeningPost(at x, y by t)

DetectBarriersToMovement(between $x1, y1$ and $x2, y2$)

ReconnoiterArea(between $x1, y1$ and $x2, y2$)

ReconnoiterRoute(from $x1, y1$ to $x2, y2$ by t)

LocateBypassOfArea(between $x1, y1$ and $x2, y2$)

ReconnoiterTerrain(between $x1, y1$ and $x2, y2$)

ReconnoiterDefilesOnRoute(from $x1, y1$ to $x2, y2$ by t)

ReconnoiterLateralRoutesAlongRoute(from $x1, y1$ to $x2, y2$ by t)

ReconnoiterApproachToRoute(from $x1, y1$ to $x2, y2$ by t)

IdentifyVehicles&Personnel(between $az1$ and $az2$)

IdentifyThreatVehicles(between $az1$ and $az2$)

MoveToMaintainContact(with target)

Hide&MaintainContact(with target)

HideFromEnemy(between bearing1 and bearing2)

Vehicle-level commands are expressed in vehicle-centered north-oriented world coordinates. Parameters typically specify where, when, how fast, and how important the task is. Typical interval between vehicle-level commands is 50 s.

Autonomous Mobility Subsystem-Level Commands

Init

E-stop

Suspend/Resume

RetroTraverse(to position, velocity, heading by t)

TurnAround(position, velocity, heading by t)

BackUp(to position, velocity, heading by t)

GoWithinCorridorTo(position, velocity, heading, right boundary, left boundary by t)

GoToRoad(position at velocity or by t)

GoOnRoadTo(position at velocity in lane by t)

GoBesideRoadTo(position at velocity, offset until t)

GoStealthyTo(position, velocity, heading by t)

GoToHillCrest(position, heading by t)

LeaveHillCrest(position, heading by t)

GoToFeature(feature, position, heading by t)

DashTo(position, velocity, heading at t)

Hide(position, heading by t)

HullDown(position, heading)

StopAt(phase line by t)

ScanTreeLine(bearing, elevation, length)

ConductSecurityHalt(position, heading at t)

Subsystem-level commands are expressed in vehicle-centered vehicle-oriented world coordinates. Parameters may specify position, velocity, heading, and timing requirements. Typical interval between subsystem-level commands is 5 s.

Primitive Level: Primitive Driver Commands

Init

E-stop

Suspend/Resume

GoTo(position, velocity, heading by t)

FollowLeadVehicle(at distance)

Primitive Level: Gaze Commands

Init

E-stop

Suspend/Resume

FixatePoint(at range, bearing, elevation)

TrackObject(at range, bearing, velocity at t)

ScanTrajectory(from range1, bearing1, elevation1 to range2, bearing2, elevation2)

Primitive-level commands are expressed in vehicle-centered vehicle-oriented coordinates. Typical interval between primitive-level commands is 500 ms.

Servo Level: Drive Commands

Init

E-stop

Suspend/Resume

GoTo(range, bearing, speed, heading by t)

Servo Level: Look Commands

Init

E-stop

Suspend/Resume

GoTo(range, bearing, speed by t)

Servo-level commands are expressed in vehicle-centered vehicle-oriented coordinates. The interval between servo commands is 50 ms.

Actuator Commands

Init

E-stop

GoTo(position at t)

GoAt(velocity at t)

ExertForce(amount at t)

Actuator commands are expressed in actuator coordinates. The interval between actuator commands is 5 ms.

DEMO III CONTROL HIERARCHY

In each BG module, commands are decomposed into approximately a 10-step plan for each of its subordinate BG modules. For each plan, an executor cycles through the plan issuing commands to the appropriate subordinate BG modules. Commands into each BG module consist of at least six elements:

1. *ActionCommand* (*ac*1): describes the action to be performed and may include a set of modifiers such as priorities, mode, path constraints, acceptable cost, and required conditions.
2. *GoalCommand* (*gc*1): describes the desired state (or goal state) to be achieved by the action. Mobility system state typically includes the position, heading, velocity, and turning rate of the system being controlled. The goal may include the name of a target or object that is to be acted upon. It may also include a set of modifiers such as tolerance.
3. *GoalTime* (*gt*1): defines the timing constraint on achieving the goal plus modifiers such as tolerance.
4. *NextActionCommand* (*ac*2): describes the planned next action to be performed plus modifiers.
5. *NextGoalCommand* (*gc*2): describes the planned next goal state to be achieved plus modifiers.
6. *NextGoalTime* (*gt*2): describes the timing constraint on achieving the next goal plus modifiers.

If we designate levels in the hierarchy by a superscript, and a node index within each level by a subscript, then input to each behavior generation (BG) process is a command data structure of the following form:

$$ac1_i^j = \text{ActionCommand plus modifiers for}$$
$$\text{BG module } i \text{ at level } j$$

$$gc1_i^j = \text{GoalCommand state plus modifiers for}$$
$$\text{BG module } i \text{ at level } j$$

$$gt1_i^j = \text{GoalTime plus modifiers for when } gc1_i^j$$
$$\text{should be achieved}$$

$$ac2_i^j = \text{NextActionCommanded plus modifiers for}$$
$$\text{BG module } i \text{ at level } j$$

$$gc2_i^j = \text{NextGoalCommand state plus modifiers for}$$
$$\text{BG module } i \text{ at level } j$$

$$gt2_i^j = \text{NextGoalTime plus modifiers for when } gc2_i^j$$
$$\text{should be achieved}$$

Figure 9.3 shows the command and plan structure for the first five levels of a Demo III XUV.

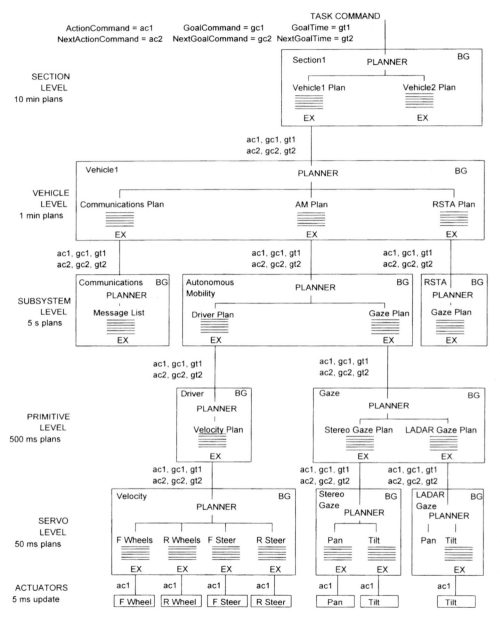

FIGURE 9.3 Command and plan structure for Demo III. Note that the plan for each BG module is generated by, and resides in, the BG module above it. For example, the AM plan for the autonomous mobility BG is generated by the vehicle-level planner. The AM plan resides in the vehicle-level BG module and is transformed into commands for the autonomous mobility BG by the vehicle-level AM executor.

Section (Level 5)

Commands to Section level BG processes would have the following form:

Section1 Command Structure

ActionCommand = $ac1_1^5$	GoalCommand = $gc1_1^5$	GoalTime = $gt1_1^5 \approx t + 10$ min
NextActionCommand = $ac2_1^5$	NextGoalCommand = $gc2_1^5$	NextGoalTime = $gt2_1^5 \approx t + 20$ min

where $\approx$ means approximately.

The planner in each section-level BG process decomposes commands into plans for each of its vehicle BG processes. Each subordinate plan is designed to have about 10 steps. For example, a section with two vehicles would have a plan for two vehicles of the following form:

Vehicle1 Plan	Vehicle2 Plan	Typical Goal Times
$ap1_1^4, gp1_1^4, gt1_1^4$	$ap1_2^4, gp1_2^4, gt1_2^4$	$gt1_i^4 \approx t + 1$ min
$ap2_1^4, gp2_1^4, gt2_1^4$	$ap2_2^4, gp2_2^4, gt2_2^4$	$gt2_i^4 \approx t + 2$ min
$ap3_1^4, gp3_1^4, gt3_1^4$	$ap3_2^4, gp3_2^4, gt3_2^4$	$gt3_i^4 \approx t + 3$ min
$ap4_1^4, gp4_1^4, gt4_1^4$	$ap4_2^4, gp4_2^4, gt4_2^4$	$gt4_i^4 \approx t + 4$ min
$ap5_1^4, gp5_1^4, gt5_1^4$	$ap5_2^4, gp5_2^4, gt5_2^4$	$gt5_i^4 \approx t + 5$ min
$ap6_1^4, gp6_1^4, gt6_1^4$	$ap6_2^4, gp6_2^4, gt6_2^4$	$gt6_i^4 \approx t + 6$ min
$ap7_1^4, gp7_1^4, gt7_1^4$	$ap7_2^4, gp7_2^4, gt7_2^4$	$gt7_i^4 \approx t + 7$ min
$ap8_1^4, gp8_1^4, gt8_1^4$	$ap8_2^4, gp8_2^4, gt8_2^4$	$gt8_i^4 \approx t + 8$ min
$ap9_1^4, gp9_1^4, gt9_1^4$	$ap9_2^4, gp9_2^4, gt9_2^4$	$gt9_i^4 \approx t + 9$ min
$ap10_1^4, gp10_1^4, gt10_1^4$	$ap10_2^4, gp10_2^4, gt10_2^4$	$gt10_i^4 \approx t + 10$ min

where

$$apk_i^j, = \text{action planned for BG module } i \text{ at level } j \\ \text{for plan step } k$$

$$gpk_i^j = \text{goal planned for BG module } i \text{ at level } j \\ \text{for plan step } k$$

$$gtk_i^j = \text{goal time planned for BG module } i \text{ at level } j \\ \text{for plan step } k$$

$$t = \text{time at which the command is scheduled to begin}$$

The GoalTimes and subgoal times shown here illustrate only order of magnitude. Plan steps need not be equally spaced in time or space. There also might be more or less than 10 steps in a plan.

Vehicle (Level 4)

Commands to vehicle-level BG processes would have the following form:

Vehicle1 Command Structure

ActionCommand = $ac1_1^4$	GoalCommand = $gc1_1^4$	GoalTime = $gt1_1^4 \approx t + 1$ min
NextActionCommand = $ac2_1^4$	NextGoalCommand = $gc2_1^4$	NextGoalTime = $gt2_1^4 \approx t + 2$ min

A vehicle with three subsystems would have a plan for each subsystem of the following form:

Autonomous Mobility Plan	RSTA Plan	Communications Plan	Typical Goal Times
$ap1_1^3, gp1_1^3, gt1_1^3$	$ap1_2^3, gp1_2^3, gt1_2^3$	$ap1_3^3, gp1_3^3, t1_3^3$	$gt1_i^3 \approx t + 5$ s
$ap2_1^3, gp2_1^3, gt2_1^3$	$ap2_2^3, gp2_2^3, gt2_2^3$	$ap2_3^3, gp2_3^3, t2_3^3$	$gt2_i^3 \approx t + 10$ s
$ap3_1^3, gp3_1^3, gt3_1^3$	$ap3_2^3, gp3_2^3, gt3_2^3$	$ap3_3^3, gp3_3^3, t3_3^3$	$gt3_i^3 \approx t + 15$ s
$ap4_1^3, gp4_1^3, gt4_1^3$	$ap4_2^3, gp4_2^3, gt4_2^3$	$ap4_3^3, gp4_3^3, t4_3^3$	$gt4_i^3 \approx t + 20$ s
$ap5_1^3, gp5_1^3, gt5_1^3$	$ap5_2^3, gp5_2^3, gt5_2^3$	$ap5_3^3, gp5_3^3, t5_3^3$	$gt5_i^3 \approx t + 25$ s
$ap6_1^3, gp6_1^3, gt6_1^3$	$ap6_2^3, gp6_2^3, gt6_2^3$	$ap6_3^3, gp6_3^3, t6_3^3$	$gt6_i^3 \approx t + 30$ s
$ap7_1^3, gp7_1^3, t7_1^3$	$ap7_2^3, gp7_2^3, gt7_2^3$	$ap7_3^3, gp7_3^3, t7_3^3$	$gt7_i^3 \approx t + 35$ s
$ap8_1^3, gp8_1^3, gt8_1^3$	$ap8_2^3, gp8_2^3, gt8_2^3$	$ap8_3^3, gp8_3^3, t8_3^3$	$gt8_i^3 \approx t + 40$ s
$ap9_i^3, gp9_1^3, gt9_1^3$	$ap9_2^3, gp9_2^3, gt9_2^3$	$ap9_3^3, gp9_3^3, t9_3^3$	$gt9_i^3 \approx t + 50$ s
$ap10_1^3, gp10_1^3, gt10_1^3$	$ap10_2^3, gp10_2^3, gt10_2^3$	$ap10_3^3, gp10_3^3, t10_3^3$	$gt10_i^3 \approx t + 60$ s

Subsystem (Level 3)

Commands to subsystem-level BG processes would have the following form:

Autonomous Mobility Command Structure

ActionCommand = $ac1_1^3$	GoalCommand = $gc1_1^3$	GoalTime = $gt1_1^3 \approx t + 5$ s
NextActionCommand = $ac2_1^3$	NextGoalCommand = $gc2_1^3$	NextGoalTime = $gt2_1^3 \approx t + 10$ s

A mobility subsystem with primitive-level driver and gaze unit controllers would have the following form:

Driver Plan	Gaze Plan	Typical Goal Times
$ap1_1^2, gp1_1^2, gt1_1^2$	$ap1_2^2, gp1_2^2, gt1_2^2$	$gt1_i^2 \approx t + 0.5$ s
$ap2_1^2, gp2_1^2, gt2_1^2$	$ap2_2^2, gp2_2^2, gt2_2^2$	$gt2_i^2 \approx t + 1.0$ s
$ap3_1^2, gp3_1^2, gt3_1^2$	$ap3_2^2, gp3_2^2, gt3_2^2$	$gt3_i^2 \approx t + 1.5$ s
$ap4_1^2, gp4_1^2, gt4_1^2$	$ap4_2^2, gp4_2^2, gt4_2^2$	$gt4_i^2 \approx t + 2.0$ s
$ap5_1^2, gp5_1^2, gt5_1^2$	$ap5_2^2, gp5_2^2, gt5_2^2$	$gt5_i^2 \approx t + 2.5$ s

$$ap6_1^2, gp6_1^2, gt6_1^2 \qquad ap6_2^2, gp6_2^2, gt6_2^2 \qquad gt6_i^2 \approx t + 3.0 \text{ s}$$
$$ap7_1^2, gp7_1^2, gt7_1^2 \qquad ap7_2^2, gp7_2^2, gt7_2^2 \qquad gt7_i^2 \approx t + 3.5 \text{ s}$$
$$ap8_1^2, gp8_1^2, gt8_1^2 \qquad ap8_2^2, gp8_2^2, gt8_2^2 \qquad gt8_i^2 \approx t + 4.0 \text{ s}$$
$$ap9_1^2, gp9_1^2, gt9_1^2 \qquad ap9_2^2, gp9_2^2, gt9_2^2 \qquad gt9_i^2 \approx t + 4.5 \text{ s}$$
$$ap10_1^2, gp10_1^2, gt10_1^2 \qquad ap10_2^2, gp10_2^2, gt10_2^2 \qquad gt10_i^2 \approx t + 5.0 \text{ s}$$

Primitive (Level 2)

Commands to primitive-level BG processes would have the following form:

Driver Command Structure

ActionCommand = $ac1_1^2$	GoalCommand = $gc1_1^2$	GoalTime = $gt1_1^2 \approx t + 0.5$ s
NextActionCommand = $ac2_1^2$	NextGoalCommand = $gc2_1^2$	NextGoalTime = $gt1_1^2 \approx t + 1.0$ s

Primitive-level plans for the servo-level BG units would have the following form:

Velocity Plan	Goal Times
$ap1_1^1, gp1_1^1, gt1_1^1$	$gt1_i^1 = t + 50$ ms
$ap2_1^1, gp2_1^1, gt2_1^1$	$gt2_i^1 = t + 100$ ms
$ap3_1^1, gp3_1^1, gt3_1^1$	$gt3_i^1 = t + 150$ ms
$ap4_1^1, gp4_1^1, gt4_1^1$	$gt4_i^1 = t + 200$ ms
$ap5_1^1, gp5_1^1, gt5_1^1$	$gt5_i^1 = t + 250$ ms
$ap6_1^1, gp6_1^1, gt6_1^1$	$gt6_i^1 = t + 300$ ms
$ap7_1^1, gp7_1^1, gt7_1^1$	$gt7_i^1 = t + 350$ ms
$ap8_1^1, gp8_1^1, gt8_1^1$	$gt8_i^1 = t + 400$ ms
$ap9_1^1, gp9_1^1, gt9_1^1$	$gt9_i^1 = t + 450$ ms
$ap10_1^1, gp10_1^1, gt10_1^1$	$gt10_i^1 = t + 500$ ms

Note that time intervals in plans become uniform at the primitive level and below.

Servo (Level 1)

Commands to servo-level BG controllers would have the following form:

Velocity Command Structure

ActionCommand = $ac1_1^1$	GoalCommand = $gc1_1^1$	GoalTime = $gt1_1^1 = t + 50$ ms
NextActionCommand = $ac2_1^1$	NextGoalCommand = $gc2_1^1$	NextGoalTime = $gt2_1^1 = t + 100$ ms

Servo-level plans for each actuator would have the following form:

Wheel Motors	Front Steer Motor	Rear Steer Motor	Goal Times
$ap1_1^0, gp1_1^0, gt1_1^0$	$ap1_2^0, gp1_2^0, gt1_2^0$	$ap1_3^0, gp1_3^0, gt1_3^0$	$gt1_i^0 = t + 5$ ms
$ap2_1^0, gp2_1^0, gt2_1^0$	$ap2_2^0, gp2_2^0, gt2_2^0$	$ap2_3^0, gp2_3^0, gt2_3^0$	$gt2_i^0 = t + 10$ ms
$ap3_1^0, gp3_1^0, gt3_1^0$	$ap3_2^0, gp3_2^0, gt3_2^0$	$ap3_3^0, gp3_3^0, gt3_3^0$	$gt3_i^0 = t + 15$ ms
$ap4_1^0, gp4_1^0, gt4_1^0$	$ap4_2^0, gp4_2^0, gt4_2^0$	$ap4_3^0, gp4_3^0, gt4_3^0$	$gt4_i^0 = t + 20$ ms
$ap5_1^0, gp5_1^0, gt5_1^0$	$ap5_2^0, gp5_2^0, gt5_2^0$	$ap5_3^0, gp5_3^0, gt5_3^0$	$gt5_i^0 = t + 25$ ms
$ap6_1^0, gp6_1^0, gt6_1^0$	$ap6_2^0, gp6_2^0, gt6_2^0$	$ap6_3^0, gp6_3^0, gt6_3^0$	$gt6_i^0 = t + 30$ ms
$ap7_1^0, gp7_1^0, gt7_1^0$	$ap7_2^0, gp7_2^0, gt7_2^0$	$ap7_3^0, gp7_3^0, gt7_3^0$	$gt7_i^0 = t + 35$ ms
$ap8_1^0, gp8_1^0, gt8_1^0$	$ap8_2^0, gp8_2^0, gt8_2^0$	$ap8_3^0, gp8_3^0, gt8_3^0$	$gt8_i^0 = t + 40$ ms
$ap9_1^0, gp9_1^0, gt9_1^0$	$ap9_2^0, gp9_2^0, gt9_2^0$	$ap9_3^0, gp9_3^0, gt9_3^0$	$gt9_i^0 = t + 45$ ms
$ap10_1^0, gp10_1^0, gt10_1^0$	$ap10_2^0, gp10_2^0, gt10_2^0$	$ap10_3^0, gp10_3^0, gt10_3^0$	$gt10_i^0 = t + 50$ ms

Actuators (Level 0)

Commands to actuators would have the following form:

Actuator i

ActionCommand$_i^0 = ac1_i^0$	GoalCommand $= gc1_i^0$	GoalTime $= gt1_i^0 = t + 5$ ms

where

$$ac1_i^0 = ap1_i^0 + kfb(gc1_i^0 - x1_i^{*0})$$

$$x1_i^{*0} = \text{predicted state of } i\text{th actuator at next sample}$$

$$gc1_i^0 = gp1_i^0$$

$$kfb = \text{feedback gain}$$

EXAMPLE DATA STRUCTURES

An example of a C++ class data structure for a command from the vehicle level to the subsystem autonomous mobility level might be:

```
/* GoToHillCrest ***************************************/
class GO_TO_HILL_CREST_CMD : public RCS_CMD_MSG
{
public:
  GO_TO_HILL_CREST_CMD(); // Constructor
  void update(CMS *); // update function.
```

```
private:
      // action modifiers
   int stealthiness;  // 1 to 100% stealthy
   double speedLimit;  // in m/s
      // GoalCommand
   double x_goal;  // desired x position on a map about 50 m away
   double y_goal;  // desired y position on a map about 50 m away
   char speedAtGoal;  // desired speed in m/s at GoalCommand (0 if stop
                      // at goal)
   double headingAtGoal;  // desired heading at GoalCommand
      // goal modifiers
   double timeToGetToGoal;  // ¯5 s for a vehicle level command
   double timeTolerance;  // ±seconds
   double goalTolerance;  // close enough radius to goal
};
```

An example of a status message from the autonomous mobility subsystem–level planner to the vehicle-level executor might be:

```
/* Status feedback ****************************************/
class AM_VEHICLE-STATUS : public RCS_STAT_MSG
{
public:
   AM_VEHICLE-STATUS ();  // Constructor
   void update(CMS *);  // update function.
private:
   boolean ExitIfPastGoal;  // task done flag
      // predicted state yd* at command GoalTime=gtl₁³
   double x_predictedAtGoalTime;
   double y_predictedAtGoalTime;
   double speed_predictedAtGoalTime;
   double heading_predictedAtGoalTime;
      // estimated time to reach GoalCommand
   double estimatedTimeToGoal;
      // predicted state at planning horizon (i.e., at last subgoal
      // yd10₁²)
   double x_predictedAtPlanHorizon;
   double y_predictedAtPlanHorizon;
   double speed_predictedAtPlanHorizon;
   double heading_predictedAtPlanHorizon;
};
```

TABLE 9.1 Planning Horizon, Replan Interval, and Executor Reaction Latency at Each Level of the 4D/RCS Hierarchy

Level		Planning Horizon	Replan Interval	Reaction Latency
1	Servo	50 ms	50 ms	5 ms
2	Primitive	500 ms	50 ms	50 ms
3	Subsystem	5 s	500 ms	200 ms
4	Vehicle	50 s	5 s	500 ms
5	Section	10 min	50 s	2 s
6	Platoon	2 h	10 min	5 s
7	Battalion	24 h	2 h	20 s

MULTILEVEL PLANNING

The 4-D/RCS architecture has seven levels of distributed, hybrid, deliberative/reactive control. There is an order-of-magnitude difference in range and resolution in space and time between successive levels. 4-D/RCS is designed to enable long-range big-picture plans for complex problems at higher levels while producing high-speed high-precision control at lower levels. Multilevel hierarchical planning makes it feasible to replan frequently at all levels because plans are limited in range and resolution at each level. Multilevel world modeling makes it possible to update the knowledge database frequently because multilevel maps are limited in range and resolution, and hierarchical symbolic data structures can be designed to represent only what is necessary for making decisions at each level. Multiresolutional representation of space limits the amount of information that must be computed for each map and the number of states in plan space that must be searched at each level. Multiresolutional representation of time limits the number of events that must be represented at each level. As a result, the world model can be extremely rich and detailed at the point of interest, yet contain relatively little information that must be updated in real time. Therefore, new plans can be generated in each node well before current plans become obsolete. In many cases, new plans can be generated in the time required to execute only one step in the current plan. Thus, action can always take place in the context of a recent plan, and feedback through the executor can close a reactive control loop using recently selected control parameters.

Suggested 4-D/RCS specifications for the planning horizon, replanning interval, and reaction latency at all seven levels are shown in Table 9.1. The planning horizon refers to the future point in time to which each level plans. Plans at each level typically have about 10 steps between the anticipated starting state and a planned goal state at the planning horizon. Thus, the planning horizon typically grows by an order of magnitude longer at each successively higher level. Reaction latency is the minimum delay through the reactive feedback loop at each level. Reaction can interrupt cyclic replanning to select an emergency plan immediately and begin a new replanning cycle based on new information. Reaction latencies at each level are determined by computational delays in updating the world model as well as the sampling frequency and computation cycle rate of the executors. The fastest servo update rate on the XUV is 200 Hz. Thus, the reaction latency at the servo level is 5 ms. The required execution cycle rate at other levels depends on the dynamics of the mechanism being controlled and the speed of the computers available.

Figure 9.4 is a block diagram of the first five levels in the 4-D/RCS architecture for Demo III. On the right, behavior generation modules decompose high-level mission commands into low-level actions. The text inside the planner indicates the planning horizon at each level. Each planner has a world model simulator that is appropriate for the problems encountered at its level. In the center of Figure 9.4, each map in the KD as a range and resolution that is appropriate for path planning at its level. Maps may have many overlays for different information, such as terrain elevation, roughness, slope, obstacles, traversability, and risk, as well as classified regions such as dirt, grass, bushes, trees, rocks, roads, mud, water, and cover (i.e., regions where overhanging foliage is high enough above the ground to permit a vehicle to drive under). At each level there are symbolic data structures and segmented images with labeled regions that describe entities, events, and situations that are relevant to decisions that must be made at that level. On the left is a sensory processing hierarchy that extracts from the sensory data stream the information needed to keep the world model knowledge database current and accurate. Maps, images, and symbolic data structures are linked by pointers as described in Chapter 7 and 8.

At the bottom of Figure 9.4 are actuators that act on the world and sensors that measure phenomena in the world. The Demo III vehicles have a variety of sensors, including a laser range imager (LADAR), stereo CCD cameras, stereo FLIR cameras, a color CCD camera, a vegetation penetrating radar, GPS (global positioning system), an inertial navigation package, actuator feedback sensors, and a variety of internal sensors for measuring parameters such as engine temperature, speed, vibration, oil pressure, and fuel level. The vehicle also carries a reconnaissance, surveillance, and target acquisition (RSTA) mission package that includes zoom lenses for CCD cameras and FLIRs, a laser range finder, and an acoustic package.

In Figure 9.4, the servo level has no map representation. The servo level deals with actuator dynamics and reacts to sensory feedback from actuator sensors. The primitive-level map has range of 5 m with resolution of 4 cm. This enables the vehicle to make small path corrections to avoid bumps and ruts during the 500-ms planning horizon of the primitive level. The primitive level also uses accelerometer data to control vehicle dynamics and prevent excessive bouncing or rollover during high-speed driving. The subsystem-level map has a range of 50 m with resolution of 40 cm. This map is used to plan about 5 s into the future to find a path that avoids obstacles and

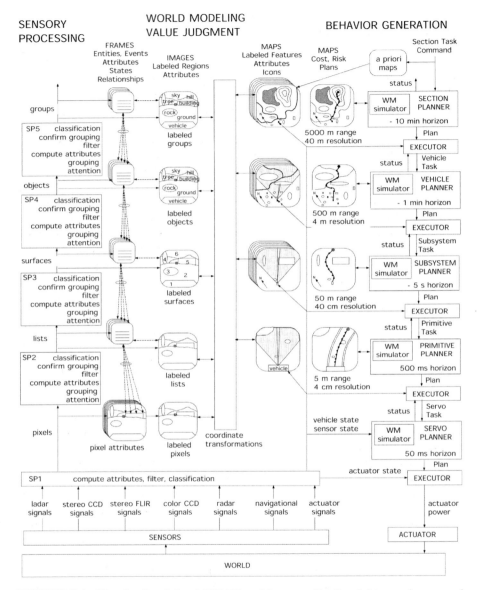

FIGURE 9.4 Five levels of the 4-D/RCS architecture. On the right are planner and executor modules. In the center are maps for representing terrain features, road, bridges, vehicles, friendly/enemy positions, and the cost and risk of traversing various regions. On the left are sensory processing functions, symbolic representations of entities and events, and segmented images with labeled regions. The coordinate transforms in the middle use range information to assign labeled regions in the entity image hierarchy on the left to locations on planning maps on the right. This causes the entity class hierarchy to be orthogonal to the BG process hierarchy.

provides a smooth and efficient ride. The vehicle-level map has a range of 500 m with resolution of 4 m. This map is used to plan paths and individual tactical behaviors about 1 min into the future, taking into account terrain features such as roads, bushes, gullies, tree lines, and other vehicles. The section-level map has a range of 5000 m with resolution of about 40 m. This map is used to plan about 10 min into the future to accomplish group tactical behaviors. Higher-level maps (not shown in Figure 9.4) are used to plan section and platoon missions lasting about 2 and 24 hours, respectively. These are derived from military maps and intelligence provided by the digital battlefield database.

At all levels (except the lowest), 4-D/RCS planners are designed to generate new plans well before current plans become obsolete. Thus, action can always take place in the context of a recent plan, and feedback through the executors can close reactive control loops using recently selected control parameters. To meet the demands of Demo III, the 4-D/RCS architecture specifies that replanning should occur within about one-tenth of the planning horizon at each level.

Executors can react to sensory feedback even faster. If the executor senses an error between its output GoalCommand and the state predicted (status from the subordinate BG planner) at GoalTime, it may react by modifying the ActionCommand so as to cope with that error. This closes a feedback loop through the executor at that level within the specified reaction latency.

The type of executor reaction depends on the size and nature of the detected error. If the error is small, the executor may simply modify its ActionCommand in a manner designed to reduce the error. For example, if the status reported from the subordinate planner indicates that the vehicle is going to arrive at the goal point late, the executor might modify its ActionCommand to speed up or delete some low-priority activities. However, if the error is out of range, the executor may select a stored emergency plan from an exception handler, substitute it for the current plan, and modify its ActionCommand and GoalCommand to its subordinate planner appropriately. For example, an event such as the discovery of an unexpected obstacle in the AM planned path (generated by the vehicle planner) may cause the AM planner to make a plan that deviates significantly from its commanded goal. In this case, the vehicle-level executor may modify its ActionCommand $ac1_1^3$ in a manner designed to buy time for the vehicle-level planner to generate a new AM plan. For example, it may command the AM level to reduce speed or stop and direct AM driving cameras or RSTA sensors to collect information about the obstacle while a new AM plan is being generated by the vehicle-level planner. All of this executor response should take place within the 500-ms reaction latency of the vehicle-level executor.

Typically, evoking an emergency plan will cause the executor to request its planner to begin a new replanning cycle immediately. As shown in Figure 9.4, the period required for replanning at the vehicle level is 5 s. The replanning period at the AM level is 0.5 s. Thus, the emergency plan evoked by the vehicle-level executor can handle the problem of what the AM level should plan to do over the next 5 s while the vehicle-level planner generates a new AM plan out to its 1-min planning horizon.

TWO KINDS OF PLANS

There are two kinds of plans that are required by the Demo III vehicles: (1) path plans for locomotion, and (2) task plans for other types of behavior. A typical path plan consists of a series of waypoints on a map. A typical task plan consists of a set of instructions or rules that describes a sequence of actions and subgoals required to complete the task. Both path plans and task plans can be represented in the form of augmented state graphs, or state tables, which define a series of planned actions (subtasks) with a desired state (subgoal) to be achieved by each action in the plan. Typically, states are represented by nodes and actions by arcs that connect the nodes. Both types of plans can be executed by the same executor mechanism.

In principle, both types of planning can be performed by searching the space of possible futures to find a desirable solution. However, path planning typically requires searching only a two-dimensional space on a map. Task planning requires searching an N-dimensional space of all possible states and actions. Searching high-dimensional spaces can be accomplished by evolutionary algorithms [Fogel99] or reinforcement learning techniques [Sutton and Barto98]. However, these methods are typically too slow for real-time use at levels where plans must be recomputed faster than once every few minutes. Therefore, real-time task planning is typically done by searching a library of schema or recipes that have been developed off-line and stored where they can be accessed by rules or case statements when conditions arise. When there is more than one recipe or schema that are appropriate to a task, each may be submitted to the world model for simulation and the predicted results evaluated by the value judgment process. The plan selector then selects the best recipe or schema for execution.

In 4-D/RCS, path planners use cost maps that represent the estimated cost or risk of being in, or traversing, regions on the map. Values represented in cost maps depend on mission priorities and knowledge of the tactical situation represented in the KD. Path planners search the cost maps for routes that have the lowest cost under a given situation. Task planners use rules of engagement, military doctrine, and case-based reasoning to select modes of operation and schema for tactical behaviors. State variables such as mission priorities and situational awareness determine cost functions and hence decisions regarding which type of behavior to select.

For example, if enemy contact is likely or has occurred, cost maps of open regions and roads will carry a high cost and regions near tree lines and under tree cover will have lower cost. In this case, path planners will

TABLE 9.2   Range and Resolution of Maps at All levels in the Demo III 4-D/RCS Architecture

	Level	Map Resolution	Map Range[a]	Function Performed
1	Servo	n/a	n/a	Actuator servo
2	Primitive	4 cm	5 m	Vehicle heading, speed
3	Subsystem	40 cm	50 m	Obstacle avoidance
4	Vehicle	4 m	500 m	Single-vehicle tactical behaviors
5	Section	40 m	10 km	Section-level tactical behaviors
6	Platoon	40 m	100 km	Platoon-level tactical behaviors
7	Battalion	400 m	1000 km	Battalion-level tactical behaviors

[a]Range is measured from the vehicle at the center of each map.

plan cautious routes near tree lines or through wooded areas, and task planners will plan behaviors designed to search for evidence of enemy activity in likely places. However, if enemy contact is unlikely, roads will have a very low cost and open regions will be less costly than wooded areas. This will cause path planners to plan higher-speed routes on the road or through open regions, and task planners to focus on issues such as avoiding local traffic. Thus, a very small amount of information, such as knowledge that enemy contact is likely or unlikely, can completely change the tactical behavior of the vehicle in a very logical, intuitive, and meaningful way.

For the Demo III program, the range and resolution of maps is limited to about 128 resolution elements from the center of the map in each direction at each level. This means that each map contains about 256×256 ($\approx 64,000$ pixels). The range and resolution of maps at all levels of the Demo III 4-D/RCS hierarchy are shown in Table 9.2. Maps at each level provide information to planners about the position, attributes, and class of entities. For example, maps at various levels may indicate the shape, size, class, and motion of objects such as obstacles and vehicles, and the location of roads, intersections, bridges, streams, woods, lakes, buildings, and towns.

In general, map range and resolution depend on velocity and the planning time horizon. For any given planning time horizon, the map range must be sufficient to guarantee that the plan will fit on the map. For different vehicle speeds, the map resolution required for planning at various levels will be different. The numbers in Table 9.2 are for a ground vehicle traveling about 10 m/s. A helicopter skimming over the ground at 100 m/s would require planning maps with an order-of-magnitude greater range and an order-of-magnitude lower resolution than that shown above. For systems with widely varying velocities, map range and resolution may be velocity dependent.

EXAMPLE SCENARIOS

We now examine two scenarios that illustrate system operation based on the
4D/RCS architecture. Figure 9.5 shows the starting conditions for the first
scenario. To make the figures in these example scenarios more easily readable,
we will use the following notation:

- The Section1 GoalCommand $gc1_1^5$ will be designated by S1.
- The Vehicle1 planned subgoals $(gp1_1^4, gp2_1^4, gp3_1^4, \ldots, gp10_1^4)$ will be designated by $(V1, V2, V3, \ldots, V10)$.
- The autonomous mobility planned subgoals $(gp1_1^3, gp2_1^3, gp3_1^3, \ldots, gp10_1^3)$ will be designated by $(A1, A2, A3, \ldots, A10)$.
- The primitive driver planned subgoals $(gp1_1^2, gp2_1^2, gp3_1^2, \ldots, gp10_1^2)$ will be designated by $(D1, D2, D3, \ldots, D10)$.

Autonomous Mobility Plan

At a point in time $t = 0$, Vehicle1 is at position V0. The platoon-Section1 executor has issued a command to Section1 to establish an overwatch position on hill S1 at about $t + 10$ min. S1 is the first subgoal in a Section1 plan that extends approximately 2 hours into the future. Upon receiving the GoalCommand S1, the Section1 planner has generated a vehicle plan for each vehicle in Section1 to achieve S1. The Vehicle1 plan consists of a series of subgoals V1 through V10. V10 is the Vehicle1 contribution to achieving the Section1 GoalCommand S1. A feature such as a road will be noticed by each planner as soon as it appears within the planning horizon and is represented on the planning map. For example, the section-level planner takes the road into account from the beginning as illustrated by the plan V1 through V10. The Vehicle1 plan calls for the road to be encountered at V3 about 3 min from the starting position at $t = 0$. The Section1 planner may specifically designate the planned subgoal V3 to be the road entry point.

Also at $t = 0$, the Section1-to-Vehicle1 executor has issued a command to Vehicle1 to Goto the first subgoal V1. The Vehicle1 planner has then generated an autonomous mobility plan consisting of a series of subgoals A1 through A10. A10 is the autonomous mobility contribution to achieving the Vehicle1 GoalCommand V1. Vehicle1 subgoals V1 through V10 define waypoints vx1 through vx10 on the ground. Autonomous mobility subgoals A1 through A10 define waypoints ax1 through ax10.

As time progresses the vehicle moves along its planned paths. As this happens, subgoals are accomplished at the beginning of the plan, and new subgoals are added to the end of the plan. Adding new subgoals at the planning horizon and deleting subgoals that have been accomplished requires replanning. In our example, replanning occurs periodically, at intervals equal to one-tenth the planning horizon. At each replanning cycle, each plan will scroll forward so that there is always a plan from the current state to a planning

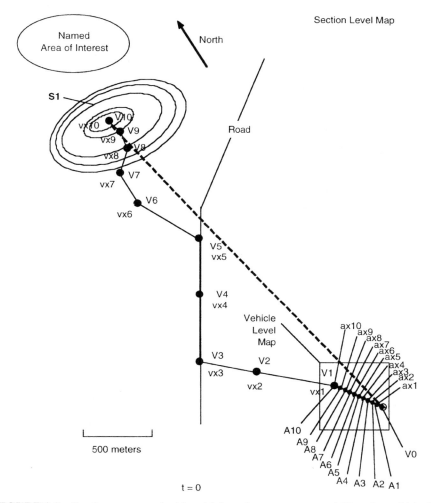

FIGURE 9.5 Section command with vehicle and autonomous mobility plans. S1 is the Section1 GoalCommand $gc1_1^5$. V1 through V10 are Vehicle1 plan subgoals plotted on a section-level map (range of 5000 m). A1 through A10 are autonomous mobility plan subgoals plotted on the vehicle-level map (range of 500 m). V0 is the vehicle current position, ax1 through ax10 are waypoints on the ground defined by the autonomous mobility plans A1 through A10, and vx1 through vx10 are waypoints on the ground defined by the vehicle plan.

horizon. At each level, replanning occurs 10 times more often than at the next higher-level.

Each BG module adds to the list of planned waypoints at its own rate, and leave a trail of historical waypoints behind. This is illustrated in Figure 9.6

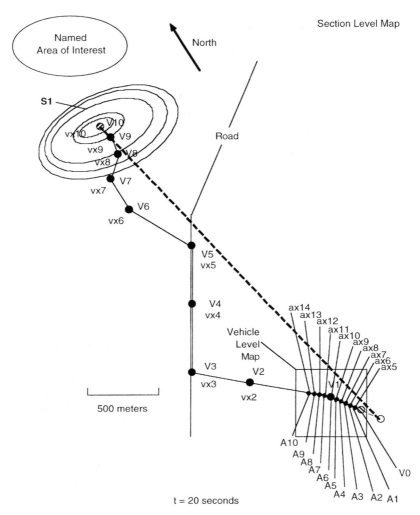

FIGURE 9.6 Replanning adds new waypoints and deletes old. Here the current ve-hicle position V0 has moved to a new location, and new waypoints ax11 through ax14 have been added as autonomous mobility subgoals A7 through A10. Previous waypoints ax5 through ax10 are scrolled to make a new plan. Passed waypoints ax1 through ax4 are deleted.

at $t = 20$ s. Figure 9.7 shows the situation at $t = 2$ min where the vehicle is approaching the road. The road first appears on the vehiclel-level map at about 500 m in the distance. At this point, the vehicle-level planner expects the road to be encountered about 1 min in the future, assuming that the vehicle drives at 10 m/s.

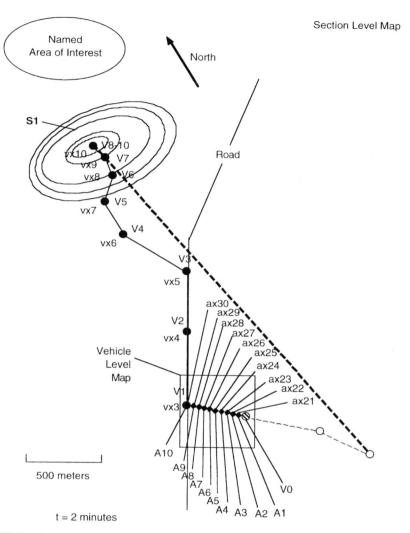

FIGURE 9.7 Replanning at $t = 2$ min. The Vehicle1 plan data structure has been scrolled. The original vx1 and vx2 waypoints have been passed; vx3 as been scrolled into V1. V9 and V10 become the first two Vehicle1 task subgoals to be pursued at the observation point S1. At this point in time, the road first appears on the vehicle-level map.

At $t = 2$ min 20 s, the transition between cross-country and on-road driving is fully incorporated by the vehicle-level planner into the autonomous mobility plan as shown in Figure 9.8. At this point, the road is still beyond the range of the LADAR or stereo, so a range image is not available. However, the road should be visible to the CCD or FLIR cameras. Thus, bearing and elevation

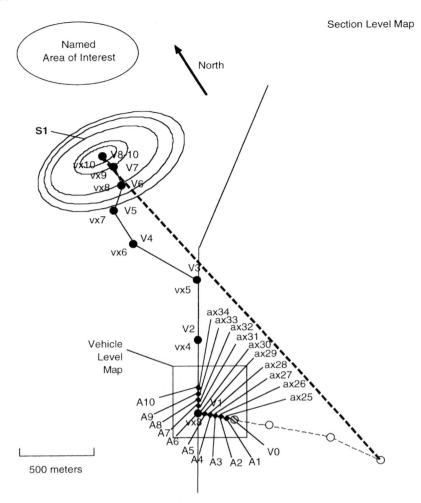

t = 2 minutes 20 seconds

FIGURE 9.8 Autonomous mobility plan generated by the vehicle-level planner for the transition from cross-country to on-road driving. The road is well within the vehicle-level planning horizon but still beyond the subsystem-level planning horizon.

data can be obtained to about 0.1° accuracy and the RSTA package can use its laser range finder to measure range at any point in the visual field.

At $t = 2$ min 55 s, the road is only about 5 s (or 50 m) away, as shown in Figure 9.9 on the section and vehicle level maps, and in Figure 9.10 on vehicle- and subsystem-level maps. At $t = 2$ min 55 s, the road is within range of the LADAR. The road now appears on the subsystem-level map and falls

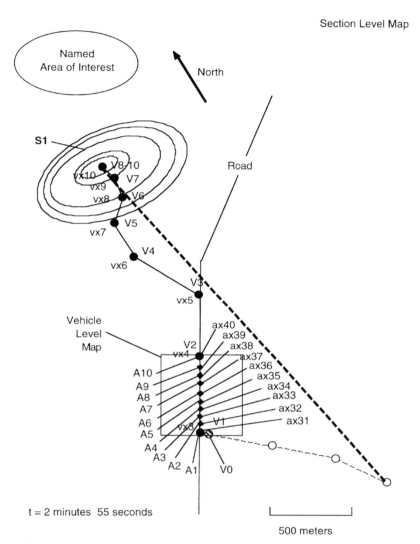

FIGURE 9.9 Autonomous mobility plan generated by the vehicle-level planner at the point where the transition from cross-country to on-road driving is only 5 s in the future. At this point, the road first comes within the planning horizon of the autonomous mobility planner.

within the planning horizon of the autonomous mobility planner, as shown in Figure 9.10. Note the change in scale from Figure 9.9. Obstacle avoidance maneuvers can be seen in the primitive driver plan on the subsystem-level map as the vehicle makes its approach to the road.

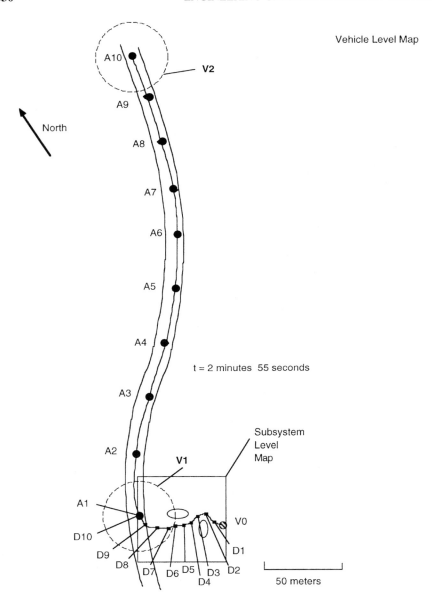

FIGURE 9.10 Primitive driver plan generated by the subsystem-level planner at the point where the road first appears within the planning horizon of the AM subsystem. The subsystem-level planner performs obstacle avoidance over the 50 m in front of the vehicle in the subsystem-level map.

At $t = 2$ min 57 s, the primitive driver plan (generated by the autonomous mobility planner) extends beyond the current autonomous mobility command A1 toward the next command A2, as shown in Figure 9.11. In both Figure 9.10 and 9.11, A1 defines the current command to the subsystem level and A2 defines the next command. As the vehicle (at V0) approaches the commanded goal A1, the distance between A1 and V0 will decrease until the vehicle-to-AM executor decides that A1 has been reached. At that time the vehicle-level executor detects a DONE condition for A1. This causes the subgoals in the autonomous mobility plan to scroll forward one step so that $A2 \rightarrow A1, A3 \rightarrow A2, A4 \rightarrow A3, \ldots$ and a new A10 is added to the end of the autonomous mobility plan. The vehicle-to-AM executor then issues the new A1 as a new current command and the new A2 as a new next command to the AM planner. At every level, the executor receives a report from its lower-level planner as to whether the current command is predicted to be achieved on time. If so, the executor simply waits. If not, the executor may issue an emergency command to the subordinate planner and request a replan by its own planner.

In the situations illustrated in Figures 9.10 and 9.11, the AM subgoal A1 coincides with the vehicle goal V1. Therefore, when the vehicle reaches A1, it also reaches V1. The section-to-vehicle executor decides independently when the DONE conditions have been satisfied for V1. This causes the vehicle plan to scroll forward one step so that $V2 \rightarrow V1, V3 \rightarrow V2, \ldots$ and a new V10 is added at the end of the vehicle plan. The section-to-vehicle executor then issues the new V1 as the new current command to the vehicle level, with the new V2 as the new next command.

In the scenario illustrated in Figures 9.5 to 9.11, each level in the system was able to carry out the plans from a higher level successfully. This, of course, is not always the case. Especially on the battlefield, unforeseen situations will arise that prevent plans from being executed as planned. In the next scenario (illustrated in Figures 9.12 through 9.18), we examine the interaction between BG processes at the vehicle and subsystem levels when unexpected obstacles are encountered.

Obstacle Avoidance

In Figure 9.12, the vehicle-level planner generates a new autonomous mobility (AM) plan about every 5 seconds with subgoals (A1, A2, A3, ... A10) about 50 meters apart (assuming 10 meters per second velocity.) The AM plan is designed to achieve the first vehicle plan subgoal V1 and look beyond to V2. The vehicle-level planner operates by connecting selected pixels (i.e., plan-graph nodes) on the vehicle-level map (500 meter range, 4 meter resolution). The cost of traversing a path between each pair of nodes is evaluated by the vehicle-level value judgment process and assigned to the arc connecting the two nodes. The vehicle-level planner computes the total cost of reaching V1 from each node and assigns this cost to each node in the AM plan-graph.

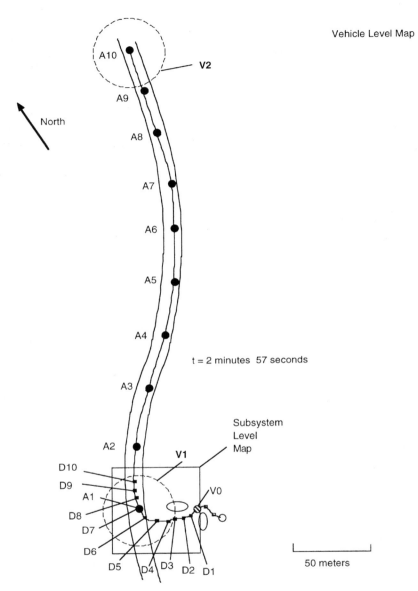

FIGURE 9.11 Primitive driver plan generated by the subsystem-level planner 2 s after the road first appears within the planning horizon of the subsystem-level planner. The primitive driver plan avoids obstacles and plans the transition between off-road and on-road driving with subgoals D5, D6, and D7.

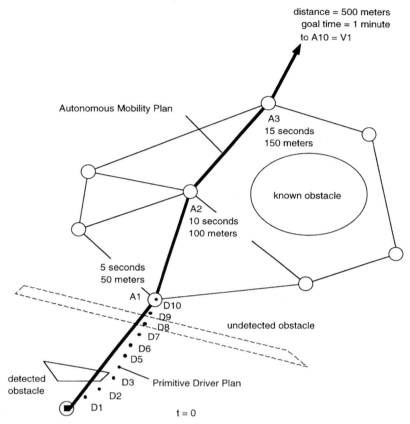

FIGURE 9.12 Plans at two levels. The first three steps in an autonomous mobility (AM) plan (generated by the vehicle-level planner) are indicated by heavy lines denoting planned actions connecting planned subgoals A1,A2,A3,...,A10. A primitive-level driver plan (generated by the subsystem-level planner) is shown as a series of dots labeled D1,D2,D3,...,D10 connecting the current vehicle position (indicated by the small black square) with the first AM subgoal A1. The additional nodes and arcs are AM subgoals and actions that were simulated and evaluated during the last vehicle-level replanning cycle but not selected as part of the AM plan.

The lowest cost path connecting the current vehicle position with V1 is selected as the AM plan. In Figure 9.12, only the first three segments of the AM plan are shown as three dark line segments connecting the AM plan nodes A1,A2,A3,... The additional nodes and thin line segments represent nodes and paths that were evaluated during the vehicle-level planning process but not selected for the AM plan.

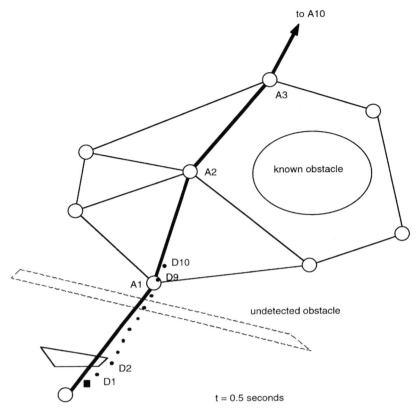

FIGURE 9.13 Replanning after a single step in the primitive driver plan. The subsystem-level planning horizon is now beyond the first AM subgoal A1. The last primitive driver subgoal D10 is now headed toward the next AM subgoal A2.

The entire AM plan covers a time interval of about 1min. It consists of about 10 AM planned actions and subgoals spaced about 5 s apart in time and about 50 m apart in distance. The last subgoal in the AM level plan (A10) is derived from the first two subgoals in the vehicle plan (V1 and V2). The vehicle-AM executor transforms the first planned subgoal A1 in the AM plan into an AM GoalCommand to be achieved at GoalTime = 5 s. The vehicle-AM executor also transforms the next planned subgoal A2 into an AM NextGoalCommand to be achieved at NextGoalTime = 10 s. These are input to the subsystem-level AM BG module.

The subsystem-level planner generates a new primitive driver plan and a primitive gaze plan (not shown in the figure) every 500 ms. Primitive driver subgoals (D1, D2, D3, . . . D10) are located about 5 m apart at a speed of 10 m/s. The primitive driver plan is the lowest cost path connecting selected nodes on the AM level map (50 m range, 40 cm resolution). Note that in Figure 9.12 the

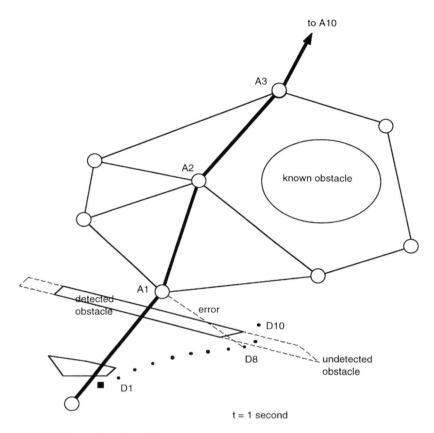

FIGURE 9.14 Replanning after two steps in the primitive driver plan. Sensors have now detected the center of the previously undetected obstacle. The subsystem-level planner has now generated a plan that by-passes what is known of the obstacle on the right. However, this creates a large error between the primitive driver goal A1 and the predicted position (D8) of the vehicle at the A1 GoalTime (t = 5 s).

primitive driver plan (D1,...,D10) deviates from the straight line of the AM plan in order to bypass a small obstacle detected by sensors. This deviation in the primitive driver plan requires no action by the vehicle-AM executor since the primitive driver plan passes through A1 at the planned time = 5 s. Note also that neither the AM plan nor the primitive driver plan take into account the undetected obstacle which cannot be observed at time t = 0.

In Figure 9.13 at t = 0.5 s, the primitive driver plan extends past the first AM subgoal A1 heading toward A2. The status feedback from the AM planner to the vehicle-AM executor indicates that D9 in the primitive driver plan is on schedule to achieve the AM subgoal A1 at t = 5 s. At t = 0.5 s, the undetected obstacle still has not been observed.

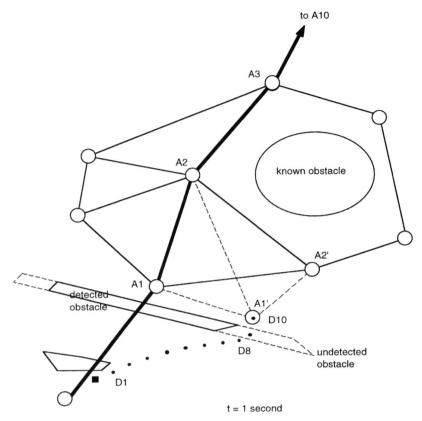

FIGURE 9.15 Execution reaction to a large predicted error. One possible reaction is for the vehicle-AM executor to generate an alternative primitive driver goal A1′ at the predicted position D10.

In Figure 9.14 at $t = 1$ s, part of the previously undetected obstacle is now observed. In response, the subsystem-level planner replans, generating an alternate primitive driver plan (D1–D10) that attempts to bypass the detected part of the obstacle. This produces status feedback from the subsystem-level planner to the vehicle-AM executor that the planned state (D8) at $t = 5$ s will be very different from A1. As a result, the vehicle-AM executor detects is a large predicted error at the GoalTime $t = 5$ s. In response to this, the vehicle-AM executor evokes an emergency plan designed for unexpected obstacles. The executor also calls for the vehicle-level planner to restart its replanning cycle, taking into account the newly detected obstacle.

There are several possible emergency plans that might be evoked. One alternative, shown in Figure 9.15, would be for the vehicle-AM executor to generate an alternative A1′ that coincides with where the primitive driver

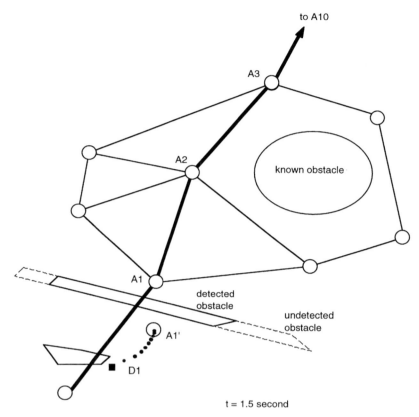

to A10

A3

known obstacle

A2

A1

detected
obstacle

undetected
obstacle

A1'

D1

t = 1.5 second

FIGURE 9.16 Another possible reaction to a large predicted error. Alternatively, the vehicle-AM executor might generate an alternative primitive driver goal A1' that will cause the subsystem-level planner to make a primitive driver plan that will bring the vehicle to a controlled stop and wait for a new AM plan from the vehicle-level planner.

plan expects to be at D10. Then, the vehicle-AM executor would request the vehicle-level planner to quickly analyze the new edges (dashed) in the vehicle-level plan graph and formulate a new AM plan. Another more conservative approach shown in Figure 9.16 would be for the vehicle-AM executor to issue an emergency halt command to the subsystem-level planner to bring the vehicle to a controlled stop and wait for the vehicle-level planner to generate a new plan before proceeding.

Assuming the first alternative was chosen, the vehicle continues along the primitive driver plan until, as shown in Figure 9.17, at $t = 3$ s, the full extent of the obstacle comes into view. At this point, the subsystem-level planner has generated a new primitive driver plan (D1–D10) around the end of the obstacle. Again, the vehicle-AM executor places a new A1' at D10 and requests the vehicle-level planner to assess the cost of the dashed arcs. As soon

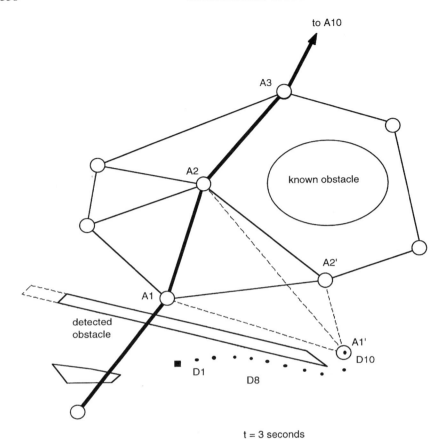

t = 3 seconds

FIGURE 9.17   Continuing with obstacle avoidance. If the execution reaction shown in Figure 9.15 is chosen, the subsystem-level planner continues planning to by-pass the obstacle on the right. At $t = 3$ s the full extent of the obstacle is detected, and the vehicle-AM executor has again placed a new A1' at D10.

as the cost of these arcs is known, the vehicle-AM executor can instruct the subsystem-level planner which of the three nodes to head for, the original A1, A2', or A2. This assessment must take into account the cost (if any) of by-passing A1. Assuming, the lowest cost path goes through A2', the primitive driver plan generated at 4.5 s passes through A1' and heads toward A2' as shown in Figure 9.18.

In Figure 9.19, replanning by the vehicle-level planner (which began anew at $t = 1$ second) has been completed at $t = 6$ s. At this point, a new AM plan has been found that passes to the right of the known obstacle. The vehicle-AM executor therefore issues a new A1 to the subsystem-level planner. Note that in this case, the new A1 is very close to the current position.

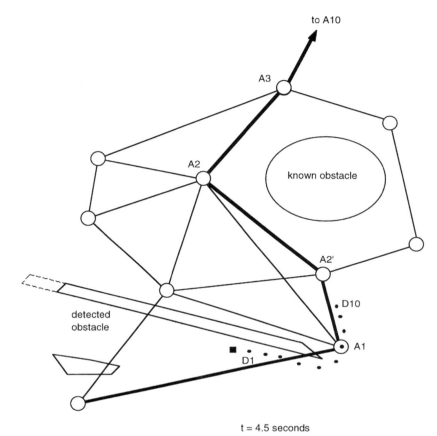

FIGURE 9.18 Continuing with obstacle avoidance. By this time, the vehicle-AM executor has determined that the lowest cost path would be to skip the original A1 and head instead for A2′.

Thus, the primitive driver plan (D1–D10) extends well beyond A1 on the way to A2.

Figure 9.20 illustrates the relative size of map resolution elements at the subsystem, vehicle, and section levels. The primitive driver plan is shown on the 50×50 m AM subsystem map. Primitive driver plan subgoals have a resolution of 0.4 m, which is the resolution of the subsystem-level map. The resolution elements in the subsystem-level map are small dots, and the squares represent 10×10 resolution elements. The AM plan is shown on the 500×500-m vehicle map. AM plan subgoals have resolution of 4 m, which is the resolution of the vehicle map. The large squares represent 10×10 resolution elements on the vehicle map. The entire vehicle-level map does not fit within the figure.

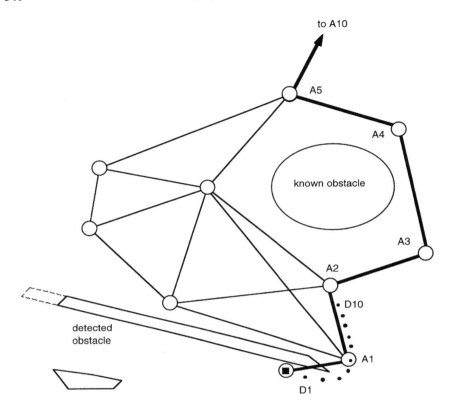

t = 6.0 seconds

FIGURE 9.19 A new AM plan. By $t = 6.0$ s, the vehicle-level planner has generated a new AM plan (after beginning replanning at $t = 1$ s when the undetected obstacle was initially encountered.) This new plan now by-passes the known obstacle on the right because this is a lower cost path to A10 from the A1′ that was defined at $t = 1$ s when its replanning cycle began. (See Figure 9.15.)

The vehicle command goal $gc1_1^4$ (shown at the upper right of Figure 9.20) is the first subgoal in the vehicle plan on the section-level map. The section-level map is 5000×5000 m with resolution elements of 40 m. The area occupied by the vehicle command goal $gc1_1^4$ is the size of a single resolution element on the section map. A GoalCommand from a higher level is typically specified only to the resolution of the higher-level planning map where it was generated. The planned paths are refined at each hierarchical level to the resolution of the planning map at that level.

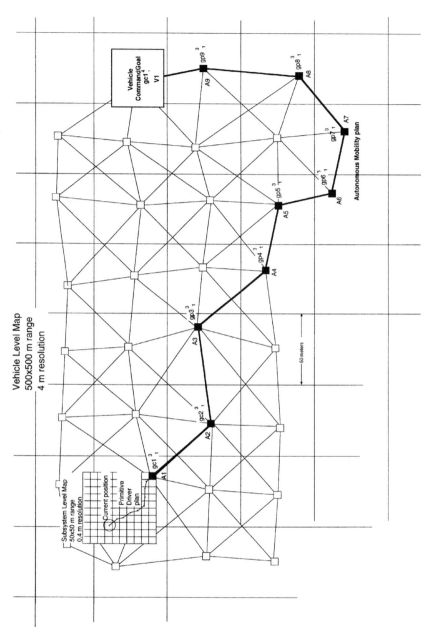

FIGURE 9.20 Subsystem- and vehicle-level planning maps. The graph consisting of small unfilled squares and connecting edges represent nodes and paths that have been simulated and evaluated on the vehicle-level map but not selected for the AM plan. The AM plan consists of filled squares connected by dark lines on the vehicle-level map. The primitive driver plan consists of connected points on the AM subsystem map. The vehicle GoalCommand $gc1_1^4$ shown at the upper right is the first step in the section plan.

341

SENSORY PROCESSING AND WORLD MODELING

The suite of sensors to be used for driving the Demo III vehicles are shown at the bottom of Figure 9.4. The sensory processing algorithms being developed at NIST integrate GPS, inertial and dead reckoning sensors, and a LADAR range imaging camera. A Kalman filter is used to compute vehicle position and orientation using data from inertial and dead reckoning sensors and a carrier phase differential GPS unit. LADAR image processing algorithms classify the terrain in front of the vehicle into five classes: clear, positive obstacle, negative obstacle, cover, and unseen. Clear terrain is open space where it is safe for a vehicle to drive. Positive obstacles are objects such as poles, walls, fences, tree trunks, buildings, and rocks of more than a given size (typically, 20 cm). Negative obstacles are ditches and gullies of more than a given width and depth (typically, 50 cm). Cover is terrain underneath objects (such as overhanging tree branches or highway overpasses) that is clear of obstacles where the vehicle can drive. Unseen is terrain that has not been viewed by the LADAR camera.

Figure 9.21*a* shows a LADAR image of a grassy field with a group of trees in the background on the left. Figure 9.21*b* shows the grassy field classified as drivable. Tree trunks and branches lower than 2 m are classified as positive obstacles, and branches higher than 2 m are classified as cover [Chang et al.99]. Since every pixel in the LADAR image has a range value, each pixel can be transformed into map coordinates. Figure 9.22 shows a map in which the path of the vehicle is shown as a dashed line. The current position of the vehicle is at the center. This map shows regions classified into four classes: (1) drivable regions are medium gray, (2) positive obstacles are dark gray, (3) cover is light gray, and (4) regions that have not been seen by the LADAR camera are white.

The map-updating algorithm is based on the concept of confidence-based mapping [Oskard et al.90]. For example, when a map pixel receives a vote for "obstacle," the pixel's obstacle confidence goes up by a given amount. When a map pixel receives a "drivable" vote, the pixel's obstacle confidence goes down. When a pixel's obstacle confidence rises above threshold, it is labeled as an obstacle on the map. The map in Figure 9.22 is 200×200 m in range with 512×512 pixels in resolution. Each map pixel is thus about 0.4 m square. As the vehicle moves, the map scrolls so that the vehicle remains in the middle.

Stereo processing algorithms developed at JPL for CCD and FLIR cameras have also been integrated onto the Demo III sensory processing system [Matthies et al.96]. Range images resulting from stereo processing can be integrated into the same map as shown in Figure 9.22 using the confidence-based mapping technique. The relative value of a vote from stereo versus a vote from LADAR is determined by the signal-to-noise ratio and false alarm probability of the two sensor systems. A radar system is also scheduled to be integrated onto the Demo III sensory processing system. Radar data will then

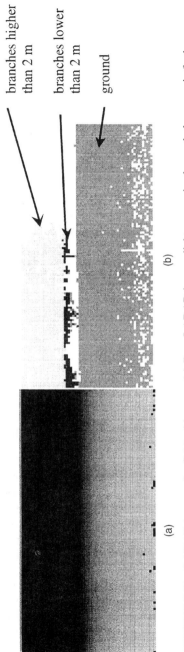

branches higher
than 2 m

branches lower
than 2 m

ground

(b)

(a)

FIGURE 9.21 LADAR image of a field with trees. (*a*) Raw LADAR image: lighter gray is closer, darker gray is farther away. (*b*) Image segmented into four classes: medium gray, clear ground; dark gray, positive obstacle; light gray, cover provided by branches higher than 2 m.

343

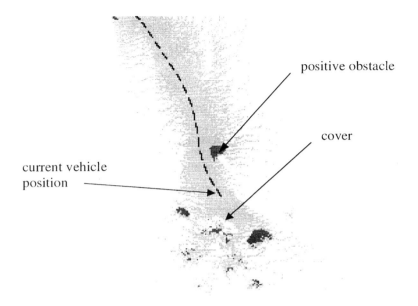

positive obstacle

cover

current vehicle
position

FIGURE 9.22 Map generated in real time by the LADAR images showing four classified regions. Drivable regions are medium gray, positive obstacles are dark gray,
cover is light gray, and regions that have not been seen by the LADAR camera are
white.

be integrated into the same map using the confidence-based update algorithm.
Figure 9.23 shows a Demo III XUV driving down a muddy dirt road at Fort
Knox, Kentucky.

Range values computed from LADAR or stereo images can be used to
generate terrain elevation maps. LADAR and stereo data collected from the
XUV in Figure 9.23 was used to generate the terrain elevation map shown
in Figure 9.24. This terrain map is updated 10 times per second. The field
of view of the LADAR is 90° wide by 20° high and is shown as a faint
outline at the top of the figure. To enable the LADAR to view a larger vertical
field it is mounted on a tilt mechanism that scans up and down about once per
second. The periodic waves in the roadway are artifacts generated by backlash
in the LADAR tilt mechanism. This will be corrected in the future. The stereo
cameras have a 40° by 40° degree field of view. To enable stereo system to have
a wider field of view, the cameras are panned side to side as the vehicle moves
forward. Terrain data from stereo are registered in the map with terrain data
from LADAR using information from pan/tilt encoders and vehicle motion
information computed from inertial sensors, odometry, and GPS.

In this example, the vehicle-level plan calls for the XUV to make a left turn
into the trees on the left side of the road as soon as an opening is detected

FIGURE 9.23 Demo III XUV driving down a muddy dirt road at Fort Knox, Kentucky. The road can be seen turning to the right about 70 m in front of the XUV. Not visible in this photo is a break in the woods about 15 m ahead of the XUV on the left. [Photo courtesy of General Dynamics Robotic Systems.]

FIGURE 9.24 Terrain elevation map generated from combined LADAR and stereo images. The pixel size in this map is 40 cm. The periodic waves in the roadway are artifacts generated by backlash in the LADAR tilt mechanism. The subsystem-level plan is depicted as a string of balls in front of the vehicle. [Photo courtesy of General Dynamics Robotic Systems.]

FIGURE 9.25 Terrain elevation map at the point where the vehicle is departing the road. This figure shows the autonomous mobility plan through a clearing in the brush on the left side of the road. At this point the vehicle is about 15 m farther down the road than in Figure 9.24. [Photo courtesy of General Dynamics Robotic Systems.]

that the vehicle can negotiate. In Figure 9.24 it can be seen that the sensors detect a break in the trees on the left at about 15 m in front of the XUV. The subsystem-level mobility planner therefore generates a plan to turn left at that point. Figure 9.25 shows the subsystem-level terrain elevation map generated at the point where the XUV is planning to depart from the road. Note that the mobility plan now extends into the woods and picks the smoothest route.

Terrain elevation maps enable the XUV to evaluate path plans on the basis of terrain elevation and roughness in addition to obstacles, cover, ground, and unseen. The planner tries to pick the smoothest path and avoid unseen areas. However, if this is not possible or too costly, the XUV slows down as it approaches rough or unseen areas. This causes the XUV to drive cautiously up to the edge of ditches until it can see the bottom. Then the terrain elevation map provides the information necessary to decide whether the ditch is too deep or steep for the XUV to cross. If so, the XUV must back up and try another path. If not, the XUV uses the elevation map to compute how best to traverse the ditch. The XUV also slows down when it cannot see over the crest of a hill. Once it can see that the ground is clear on the other side, it accelerates to a speed that is appropriate for mission priorities and terrain roughness.

In addition to terrain elevation, color CCD camera images are being processed to classify regions in the image based on color and texture. Regions can be classified as grass, dirt, rocks, roads, trees, and walls. All pixels have range values so that pixel classes can be overlaid on the map. As new image processing algorithms are developed, additional map overlays will be added for cost value attributes such as traversibility, risk, and visibility from enemy positions. This will enable the path planners to generate driving behavior that is appropriate for the tactical situation.

Each Demo III vehicle also carries a RSTA subsystem that contain a high-resolution camera, a FLIR, and a laser range finder. A separate image processing system processes data from the RSTA subsystem. In the future, the RSTA information will be incorporated into maps at the vehicle, section, and platoon levels of the 4D/RCS architecture so that longer-range tactical behaviors can be generated.

SUMMARY AND CONCLUSIONS

The Army Demo III experimental unmanned ground vehicle project described in this chapter is only one example of the types of intelligent systems that are currently under development. There are many others. The Navy has a significant unmanned undersea vehicles program to map the ocean bottom and collect information about what goes on under the sea. The Air Force has numerous unmanned air vehicle programs and plans for several more. DARPA has major research programs in sensors and autonomous systems. In the commercial world, intelligent manufacturing systems are having a profound influence on how goods are produced. Computer systems are ubiquitous in business. Electronic commerce is becoming a significant factor. Universities around the world have major research programs in robotics and intelligent systems. Computer technology is enabling systems that mimic the massive parallelism of the brain. New insights into how to design and build intelligent systems are being published in hundreds of journal articles every month.

We are beginning to understand how to design sensory processing algorithms that can appreciate what is significant in the world and focus attention on what is important to behavioral goals and objectives. We are beginning to see how to construct internal world models that demonstrate situational awareness, including self awareness. We are beginning to understand how to represent knowledge about space and time, entities and events, and relationships that have meaning and worth. We are beginning to understand how to design planners with intentions and motives and how to build decision-making processes that take worth into consideration. We are beginning to understand how to build reasoning processes that exhibit elements of imagination and insight. In short, we are beginning to see the faint outlines of mind emerging from within the intelligent systems that we build.

Of course, we are still a very long way from building a functional equivalent of the human mind. Even if we already had a fully developed theoretical model of the brain, to duplicate the computational power of the brain we still need at least three, and perhaps six orders of magnitude increase in computational power. We are further still from understanding the processes of mind that operate within the brain. Understanding how a computer works is not the same as understanding the processes running on the computer. The processes that give rise to mind are sophisticated far beyond our understanding. Nevertheless, a great deal is known and more is being learned at a rapid pace. Progress is rapid and there appear to be no fundamental barriers standing in the way. The time may have come to consider whether intelligent and cognitive systems research should be promoted to the status of legitimate science, in the same class with physics, chemistry, astronomy, and biology. It may be time to move beyond toy robot experiments in university labs, and even beyond specific military/commercial applications, to serious scientific research. As we ponder this question, we must consider whether the engineering of mind is a goal that ought to be pursued. What would be the consequences if we really could engineer mind? We address these issues in the next chapter.

10 Future Possibilities

Sybil of Delphi Unrolling a Parchment Scroll and Apparently Turning Towards the Crowd to Read It by Michelangelo, from the Sistine Chapel in the Vatican. [Courtesy of the Vatican Museums.]

The essence of higher intelligence is the ability to look into the future, to accurately predict what will happen, to recognize what is important, and to act so as to minimize cost and risk, and maximize benefit. The ability to predict the future is extremely valuable. It enables the intelligent system to avoid danger, anticipate trouble, prepare for adversity, and take advantage of opportunities that require preparatory or preemptive action.

Predicting the future is difficult.* It requires the ability to acquire, maintain, and use a world model that accurately reflects the state of the world and properly represents the important dynamic and causal relationships that make things happen in the world. The accuracy of the prediction depends on the accuracy of the model. A perfect world model would be able to predict with complete accuracy. Of course, no model of the natural world can ever be perfect. In a natural environment, most of the relevant variables cannot be measured precisely, and many are not even known. Many processes in nature are chaotic. In chaotic systems, a tiny difference in state at a time t can lead to dramatically different states at time $t + T$, where T is an arbitrary interval. If the world were completely chaotic, planning for the future would be impossible, because prediction would be impossible. In a chaotic world, all behavior would be reactive.

Fortunately, the world is not completely, or even mostly, chaotic. The world contains many regularities. There are physical laws that make many events quite predictable over long periods into the future. Eclipses of the sun and moon can be predicted centuries in advance. The seasons are predictable and calendars can be constructed that are reliable for planting and harvesting crops. Light travels in a straight line. The earth is a globe. Gravity causes objects to fall toward the center of the earth. The horizon makes a right angle with the gravity vector. The sun comes up every morning and sets every evening. The world can be classified into animal, vegetable, and mineral. Animals and people behave in characteristic ways in a variety of circumstances: The thigh bone is connected to the hip bone. Life begets life.

Objects and events have attributes and states that can be measured. Objects have size, shape, color, mass, temperature, position, orientation, and velocity. Events have duration, magnitude, frequency, phase, and location. Many object attributes are invariant with translation and rotation. Some are invariant with illumination and temperature. Many attributes and states can be predicted over time using differential equations or rules of logic. Of course, most of the regularities in the world are approximate and subject to perturbation by a wide variety of influences. Light bends where the index of refraction changes or where the gravity field is intense. The length of the day varies with the time of year. Clouds may obscure the rising and setting of the sun. The weather depends on many factors other than the calendar date. People do not always behave as expected. Regularity is intermixed with chaos. Quiet afternoons are interrupted by thunderstorms. Long periods of stability are occasionally shat-

---

*"Prediction is very hard, especially when it's about the future."—Yogi Berra [Time99].

tered by earthquakes or fires. Periods of plenty are interspersed with famine, drought, and plague. Sickness, injury, violence, and death often occur unexpectedly. Predictions are only as good as the assumptions in our model. When our models are wrong, our predictions fail. When our models are refined, our predictive powers improve.

To deal with apparently chaotic events in the world, ancient and primitive cultures adopted world models filled with myths of gods and devils. Priests and seers could always interpret plagues, droughts, floods, earthquakes, and wars as expressions of the gods' disapproval or displeasure, or as collateral fallout from cosmic encounters between deities. Thus, primitive models could explain everything, but they predicted very little. The behavior of the gods is very capricious. Modern models of the world are based on scientific laws and theories of physics, chemistry, mathematics, astronomy, geology, and biology. Scientific models both explain and predict events and relationships with great precision—at least in those situations where the relevant variables are known and accurately represented in the models. Scientific models predict relationships between force, mass, and acceleration. They predict the behavior of electromagnetic waves. They predict the properties of materials and the outcome of processes. Meterological models can predict the weather and the paths of hurricanes (with a degree of uncertainty that grows with time). Good predictions can be made for population growth under various environmental conditions. Statistical predictions can be made for stocastic events like earthquakes and highway traffic deaths.

TECHNOLOGY FORECASTS

Technology forecasting over periods of more than a decade is notoriously inaccurate. In 1899, Charles Duell, Commissioner of the U.S. Patent Office, recommended that the patent office be closed because everything that could be invented had already been invented. In 1932, Albert Einstein said "There is not the slightest indication that [nuclear] energy will ever be attainable." In 1943, Thomas J. Watson and John von Neumann estimated that the total world market for computers might be five. In 1981, Bill Gates predicted that no one would ever need more than 640K of RAM.

In the early days of artificial intelligence research, it was predicted that the most difficult problems would lie in such high-level intellectual activities as mathematical reasoning. But it turns out that computers are quite good at proving mathematical theorems and can play intellectual games such as chess and bridge at the level of human experts. On the other hand, the ability of computers to understand simple stories is nowhere near that of an average 6-year-old. Computer-controlled hands and fingers cannot tie a shoe with anything like the dexterity of a 4-year-old. The ability of machine vision systems to track objects moving through the woods under various lighting conditions is far inferior to that of a cat or bird.

Particularly in the fields of robotics and artificial intelligence there have been many naive and extravagant claims, and many unfortunate predictions. The Society of Manufacturing Engineers predicted during the early 1980s that by the year 1996, almost all manufacturing jobs would be done by robots, and very few human workers would be required in the labor force. Herbert Dreyfus has made a literary career out of ridiculing erroneous predictions by authority figures in artificial intelligence. The fallacy of the first step has lured more than one expert into excessively optimistic prognostications. Numerous blind alleys have been followed, and some are still being pursued. There have been many surprises. One of the biggest has been that rudimentary skills of perception, manipulation, communication, and locomotion are far more complicated than almost anyone predicted.

Of course, not all technology forecasts have been overly optimistic. In many areas, computer capabilities have exceeded expectations. The amount of computational power per unit of size, weight, and cost has far surpassed even the wildest predictions. The power of computers to solve equations and model complex phenomena such as nuclear explosions is far superior to that of the human brain. Computers are as good or better than humans at doing mathematics and logic, executing formal rules, and remembering detailed information. Today, computers can fly airplanes, pilot ships, make reservations, control traffic, maintain inventory, operate machine tools, sequence production lines, oversee medical laboratory testing, launch rockets, and map the human genome. There are many practical applications of artificial intelligence, robotics, and computer-controlled machines in manufacturing, aerospace, and military weapons systems. There are thousands of experiments under way in academic, corporate, and government research laboratories. There are large research and development programs funded by the government, many for military applications.

THE FUTURE OF INTELLIGENT SYSTEMS

To predict what the future will bring in the field of intelligent systems requires a model of the world that accurately reflects the current state of the art and probable future developments in the cognitive and neurosciences, computer science, solid-state physics, electronics, computer and software engineering, and intelligent manufacturing research. It requires an imagination that can predict the impact of future discoveries in these areas and a value judgment system that can evaluate the costs and benefits of each possible result. It requires a reliable estimate of the set of possible actions, strategies, and policies that might be adopted in the future, a wise evaluation of the probable results of each series of possible actions, and a lucky guess of what course of action will be followed by decision makers in society.

Needless to say, the space of possible futures is very large and grows geometrically with extending time horizons. Over a period of decades, many possibilities simply cannot be anticipated. No one can predict new discoveries before they are discovered. One can only assume that there will be new discoveries, and these will bring new capabilities, and these will lead to even further discoveries. It also is impossible to explore all conceivable policies or predict what policymakers will do under the pressures of future events. Clearly, future developments depend heavily on the level of funding for research, which in turn depends on what costs and benefits are perceived by decision makers. Often, events such as economic recession, political disputes, natural disasters, war, and other emergencies arise and overcome the best-laid plans.

Although obviously no one can anticipate everything, one can make conservative assumptions such as that present trends in science and technology will continue for two or three more decades. One can conservatively assume that there will be no fundamental new discoveries, only incremental improvements on current technology. One can assume that financial investments in technology development will be conservative (i.e., less than commensurate with the real potential benefits). Using only these conservative assumptions, we predict that the science of intelligent systems will develop within the next two decades to the point that it can support the development of a computational theory of mind. If this occurs, we further predict that it will have a profound impact on science, economics, national security, and human well being.

IMPLICATIONS FOR SCIENCE

All of science revolves around three fundamental questions:

1. What is the nature of matter and energy?
2. What is the nature of life?
3. What is the nature of mind?

Over the past 300 years, research in the physical sciences, have produced a wealth of knowledge about the nature of matter and energy, both on our own planet and in distant galaxies. We have developed mathematical models that enable us to understand at a very deep level what matter is, what holds it together, and what gives it its properties. Our models of physics and chemistry can predict with incredible precision how matter and energy will interact under an enormous range of conditions. We have a deep understanding of what makes the universe behave as it does. Our knowledge of the physical universe includes precise mathematical models that stretch over time and space from the scale of quarks to the scale of galaxies.

Over the past half century, the biological sciences have produced a revolution in knowledge about the nature of life. We have developed a wonderfully powerful model of the molecular mechanisms of life. The first draft of the human genome has been published. We may soon understand how to cure cancer and AIDS. We are witnessing an explosion in the development of new drugs and new sources of food. Within the next century, biological sciences may eliminate hunger, eradicate most diseases, and discover how to slow or even reverse the aging process.

Yet, of the three fundamental questions of science, the most profound may be "What is mind?" Certainly, this is the question that is most relevant to the fundamental nature of human beings. We share most of our body chemistry with all living mammals. Our DNA differs from that of chimpanzees by only a tiny percentage of the words in the genetic code. Even the human brain is similar in many respects to the brains of apes. Who we are, what makes us unique, and what distinguishes us from the rest of creation lies not in our physical elements, or even in our biological makeup, but in our minds.

It is the mind that sharply distinguishes the human race from all other species. It is the mind that enables humans to understand and use language, to manufacture and use tools, to tell stories, to compute with numbers, and reason with rules of logic. It is the mind that enables us to compose music and poetry, to worship, to develop technology, and organize political and religious institutions. It is the mind that enabled humans to discover how to make fire, to build a wheel, to navigate a ship, to smelt copper, refine steel, split the atom, and travel to the moon.

The mind is a process that emerges from neuronal activity within the brain. The human brain is arguably the most complex structure in the known universe. Compared to the brain, the atom is an uncomplicated bundle of mass and energy that is easily studied and well understood. Compared to the brain, the genetic code embedded in the double helix of DNA is relatively straightforward. Compared to the brain, the molecular mechanisms that replicate and retrieve information stored in the genes are quite primitive. One of the greatest mysteries in science is how the computational mechanisms in the brain generate and coordinate images, feelings, memories, urges, desires, conceits, loves, hatreds, beliefs, pleasures, disappointment, and pain that make up human experience. The really great scientific question is: What causes us to think, imagine, hope, fear, dream, and act like we do? Understanding the nature of mind may be the most challenging problem, not only in science, but in all of human experience.

We have only just entered an era in history in which technology is making it possible to seriously address the question of the nature of mind. Prior to about 125 years ago, inquiry into the nature of mind was confined to the realm of philosophy. During the first half of the twentieth century, the study of mind expanded to include neuroanatomy, behavioral psychology, and psychoanalysis. The last 50 years have witnessed an explosion of knowledge in neuroscience and computational theory. Much is now understood about how the brain works, and more is being learned at a rate faster than almost anyone can comprehend. Research on mental disease and drug therapy has led to a wealth of knowledge about the role of various chemical transmitters in the mechanisms of neurotransmission. Single-cell recordings of neural responses to different kinds of stimuli have shown much about how sensory information is processed and muscles are controlled. The technology of brain imaging is now making it possible to visually observe where and when specific computational functions are performed in the brain. Researchers can literally see patterns of neural activity that reveal how computational modules work together during the complex phenomena of sensory processing, world modeling, value judgment, and behavior generation. It has become possible to visualize what neuronal modules in the brain are active when people are thinking about specific things and to observe abnormalities that can be directly related to clinical symptoms [Carter98].

Prior to brain imaging technologies, experimental procedures for studying the brain were confined to very indirect and ambiguous methods. Simple stimulus–response behavioral experiments are inadequate for investigating internal states or intentions. Psychological testing cannot unambiguously attribute experimental results to computational processes in the brain. Studies of behavioral abnormalities caused by lesions are laborious, often ambiguous, and always lacking in controls. Single-cell recordings can study only a few neurons at a time and can be obtained only through intrusive surgical procedures that rarely are appropriate for human subjects. Electroencephalograms (EEG) are unable to pinpoint brain activity to precise locations. The recent imaging technologies of positron emmision tomography (PET) and functional magnetic resonance imaging (fMRI) have begun to overcome these deficiencies. Modern brain imaging provides a clear high-resolution picture of which computational modules are activated and how they interact over time during complex intellectual tasks preformed by human subjects who are able to report what they are thinking as the task proceeds [Carter98].

Today, progress in the neurosciences is rapid. The 1990s, designated the "Decade of the Brain," produced an enormous expansion of understanding of the molecular and cellular processes that enable computation in the neural substrate. As the fields of brain research and intelligent systems engineering converge, the probability grows that we may be able to construct what Edelman calls a "conscious artifact" [Edelman99a]. Such a development would provide answers to many long-standing scientific questions regarding the relationship between the mind and the body. At the very least, building artificial models of the mind would provide new insights into mental illness, depression, pain, and the physical bases of perception, cognition, and behavior. It would open up new lines of research into questions that hitherto have not been amenable to scientific investigation. For example, we may be able to understand and describe intentions, beliefs, desires, feelings, and motives in terms of computational processes with the same degree of precision that we now can apply to the exchange of energy and mass in radioactive decay, or to the sequencing of amino acid pairs in DNA. We may discover whether humans are unique among the animals in their ability to have feelings. We may know to what extent we alone have the ability to experience pain, pleasure, love, hate, jealousy, pride, and greed. We may learn whether artificial minds can appreciate beauty and harmony, or comprehend abstract concepts such as truth, justice, meaning, and fairness. Can silicon-based intelligence exhibit kindness or show empathy? Can machines pay attention, be surprised, or have a sense of humor? Can they feel reverence? Might they worship God? Could they be agnostic? Might they become religious zealots?

Functional imaging of the brain may be the last fundamental technological advance that is required before a computational theory of mind can mature into a widely accepted scientific theory. If this is true, no additional fundamental technological breakthrough will be required before truly intelligent, fully conscious machines can be designed and built. If this is true, all that is necessary is for current trends in cognitive neuroscience and computational hardware and software to continue for another two or three decades.

Clearly, the computing power to build an artificial mind is emerging. Since computers were invented about a half century ago, the rate of progress in computer technology has been astounding. Since the early 1950s, computing power has doubled about every three years. This is a compound growth rate of a factor of 10 per decade, a factor of 100 every two decades. This growth rate shows no sign of slowing, and in fact, may be accelerating. Today, a typical personal computer costing less than $1000 has more computing power than a Cray I, the top-of-the-line supercomputer only two decades ago. One giga-op (1 billion operations per second) single-board computers are now on the market. There appears to be no theoretical limit that will slow the rate of growth in computing power for at least the next few decades. This means that within 30 years a relatively inexpensive network of 10 single-board computers could have computational power approaching 10^{13} operations per second. Properly utilized, this could rival the computational power of the human brain.

The brain contains around 10^{11} neurons, most of which receive information from thousands of inputs and send messages to hundreds of outputs with a bandwidth of about 100 Hz. To precisely model all the synaptic computations in the brain might require as many as 10^{16} operations per second. However, functional equivalence to the brain can almost certainly be achieved at much lower computational rates. The biological brain is handicapped by the fact that neurons are noisy and unreliable computing devices. The brain achieves precision and reliability through redundancy. Neuronal outputs are averaged over populations of thousands of neurons to arrive at a precise and reliable value for each computational variable [Georgopoulos95]. This suggests that if we really understood the algorithms being computed by the brain, we might achieve functional equivalence to the neural substrate with two or three orders of magnitude lower computational rates, perhaps only 10^{13} operations per second [Moravec98]. Thus, it appears likely that within 30 years, the computing power will exist to build machines that are functionally comparable to the human brain.

In the field of software engineering, progress is also rapid. After many years of disappointing results, productivity in software development is making rapid strides. Much has been learned about how to write code for software agents and build complex systems that process signals, understand images, model the world, reason and plan, and control complex behavior. Despite many false starts and overly optimistic predictions, artificial intelligence, intelligent control, intelligent manufacturing systems, and smart weapons systems have begun to deliver solid accomplishments. We know how to build systems that seek goals, that reason and plan, and that build and use a rich and dynamic world model to predict the future and optimize behavior. We are learning how to build systems that learn from experience as well as from teachers and programmers. We understand how to use computers to measure attributes of objects and events in space and time. We know how to extract information, recognize patterns, detect events, represent knowledge, classify and evaluate objects, events, and situations. We know how to build internal representations of objects, events, and situations, and how to produce computer-generated maps, images, movies, and virtual reality environments. We have algorithms that can evaluate cost and benefit, make plans, and control machines. We have engineering methods for extracting signals from noise. We have solid mathematical procedures for making decisions amid uncertainty. We are developing new manufacturing techniques to make sensors tiny, reliable, and cheap. Special-purpose integrated circuits can be designed to implement neural networks or perform parallel operations such as are required for low-level image processing. We know how to build human–machine interfaces that enable close coupling between humans and machines. We are developing vehicles that can drive without human operators on roads and off. We are discovering how to build controllers that generate autonomous tactical behaviors under battlefield conditions.

To be sure, the engineering of mind represents an enormously complex challenge. Much more than raw computing power and a grab bag of algorithms is required to create real intelligence. We must be able to integrate perception, cognition, and control into a coherent system architecture that can blend planning and reactive behavior. We must be able to integrate images with symbolic knowledge in a system that can react to real-time situations. We must be able to integrate reason and emotion into a system that makes wise decisions in a world filled with uncertainties. We must be able to integrate software and hardware systems into computing modules and data structures that are functionally equivalent to those in the brain.

The technical challenge is enormous. We don't yet have an engineering procedure for designing systems that mimic how that brain reasons about space and time. We don't know exactly what it might mean for computers to feel emotions or experience conscious insight. We still do not fully understand the fundamental nature of the computational mechanisms in the brain that give rise to the phenomena of mind. However, we are not without considerable understanding in these areas. The field of intelligent systems architecture development is extremely active. There are several attractive hypotheses that could mature into a formal theoretical framework. RCS is only one of these. There are many architectural approaches to building intelligent systems, and many software engineering tools for implementing them (see the discussion in Chapter 5). There are machine vision systems that can recognize faces, decipher handwritten messages, and describe scenes. There is commercial software available for recognizing spoken and written words. There is commercial software for animation and visualization of complex environments. There are software tools for simulation of large complex systems such as factories, transportation systems, and natural ecological systems. Software packages are available for production planning and inventory management. There are software tools for designing and using expert systems, neural nets, genetic algorithms, finite-state automata, Petri nets, object-oriented programming techniques, and many more. Many of these are already commercial products with thousands of users. Examples include Control Shell, MatLab, LabView, ProEngineer, Rapide, UML, Delmia, CORBA, JAVA, and Expert Systems shells.* Software libraries such as Linux and RCS [Shackelford99, Huang et al.2000] are available as freeware on the Internet.

THE DARK SIDE

Just because something is possible does not mean that it would be wise or good. Several have recently raised the question whether research in machine intelligence is ethically justified. On the negative side—what could go wrong? What if intelligent machines evolve beyond human intelligence? What if humans become dependent on their robot slaves? What if humans let machines make more of their decisions for them simply because machine-made decisions produce better results? [Kaczynski95, Moravec98, Kurzweil99, Joy2000]. What if artificial minds turn evil? Might intelligent machines develop malice toward humans or form their own political agenda? Might robots turn against their human masters and eventually eradicate the human race as described in Capek's play, *R.U.R. (Rossum's Universal Robots)* [Capek23]. In contemporary jargon, what if intelligent machines join the Dark Side?

The fact that many well-informed people consider these to be serious questions indicates the depth of emotional resistance to efforts to focus serious scientific attention on intelligent systems research. Since the publication of Frankenstein, science fiction has portrayed the attempt to build intelligent machines as a mad urge to play God. Stories such as the Golem, *R.U.R.*, and many others have predicted dire consequences from attempts to create creatures with artificial brains. Even the robot-friendly plot of *Star Wars* includes as a central character Darth Vadar, the commander of a Death Star with legions of robot warriors intent on the destruction of human civilization.

We believe that these kinds of fears are largely unfounded under any plausible scenario. It will be a very long time before robots pose any serious eco-

---

*Commercial products identified in this book are not intended to imply recommendation or endorsement, nor is it intended to imply that the products identified are necessarily the best available for the purpose. See the references for Web sites for these products.

nomic, political, or military threat to the human race. Any plausible danger to humanity from intelligent machines lies far beyond any rational planning horizon. For the foreseeable future, surely for the next century, the real danger to humans comes not from robots, but from other humans. The scourge of poverty, the existence of tyrannical governments, the threat of racial and religious conflict, and the potential for terrorist organizations wielding weapons of mass destruction represent a much more profound and immediate threat to human lives than do rebellious robots or machines usurping critical functions and institutions. People are starving and dying and being oppressed at the hands of their fellow humans in many places throughout the world right now, and it seems likely that this will continue for decades, if not centuries. Most of the large cities in the world are riddled with slums where people live in neighborhoods that are infested by vermin and rife with disease. Genocide, ethnic cleansing, religious and tribal warfare, crime, and violence are widespread. By comparison, the threat of intelligent machines taking over society is extremely remote. Human endangerment from rebellious robots is in a category with the threat of a comet or asteroid colliding with Earth and causing it to explode. Such an event is easy to imagine, but quite unlikely to occur in our lifetime or even in the next 10,000 years. Such scenarios make for thrilling plots in books and movies but should not rank high on anyone's list of real things to worry about. The possibility of a robot rebellion should certainly not be a factor in scientific policy decisions. The potential real benefits to be derived from building intelligent machines far outweigh any conceivable dangers of robots run amuck.

FEAR OF AUTOMATION

For many, fear of intelligent machines comes not from science fiction but from concerns about job security. People of all backgrounds and persuasions have deep reservations about the potential effect of intelligent robots on the labor force. A widespread popular belief is that as machines become more intelligent, jobs for human workers will be eliminated, and eventually, unemployment will become widespread. The fear of automation has haunted the labor force ever since the Luddite riots against the spinning jenny. Throughout the entire industrial revolution, workers of all kinds—from coal miners, factory workers, butchers, bakers, and candlestick makers to soldiers and astronauts—all fear that intelligent machines might take away their jobs. The fear is that eventually there will be no need for human workers at all. Machines will do all the work and ordinary people will all be unemployed.

This line of reasoning stems from the dual assumptions that (1) jobs are created because there is work to be done, and (2) there is a limited amount of work that needs doing. Therefore, whenever work is done by machines, there must be less work available for humans. This popular belief has been echoed in the writings of numerous social philosophers over the last two centuries and has been dramatized in countless movies and science fiction novels such as *Player Piano* [Vonnegut74]. It has been reinforced by organized labor's concern for job security, and exacerbated by sensational news headlines of layoffs by major corporations. Downsizing has become the hallmark of success in the modern corporate world. As productivity has increased, competition in the market place has pressured industry to reduce labor costs. People fear that if intelligent machines do more and more of the work, fewer human workers will be needed [Wilson96]. As productivity improves, downsizing will continue, and sooner or later everyone may be laid off. Fortunately, this widespread popular belief is completely wrong (at least for the economy as a whole). Although technology advances do produce some dislocations as specific job skills become obsolete, overall productivity growth has always created more jobs than are lost [Mankiw95, Denison85, Council of Economic Advisers87–99, Appelbaum and Schettkat95].

It is simply not true that jobs are created by work to be done. If it were, poverty and unemployment could not coexist, because anywhere there is poverty there is plenty of work to be done. In poor regions of the world, more food needs to be produced and distributed. New homes need to be built. Roads, bridges, schools, and hospitals need to be constructed. People need education and medical care. If jobs were created by work that needs to be done, the very existence of poverty would produce full employment automatically. The fact is that unemployment is chronic and epidemic in poor countries is prima fascia evidence that need does not produce jobs. In fact, just the opposite. In most cases, the greater the poverty, the higher the rate of unemployment. This is not because there is nothing that needs doing, but because those that are in need have neither the resources to produce what they need nor the income to buy what they need.

The fact is, unemployment is negatively correlated with productivity. Unemployment is generally low where productivity is high, and high where productivity is low. Those places on Earth with the lowest levels of unemployment tend to be those with the most advanced automation. This is because productivity increases the production of wealth (i.e., goods and services that can be sold at a profit). This produces revenue that is distributed as income to owners in the form of dividends and rent and to workers in the form of wages and salaries. Income generates consumer demand. The sale of products generates the revenue needed to hire workers. Thus, even though productivity gains from automation and intelligent machines may eliminate the need for some jobs, the net effect is to create more wealth. And wealth creation is what generates profits for employers and jobs for workers. The bottom line is that lower costs and increased sales of goods and services create many more employment opportunities than are ever lost to automation. This certainly has been the experience in all Western economies over the last 200 years. Despite widely publicized downsizing of American industry and simultaneous growth in productivity, the United States, is currently experiencing the lowest unemployment in decades.

The fact is jobs are not created by work that needs doing, but by profitable enterprises that can afford to pay workers. Jobs are created when employers generate sufficient revenue to meet a payroll. Revenue is produced when wealth is created (i.e., when goods and services are produced to meet demand generated by consumers with money to spend). Jobs tend to be readily available in places where wealth is being created or consumed. Of course, wealth cannot be consumed unless it is first created. Therefore, all jobs depend on the creation of wealth, and the more wealth is created, the more employment opportunities are created.

IMPLICATIONS FOR ECONOMIC PROSPERITY

The fact is that intelligent machines can and do create wealth; and as they become more intelligent they will create more wealth. Intelligent machines will have a profound impact on the production of goods and services. Until the invention of the computer, economic wealth (i.e., goods and services) could not be generated without a significant amount of human labor [Mankiw92]. This places a fundamental limit on average per capita income. Average income cannot exceed average worker productive output. However, the introduction of the computer into the production process is enabling the creation of wealth with little or no human labor. This removes the limit to average per capita income. It will almost certainly produce a new industrial revolution [Toffler80].

The first industrial revolution was triggered by the invention of the steam engine and the discovery of electricity. It was based on the substitution of mechanical energy for muscle power in the production of goods and services. The first industrial revolution produced an explosion in the ability to produce material wealth. This led to the emergence of new economic and political institutions. A prosperous middle class based on industrial production and commerce replaced aristocracies based on slavery. In all the thousands of centuries prior to the first industrial revolution, the vast majority of humans existed near the threshold of survival, and every major civilization was based on slavery or serfdom. Yet, a little more than 200 years after the beginning of the first industrial revolution, slavery has almost disappeared, and a large percentage of the world's population lives in a manner that far surpasses the wildest utopian fantasies of former generations.

There is good reason to believe that the next industrial revolution will change human history at least as profoundly as the first. The application of computers to the control of industrial processes is bringing into being a new generation of machines that can create wealth largely, or completely, unassisted by human beings. The next industrial revolution, sometimes referred to as the *robot revolution*, has been triggered by the invention of the computer. It is based on the substitution of electronic computation for the human brain in the control of machines and industrial processes. As intelligent machine systems become more and more skilled and numerous in the production process, productivity will rise and the cost of labor, capital, and material will spiral downward. This will have a profound impact on the structure of civilization. It will undoubtedly give rise to new social class structures and new political and economic institutions [Albus76].

THE ROLE OF PRODUCTIVITY

The fundamental importance of productivity on economic prosperity can be seen from the following equation:

$$\text{output} = \text{productivity} \times \text{input} \tag{10.1}$$

where

$$\text{input} = \text{labor} + \text{capital} + \text{raw materials}$$

productivity = efficiency by which the input of labor, capital, and raw material is transformed into output product

Productivity is a function of knowledge and skill (i.e., technology). Growth in productivity depends on improved technology. The rapid growth in computer technology has produced an unexpectedly rapid increase in productivity that has confounded predictions of slow economic growth made by establishment economists only a decade ago [Symposia88, Bluestone and Harrison2000]. In the future, the introduction of truly intelligent machines could cause productivity to grow even faster. Given only conservative estimates of growth in computer power, unprecedented rates of productivity growth could become the norm as intelligent machines become pervasive in the productive process.

Intelligent systems have the potential to produce significant productivity improvements in many sectors of the economy, in both the short and long term. Already computer-controlled machines routinely perform economically valuable tasks in manufacturing, construction, transportation, business, communications, entertainment, education, waste management, hospital and nursing support, physical security, agriculture and food processing, mining and drilling, and undersea and planetary exploration.

Manufacturing

Intelligent manufacturing systems have demonstrated the ability to dramatically reduce cost and improve quality of automobiles, trucks, airplanes, railroads, ships, appliances, furniture, clothing, chemicals, materials, medicines, recreational goods, food products, electronics, optics, construction equipment, mining and drilling equipment, farm machinery, and weapons systems. Intelligent machine controllers based on inexpensive personal computers can make it possible to automatically generate and execute programs for lifting, positioning, cutting, joining, machining, forming, finishing, and assembly operations in factories and shops throughout the world. On-line monitoring of material flow and machine availability can enable real-time optimization of production planning and scheduling. Advanced sensory perception systems can enable in-process inspection and testing. Intelligent adaptive control systems can increase the efficiency and reduce the cost of plants, processes, and machines [IMTR98].

Dramatic productivity improvements have already occurred in computer-aided design, computer-aided manufacturing, computer-aided process planning, material resource planning, inventory control, and computer-aided production scheduling. Further productivity gains are to be expected from electronic commerce. Merchandizing, contracting, purchasing, requests for proposals, bidding, maintenance, and servicing via the Internet is a high-growth sector and will remain so for many years. Many new opportunities will emerge for new ideas to enter the marketplace and for new entrepreneurs to achieve success.

Construction

Intelligent systems technology for construction has the potential to improve productivity, reduce cost, and increase quality in the construction of factories, plants, high-rise buildings, homes, highways, bridges, tunnels, port facilities, and sewer, water, electricity, and gas utilities. Studies of the construction industry have suggested that significant cost benefits can be achieved from integrating computer design data with on-site measurements to enable real-time planning, scheduling, and inventory tracking [Kent95]. Intelligent measurement systems can assure that construction tolerances are met and parts fit together without modification. Intelligent planning systems can assure that parts, tools, and materials arrive at the right place, at the right time, in the right order so that construction operations flow smoothly without delays. Complete "as-built" records can be kept for future reference to improve productivity of building operations, maintenance, repair, modification, and demolition.

Computer-aided design and engineering tools have already had an enormous impact on how buildings and structures are designed and engineered. Computer-assisted controls for construction machinery are increasing safety and productivity of human operators. In the future, intelligent machines will enable entirely new construction techniques that today are impractical, such as computer-controlled painting and custom carving and sculpting of buildings and rooms.

Transportation Safety and Efficiency

Intelligent systems technologies are about to have a revolutionary impact on automobiles, trucks, and highways around the world. For more than a decade in Europe, Japan, and the United States, serious research efforts have been directed toward intelligent vehicles and highway systems [Conference on Intelligent Vehicles98]. In Germany, experimental vision-guided automatic automobiles are undergoing regular tests driving the streets and highways in traffic at normal speeds, almost completely without human assistance. A sedan has been driven from Munich to Copenhagen under automatic control over 95% of the time [Dickmanns et al.94]. In the United States, an automobile under computer control drove from Washington, DC to Los Angeles with only occasional human intervention [Thorpe et al.97]. For several years in Japan, drivers have been able to view images of detailed street maps with directory assistance, on-line traffic information, and voice input/output. Directory assistance systems using GPS are now available on luxury automobiles in the United States.

In the future, advanced cruise control with collision avoidance and run-off-road detection will enhance safety by alerting drivers that have gone to sleep or whose attention has wandered. Automatic lane following and distance keeping will improve throughput and increase safety on congested freeways. In areas where construction is restricted by the availability of space, this could reduce the need for building additional lanes of roadway. Intelligent systems could also improve airline safety, and prevent most rail and ship collisions. A large percentage of accidents are caused by human operator errors. Most of these could be prevented by intelligent systems technology.

Business

Intelligent systems technologies have become essential elements of business management. Computers are used for design, planning, scheduling, word processing, budgeting, financial services, marketing, and customer services. Intelligent machines connected by the Internet are performing many kinds of business services and electronic commerce, including retail sales and distribution. Automatic teller machines are among the fastest-growing types of commercial robots.

Communications

Intelligent systems technologies have already had a profound impact on the communications industries. The Internet, satellite and cable television, and cellular phone systems are creating a communications network that rivals in bandwidth, complexity, and sophistication the network of neural fibers that interconnect various parts in the human brain. The potential for productivity improvements from these new technologies are beyond our ability to predict.

Waste Management

Intelligent machines could make an important contribution to environmental preservation. Intelligent machines will be able to collect trash and sort it for recycling. Intelligent machines will play a major role in cleaning up nuclear and toxic waste sites and performing tasks that are hazardous for human workers.

Hospital and Nursing Support

Intelligent fetch and carry robots are already in everyday use in hospitals [Evans and Krishnamurthy89]. Expert systems are used routinely for patient monitoring, diagnostic aids, and record keeping. Intelligent systems can be used in laboratory analysis, drug manufacturing, and prescription filling. These applications have the potential to reduce costs and improve patient care.

In-home Patient Services

Intelligent systems technology will enable a wide variety of nursing and rehabilitation services, including in-home elder care. Intelligent robotic systems could be developed for lifting and positioning invalid and infirm patients in and out of bed, on and off of the toilet, in and out of the bath. Intelligent systems can provide mobility, food preparation, physical therapy, security and health monitoring, telecommuting for work and shopping, and entertainment and education for the home patient. This could reduce the cost and improve the life of handicapped and elderly persons who prefer to remain at home rather than be institutionalized in nursing care facilities. There are about 100,000 persons that enter nursing homes every month in the United States, at an average cost of about $2500 per month. It thus can be estimated that delaying the average date of entry of the elderly to nursing homes by only one month might save the country about $3 billion per year.

Physical Security

Intelligent systems will enable advanced security systems for the detection and tracking of intruders with a minimum of false alarms. A problem with current home security systems is the high incidence of false alarms. A problem with commercial security cameras is that they cannot have both wide coverage and high resolution. Images from security cameras are often too blurry to make positive identification of suspects. Intelligent systems could solve both of these problems. Intelligent security systems could have multiple levels of alarm and could automatically track and zoom in on faces of persons that behave suspiciously.

Agriculture and Food Processing

Robotics and intelligent machine systems have begun to enter the field of agriculture and food-processing industries. Computer-controlled tractors can have a significant impact on productivity during plowing, planting, tilling, and harvesting operations. Intelligent robots can perform many of the operations required to process, package, store, and distribute food. The application of intelligent systems technologies to farming the oceans has not yet begun to be explored.

Mining and Drilling

Intelligent machines will be able to improve productivity and safety dramatically in underground mines and undersea operations. Three-fifths of Earth's surface is too deep beneath the oceans for mining or drilling operations using conventional techniques. This means that most of Earth's mineral resources and many oil fields have never been touched. Intelligent undersea robots offer significant potential for deep-sea mining and drilling.

Space and Undersea Exploration

Intelligent machines will play a major role in exploration of the seabed and the solar system. Outer space, planetary surfaces, and the bottom of the ocean all share the characteristic that manned exploration is extremely expensive and hazardous. Intelligent systems and robotics promise to reduce the cost and risk and increase the amount of knowledge that can be gathered from space and undersea exploration.

IMPACT ON ECONOMIC THEORY AND POLICY

It seems clear that intelligent systems technology will have a profound impact on productivity and hence economic prosperity. At the very least, it seems likely that this will bring about revisions in economic theory to give a more prominent role to productivity as a factor in production. Productivity may someday be treated as a primary system parameter that can be controlled directly through investment policy rather than as an exogenous variable that is beyond human influence [Solow70,94]. Productivity may no longer be computed as a residual (i.e., what is left over when everything else is accounted for), but assume a more critical role as *the* primary engine of production. The causal effect of productivity on wealth production and that of investment on productivity could become the primary focus of economic theory and policy development in the twenty-first century [Romer90,94, Grossman and Helpman94, Caballero and Jaffe93].

It may be that new economic theories based on abundance will emerge to replace current theories based on scarcity. New economic institutions and policies may arise to exploit the wealth-producing potential of large numbers of intelligent machines. As more wealth is produced without direct human labor, the distribution of income may shift from wages and salaries to dividends, interest, and rent. As more is invested in ownership of the means of production, more people may derive a substantial income from ownership of capital stock. Eventually, some form of people's capitalism may replace the current amalgam of capitalism and socialism that is prevalent in the industrialized world today [Albus76, Kelso and Hetter67].

IMPORTANCE FOR MILITARY STRENGTH

Intelligent systems technologies certainly have the potential to revolutionize the art of war. The eventual impact on military science may be as great as the invention of gunpowder, the airplane, or nuclear weapons. Intelligent weapons systems are already highly advanced. Cruise missiles, smart bombs, and unmanned reconnaissance aircraft have been deployed and used in combat with great effect. Unmanned ground vehicles and computer-augmented command and control systems are currently being developed and deployed. Unmanned undersea vehicles are patrolling the oceans collecting data and gathering intelligence. These are but the vanguard of a whole new generation of military systems that will become possible as soon as intelligent systems engineering becomes a mature discipline [Gourley2000].

In future wars, unmanned air vehicles, ground vehicles, ships, and undersea vehicles will be able to outperform manned systems. Many military systems are limited in performance because of the inability of the human body to tolerate high levels of temperature, acceleration, vibration, or pressure, or because humans need to consume air, water, and food. A great deal of the weight and power of current military vehicles is spent on armor and life support systems that would be unnecessary if there were no human operators on board. A great deal of military tactics and strategy are based on the need to minimize casualties and rescue people from danger. This would become unnecessary if warriors could remain out of harm's way.

Intelligent military systems will significantly reduce the cost of training and readiness. Compared to humans, unmanned vehicles and weapons systems will require little training or maintenance to maintain readiness. Unmanned systems can be stored in forward bases or at sea for long periods of time at low cost. They can be mobilized quickly in an emergency, and they will operate without fear under fire, the first time and every time. Intelligent systems also enable fast and effective gathering, processing, and displaying of battlefield information. They can enable human commanders to be quicker and more thorough in planning operations and in replanning as unexpected events occur during the course of battle. In short, intelligent systems promise to multiply the capabilities of the armed forces while reducing casualties and hostages and lowering the cost of training and readiness [Maggart and Markunas2000].

IMPORTANCE FOR HUMAN WELL-BEING

In the long run and most important of all, the development of intelligent machines could lead to a golden age of prosperity, not only in the industrialized nations, but throughout the world. Despite the explosion of material wealth produced by the first industrial revolution, poverty persists and remains a major problem throughout the world today. Poverty causes hunger and disease. It breeds ignorance, alienation, crime, and pollution. Poverty brings misery, pain, and suffering. It leads to substance abuse. Particularly in the third world, poverty may be the biggest single problem that exists, because it causes so many other problems. Yet there is a well-known cure for poverty. It is wealth.

Wealth: goods and services that people desire

Poverty: lack of wealth

Wealth is difficult to generate (which is why it is so hard to become wealthy). Producing wealth requires labor, capital, and raw materials—multiplied by productivity. The amount of wealth that can be produced for a given amount of labor, capital, and raw materials depends on productivity. The level of productivity that exists today is determined by the current level of knowledge embedded in workers' skills, management techniques, tools, equipment, and software used in the manufacturing process.

Productivity is the product of technology. As intelligent systems become widespread, productivity will grow and the rate of wealth production will increase. Intelligent machines in manufacturing and construction will increase the stock of wealth and reduce the cost of material goods and services. Intelligent systems in health care will improve services and reduce costs for the sick and elderly. Intelligent systems could make quality education available to all. Intelligent systems will make it possible to clean up and recycle waste, reduce pollution, and create environmentally friendly methods of production and consumption.

The potential impact of intelligent machines is magnified by that fact that technology has reached the point where machines cannot only create wealth, but can build other machines that create wealth. In other words, intelligent machines have begun to exhibit a capacity for self-reproduction. John von Neumann[66] was among the first to recognize that machines can possess the ability to reproduce. Using mathematics of finite-state machines and Turing machines, von Neumann developed a theoretical proof that machines can reproduce. Over the past two decades the theoretical possibility of machine reproduction has been demonstrated empirically (at least in part) in the practical world of manufacturing. For example, computers are routinely involved in the processes of manufacturing computers. Computers are indispensable to the process of designing, testing, manufacturing, programming, and servicing computers. On a more global scale, intelligent factories build components for intelligent factories.

At a high level of abstraction, many of the fundamental processes of biological and machine reproduction are similar. Both require the availability of energy and a suitable environment in which the reproductive processes can take place. Both require an elaborate infrastructure to support the creation of the next generation from the previous. Both require agents, tools, materials, mechanisms, a source of energy, and protection from disturbances and contamination. Both require the storage, use, and transmittal of information that specifies the materials and the processes required to assemble constituent parts into finished products.

Both biological and machine reproduction are controlled by knowledge that specifies the processes that create new members of the species. In biological reproduction, the knowledge required to build a biological creature is embedded in the DNA sequence. This knowledge controls the molecular and cellular processes embodied in entities such as RNA, mitochondria, energy sources, and membrane mechanisms that carry out the transport and assembly of nucleic acids into the proteins and other building materials necessary for reproduction at the cellular level. Knowledge embedded in DNA enables protein to be manufactured and cells to grow, reproduce, differentiate, and assemble themselves into an embryo and then a fetus within the protective environment of the womb. When the time has come, knowledge embedded in the reflexes of the mother's body expels the fetus into the world as a baby. The newborn infant then depends on knowledge located in two places: first, the knowledge embedded in the wiring diagram and preprogrammed behavioral responses specified by the genes of the infant; and second, the knowledge embedded in the maternal skills of the mother that generate the behavior required to rear the infant successfully into an adult capable of reproduction. As the child grows and matures, its brain makes new connections and modifies or prunes old connections, through a variety of learning mechanisms.

In manufacturing, the knowledge required to create a computer or a machine tool is embedded in the design, process plans, and programs that specify the manufacturing process. Knowledge embedded in the skills of workers and managers enable the transformation of specifications into decisions that control the production process. Knowledge embedded in the design and specifications for the construction of a manufacturing plant control the assembly of the factory into a place where computers and machine tools can be manufactured. Eventually, the machines and computers built in the first-generation plant are used to build the next generation of machines and computers in the second-generation plant. Thus, both biological and machine reproduction involve mechanisms that are fabricated and controlled by information that specifies the mechanical structure, the sensors, actuators, communications network, and computational architecture necessary to close the reproductive loop.

Some might object to a comparison between biological and machine reproduction on the grounds that the processes of manufacturing and engineering are fundamentally different from the processes of biological reproduction and evolution. Certainly, there are many differences between biological and machine reproduction. Machine reproduction does not require sex or involve carbon-based organisms. It takes place on a vastly different scale with entirely different materials and processes. It is not nearly as autonomous. At least for the present, machine reproduction requires a great deal of interference by human workers. But the comparison is not farfetched.

And the results can be quite similar. Both biological and machine reproduction can produce populations that grow exponentially. In fact, machine reproduction can be much faster than biological. Intelligent machines can flow from a production line at a rate of many per hour. Perhaps more important, machines can evolve from one generation to the next much faster and more efficiently than can biological organisms. Biological organisms evolve by a Darwinian process, through random mutation and natural selection. Intelligent machines evolve by a Lamarckian process, through conscious design improvements under selective pressures of the marketplace. In the machine evolutionary process, one generation of computers often is used to design and manufacture the next generation of computers that is more powerful and less costly than itself. Significant improvements can occur in a very short time between one generation of machines and the next. As a result, intelligent machines are evolving extremely fast relative to biological species. Improved models of computer systems appear every few months to vie with each other in the marketplace. Those that survive and are profitable are improved and enhanced. Those that are economic failures are abandoned. Entire species of computers evolve and are superseded within a single decade.

New tools for evolving more powerful software have also begun to emerge. Advanced software tools are being developed to assist human software engineers in improving the effectiveness and efficiency of new generations of software. Genetic algorithms and evolutionary programming techniques are active fields of research [Holland75, Fogel99]. As a result, each generation of software is more powerful, cost effective, and intelligent than the generation before. Computer software evolution may be less complex than biological evolution, but it is many times faster. New and improved software releases appear on the market every few months.

Both biological and machine reproduction are subject to evolutionary pressures that tend to reward success and punish failure. In biological evolution, selective pressures are applied by competition between individuals for food, territory, and desirable sexual partners, and between species for food and territory. Those individuals or species that are more successful in acquiring food, territory, and breeding privileges are more likely to survive and reproduce than are those that are less successful. In manufacturing, selective pressures are applied by competition between products in the marketplace. Those products that are higher quality and lower cost are more likely to succeed than those that are lower quality and higher cost. Each step of the manufacturing process is subject to improvement through new technology. As improvements are made, each generation of intelligent system can be produced more efficiently and at lower cost than the one before. Each new generation of computers is improved, both in terms of cost per unit of computation and cost per unit of memory.

The ability of intelligent systems to reproduce and evolve will have a profound effect on the capacity for wealth production. As intelligent machines reproduce, they multiply their numbers. This leads to an exponential increase in the intelligent machine population. Since intelligent machines can produce wealth, this implies that exponential growth in per capita wealth is possible. As intelligent machines evolve, they increase their effectiveness and efficiency. Thus, each new generation of intelligent machines can increase productivity (i.e., the efficiency of the productive process). This leads to growth in the exponent of the rate of growth in per capita wealth. With each new generation, goods and services become exponentially less expensive and more plentiful.

THE PROSPECTS FOR TECHNOLOGY GROWTH

It is sometimes argued that technology, and therefore productivity, cannot grow forever because of the law of diminishing returns. It is argued that there must be a limit to everything, and therefore, productivity cannot grow indefinitely. Whether this is true in some abstract sense is an interesting philosophical question. Whether it is true in any practical sense is clear. It is not. From the beginning of human civilization until now, the more that is known, the easier it is to discover new knowledge. This has always been true and remains true today. There is nothing to suggest that knowledge will be subject to the law of diminishing returns in the future, at least not in the foreseeable future. Most of the scientists that have ever lived are alive and working today. Scientists and engineers today are better educated and have better tools with which to work than ever before. In the neuro- and cognitive sciences, the pace of discovery is astonishing. The same is true in computer science, electronics, manufacturing, and many other fields as well. Today, there is an explosion of new knowledge in almost every field of science and technology.

There is certainly no evidence that we are nearing a unique point in history where progress will be limited by an upper bound on what there is to know. There is no reason to believe that such a limit even exists, much less that we are approaching it. On the contrary, there is good evidence that the advent of intelligent machines has placed us on the cusp of an S-curve where productivity can grow exponentially for many decades. Productivity growth is directly related to growth in knowledge. Growth in knowledge is dependent on the amount and effectiveness of investment in research, development, and education. This suggests that, given adequate investment in technology, productivity growth could return to 2.5%, which is the average for the twentieth century. With higher rates of investment, productivity growth could conceivably rise to 4%, which is the average for the 1960–1968 time frame, or even above 10%, which occurred during the period between 1939 and 1945 [Samuelson and Nordhaus89].

If such productivity growth were to occur, society could afford to improve education, clean up the environment, and adopt less wasteful forms of production and consumption. Many social problems that result from slow economic growth, such as poverty, disease, and pollution, would virtually disappear. At the same time, taxes could be reduced, social security benefits increased, and health care and a minimum income could be provided for all. The productive capacity of intelligent machines could generate sufficient per capita wealth to support an aging population without raising payroll taxes on a shrinking human labor force. Over the next three decades, intelligent machines might provide the ultimate solution to the social security and Medicare crisis. Benefits and services for an aging population could be expanded continuously, even in countries with stable or declining populations.

ENDING THE THREAT TO EMPLOYMENT

The scenario suggested above would become more attractive and the fears of intelligent machines less intense if jobs and income were less intimately coupled. For example, the one sector of society that harbors no fear of unemployment because of intelligent machines consists of homemakers [i.e., housewives (and the occasional house husband)]. This is undoubtedly due to the fact that homemakers typically are not paid for their labor. Therefore, intelligent machines pose no economic threat. No one views automatic washing machines and dishwashers as a threat to employment. On the contrary, household automation offers the prospect of emancipation from tedious chores. If intelligent machines can perform housework, homemakers lose no income. Instead, they gain freedom to pursue other more rewarding activities—rearing children, working, shopping, sewing, cooking, hobbies, sports, and volunteer activities.

One way that jobs and income could become less closely coupled would be if ownership of the means of production became more widely distributed among the population. For example, if the average worker had a significant ownership position in profitable companies, income from stock dividends and capital gains would begin to supplement income from wages and salaries. As more and more income is derived from ownership of the means of production, many workers would voluntarily leave the labor force to pursue other interests. This would shrink the labor pool and create job openings for those who desire to continue working. Thus, the demand for labor would grow while the threat of unemployment subsided. If everyone held a significant ownership position in the means of production, productivity gains from technology development would no longer threaten jobs. Productivity growth through advanced technology would appear much less frightening. Fear of unemployment might give way to hope for the potential benefits of increased wealth production. Under such conditions, technology development directed toward productivity growth might become a primary goal of economic policy.

Eventually, ownership and control of intelligent machines could become distributed widely enough that dividends and rent would replace wages and salaries as the primary source of personal income. If everyone owned a minimum amount of stock, as soon as dividends on that minimum rises above a subsistence level, poverty would simply disappear. Everyone would be financially independent by virtue of having an ownership stake in the economic system. Productivity growth would then become attractive because higher productivity would mean greater income for everyone. A number of other benefits might also accrue, including improved medical care, more widespread education, a cleaner and safer environment, increased political stability, strengthened democratic institutions, and greater general welfare.

WIDESPREAD OWNERSHIP

Individual stock ownership of the means of production is a growing trend in the United States. Already 44% of the U.S. population owns stocks or bonds of some type, up from 32% only a decade ago [U.S. Census Bureau99]. There are good reasons to believe that this trend will continue. Efforts are under way to expand Individual Retirement Accounts and other forms of private savings and investment mechanisms. Proposals have been made to invest part of the revenue from social security payroll taxes in stocks and bonds. If widespread ownership of intelligent machines ever becomes a high-priority national goal, a number of other more aggressive approaches might be explored. For example, tax incentives might be devised to subsidize individual ownership of stocks and bonds, similar to how homeownership is subsidized today or how farm ownership was subsidized in the past. Incentives might be increased for employee stock ownership plans [Kelso and Hetter67]. The budget surplus might be devoted to giving every citizen vouchers to be deposited in Individual Retirement Accounts. It is even possible for the Federal Reserve or the U.S. Treasury to create the equivalent of a national mutual fund and distribute ownership credits directly to the general population [Albus76,94].

If it were possible to agree that an ownership economy (i.e., an economy where everyone owns enough stock to provide at least a subsistence income from dividends and capital gains) is a desirable national goal, there are many possible approaches that could be devised to move in that direction. Once that occurs, the stage would be set for a dramatic increase in wealth production through intelligent machine systems.

SUMMARY AND CONCLUSIONS

We are at a point in history where science has good answers to questions such as, "What is the universe made of?" and "What are the fundamental mechanisms of life?" There exists a wealth of knowledge about how our bodies work. There are solid theories for how life began and how species evolved. However, we are just beginning to acquire a deep understanding of how the brain works and what the mind is.

We know a great deal about how the brain is wired up and how neurons compute various functions. We have a good basic understanding of mathematics and computational theory. We understand how to build sensors, process sensory information, extract information from images, and detect entities and events. We understand the basic principles of attention, clustering, classification, and statistical analysis. We understand how to make decisions in the face of uncertainty. We know how use knowledge about the world to predict the future, to reason, imagine, and plan actions to achieve goals. We have algorithms that can decide what is desirable, and plan how to get it. We have procedures to estimate costs, risks, and benefits of potential actions. We can write computer programs to deal with uncertainty and compensate for unexpected events. We can build machines that can parse sentences and extract meaning from messages, at least within the constrained universe of formal languages.

There is also much about the brain that we do not yet understand, and there are many processes of mind that we have not been able to duplicate with computers. Our brains can integrate information from millions of sensors into a rich mosaic of sensations, feelings, and knowledge that is far beyond anything we know how to duplicate in computers. Our central nervous system controls millions of muscles in producing behavior that is far more complex than that of any artificial control system.

The human mind can understand what is possible and predict what can be achieved through purposeful actions. The mind can imagine how the world might be changed, for better or worse, through our own actions or inaction. The mind can even imagine things that have never existed before. We can imagine how world might be changed for the better if intelligent system theory were applied to economic and social institutions and policies. We can imagine how intelligent systems principles might be applied to economic policy, to more effectively control inflation, stimulate investment, and promote economic growth. We can speculate about what rates of productivity growth are physically possible and under what circumstances. We can hypothesize how investment, savings, and inflation might be manipulated to meet societal goals. We can dream about how to determine what socioeconomic goals are most desirable and plan how best to achieve them.

We have mentioned only a tiny fraction of the possible futures. There are many possibilities that we have not discussed and many more that cannot even be imagined because the enabling discoveries have not yet occurred. Even under very conservative assumptions, the possibilities that can be generated from simple extrapolations of current trends are very exciting. We are at a point in history where some of the deepest mysteries are being revealed. We are discovering how the brain processes information, how it represents knowledge, how it makes decisions and controls actions. We are beginning to understand what the mind is. We will soon have at our disposal the computational power to emulate many of the functional operations in the brain that give rise to the phenomena of intelligence and consciousness. We are learning how to organize what we know into an architecture and methodology for designing and building truly intelligent machines, and we are developing the capacity to test our theories experimentally. As a result, we are at the dawning of an age where the engineering of mind is feasible.

In this book we have suggested one approach to the engineering of mind that we believe is promising. We have proposed a method for knowledge representation that can support an intelligent world model. We have described a world modeling system that can compute what to expect and predict what is likely to result from contemplated actions. We have outlined a behavior generating system that can choose what it intends to do from a wide variety of options and can focus available resources on achieving its goals. We have proposed a sensory processing system that can perceive what is happening, both in the outside world and inside the system itself. We have suggested a value judgment system that can distinguish good from bad and decide what is desirable. We have outlined a reference model architecture for organizing the foregoing functions into a truly intelligent system. We have hypothesized that in the near future it will become possible to engineer sentient, caring, feeling machines with intentions and motives that use reason and logic to devise plans to accomplish their objectives. And we have suggested some possible scenarios that might result from continued technological developments in the domain of mind engineering.

We believe that the engineering of mind is an enterprise that will prove at least as technically challenging as the Apollo program or the Human Genome project. And we are convinced that the potential benefits for humankind will be at least as great, perhaps much greater. Understanding of the mind and brain will bring major scientific advances in psychology, neuroscience, and education. A computational theory of mind may enable us to develop new tools to cure or control the effects of mental illness. It will certainly provide us with a much deeper appreciation of who we are and of our place in the universe.

Finally, understanding of the mind and brain will enable the creation of a new species of intelligent machine systems that can generate economic wealth on a scale hitherto unimaginable. Within a century, intelligent machines could create the wealth needed to provide food, clothing, shelter, education, medical care, a clean environment, and physical and financial security for the entire world population. Intelligent machines may eventually generate the production capacity to support universal prosperity and financial security for all human beings. Thus, the engineering of mind is much more than the pursuit of scientific curiosity. It is more even than a monumental technological challenge. It is an opportunity to eradicate poverty and usher in a golden age for all humankind.

References

Adams, M., Deutsch, O., and Harrison, J. (1985). A hierarchical planner for intelligent systems, *Proceedings of the SPIE Conference on Applications of Artificial Intelligence*, April.

Adelson, E. H., and Movshon, J. A. (1982). Phenomenal coherence of moving visual patterns, *Nature, 30*, pp. 523–525.

Advanced Technology and Research Corporation (1993). Stamp distribution network, *USPS Contract 104230-91-C-3127, Final Report*, ATR, Burtonsville, MD.

Albus, J. S. (1971). A theory of cerebellar function, *Mathematical Biosciences, 10*, pp. 25–61.

Albus, J. S. (1972). *Theoretical and Experimental Aspects of a Cerebellar Model*, Ph.D. Thesis, University of Maryland, College Park, MD.

Albus, J. S. (1975a). A new approach to manipulator control: the cerebellar model articulation controller (CMAC), *Transactions of the ASME Journal of Dynamic Systems, Measurement, and Control*, September, pp. 220–227.

Albus, J. S. (1975b). Data storage in the cerebellar model articulation controller (CMAC), *Transactions of the ASME Journal of Dynamic Systems, Measurement, and Control*, September, pp. 228–233.

Albus, J. (1976). *Peoples' Capitalism: The Economics of the Robot Revolution*, New World Books, Kensington, MD. See also Peoples' Capitalism Web page, *http://www.peoplescapitalism.org*.

Albus, J. (1979). Mechanisms of planning and problem solving in the brain, *Mathematical Biosciences, 45*, pp. 247–293.

Albus, J. (1981). *Brains, Behavior, and Robotics*, BYTE/McGraw-Hill, Peterborough, NH.

Albus, J. S., McLean, C., Barbera, A., Fitzgerald, M. L. (1982). An architecture for real-time sensory-interactive control of robots in a manufacturing environment, *Proceedings of the 4th IFAC/IFIP Symposium on Information Control Problems in Manufacturing Technology*, Gaithersburg, MD.

Albus, J. S. (1988). System description and design architecture for multiple autonomous undersea vehicles, *NIST Technical Note 1251*, National Institute of Standards and Technology, Gaithersburg, MD, September.

Albus, J. S. (1991). Outline for a theory of intelligence, *IEEE Transactions on Systems, Man and Cybernetics, 21*(3), pp. 473–509.

Albus, J. S. (1993). A reference model architecture for intelligent systems design, in *An Introduction to Intelligent and Autonomous Control* (P. J. Antsaklis, and K. M. Passino, Eds.).

Albus, J. S. (1994). Peoples' Capitalism Web page, *http://members.aol.com/jsalbus*.

379

Albus, J. (1995). The NIST Real-Time Control System (RCS): an application survey, *Proceedings of the AAAI 1995 Spring Symposium Series*, Stanford University, Menlo Park, CA, March 27–29.

Albus, J. S. (1997). The NIST Real-Time Control System (RCS): an approach to intelligent systems research, *Journal of Experimental and Theoretical Artificial Intelligence*, 9, pp. 157–174.

Albus, J. (1998). 4D/RCS: a reference model architecture for Demo III, Version 0.1, *NISTIR 5994*, National Institute of Standards and Technology, Gaithersburg, MD.

Albus, J. S. (1999). 4-D/RCS: a reference model architecture for Demo III, *Proceedings of the SPIE Vol. 3693 AeroSense Session on Unmanned Ground Vehicle Technology*, Orlando, FL, April 7–8.

Albus, J. S. (2000). 4-D/RCS reference model architecture for unmanned ground vehicles, *Proceedings of the IEEE International Conference on Robotics and Automation*, San Francisco, April 24–27.

Albus, J. S., and Hong, T. (1990). Motion, depth, and image flow, *Proceedings of the 1990 IEEE International Conference on Robotics and Automation*, Cincinnati, OH.

Albus, J., and Meystel, A. (1995). A reference model architecture for design and implementation of semiotic control in large and complex systems, in *Architectures for Semiotic Modeling and Situation Analysis in Large Complex Systems, Proceedings of the 1995 ISIC Workshop*, Monterey, CA, pp. 33–45.

Albus, J., and Meystel, A. (1996). A reference model architecture for design and implementation of intelligent control in large and complex systems, *International Journal of Intelligent Control and Systems*, 1(1), pp. 15–30.

Albus, J. S., Barbera, A. J., Fitzgerald, M. L., and Nashman, M. (1981). Sensory interactive robots, *Proceedings of the 31st General Assembly, International Institution for Production Engineering Research (CIRP)*, Toronto, Ontario, Canada, September.

Albus, J. S., McCain, H. G., and Lumia, R. (1987). NASA/NBS standard reference model for telerobot control system architecture (NASREM), *NIST Technical Note 1235*; 1989 ed., National Institute of Standards and Technology, Gaithersburg, MD, April (supersedes *NBS Technical Note 1235*, July 1987).

Albus, J. S., McLean, C., Barbera, A., Fitzgerald, M. L. (1982). An architecture for real-time sensory-interactive control of robots in a manufacturing environment, *Proceedings of the 4th IFAC/IFIP Symposium on Information Control Problems in Manufacturing Technology*, Gaithersburg, MD.

Albus, J., Meystel, A., and Uzzaman, S. (1993). Nested motion planning for an autonomous robot, *Proceedings of the IEEE Conference on Aerospace Systems*, Westlake Village, CA, May 25–27.

Albus, J., Lacaze, A., and Meystel, A. (1995). Theory and experimental analysis of cognitive processes in early learning, *Proceedings of the IEEE International Conference on Systems Man and Cybernetics*, 4, pp. 4404–4409.

Albus, J., Lacaze, A., and Meystel, A. (1997). Multiresolutional planning with minimum complexity, *Proceedings of the 1997 International Conference on Intelligent Systems and Semiotics*, Gaithersburg, MD, pp. 151–156.

Allen, J., and Koomen, J. (1983). Planning using a temporal world model, *Proceedings of International Joint Conference on Artificial Intelligence*, 8, pp. 741–747.

Appelbaum, E., and Schettkat, R. (1995). Employment and productivity in industrialized economies, *International Labor Review*, 134(4–5).

Arbib, M. (1972). *The Metaphorical Brain*, Wiley, New York.

Arbib, M. (1992). Schema theory, pp. 1427–1443 in *The Encyclopedia of Artificial Intelligence*, 2nd ed., (S. Shapiro, Ed.), MIT Press, Cambridge, MA.

Arbib, M. (1995). Schema theory, pp. 830–834 in *The Handbook of Brain Theory and Neural Networks* (M. Arbib, Ed.), MIT Press, Cambridge, MA.

Arkin, R. C. (1986). Path planning for a vision-based autonomous robot, *Proceedings of the SPIE Conference on Mobile Robots*, Cambridge, MA.

Arkin, R. C. (1990). Integrating behavioral, perceptual, and world knowledge in reactive navigation, *Robotics and Autonomous Systems*, 6, pp. 105–122.

Arkin, R. C. (1998). *Behavior-Based Robotics*, MIT Press, Cambridge, MA.

Ashby, R. (1952). *Design for a Brain*, Chapman and Hall, London.

Ashby, R. (1958). *Introduction to Cybernetics*, Wiley, New York.

Ashby, R. (1981). In *Mechanisms of Intelligence: Ross Ashby's Writings on Cybernetics* (R. Conant, Ed.), Intersystems Publications, Seaside, CA.

Ballard, D. (1997). *An Introduction to Natural Computation*, MIT Press, Cambridge, MA.

Barbera, A. J., Albus, J. S., and Fitzgerald, M. L. (1979). Hierarchical control of robots using microcomputers, *Proceedings of the 9th International Symposium on Industrial Robots*, Washington, DC, March 13–15.

Barbera, A. J., Fitzgerald, M. L., Albus, J. S., and Haynes, L. S. (1984). RCS: the NBS real-time control system, *Proceedings of the Robots 8 Conference and Exposition*, Detroit, MI, June 4–7.

Barlow, H. (1972). Single units and sensation: a neuron doctrine for perceptual psychology? *Perception*, *1*, pp. 371–394.

Barlow, H. (1994). What is the computational goal of the neocortex?, pp. 1–22 in *Large Scale Neuronal Theories of the Brain* (C. Koch and J. Davis, Eds.), MIT Press, Cambridge, MA.

Barto, A. (1992). Reinforcement learning and adaptive critic methods, pp. 469–491 in *Handbook of Intelligent Control* (D. White and D. Forge, Eds.), Van Nostrand Reinhold, New York.

Beakley, B., and Ludlow, P. (1992). *The Philosophy of Mind: Classical Problems/Contemporary Issues*, MIT Press, Cambridge, MA.

Beer, S. (1995). *Brain of the Firm*, Wiley, New York.

Bellman, R. (1957). *Dynamic Programming*, Princeton University Press, Princeton, NJ.

Bernus, P., Mertins, K., and Schmidt, (1998). *Handbook on Architectures of Information Systems*, Springer-Verlag, New York.

Biederman, I. (1990). Higher-level vision, in *Visual Cognition and Action*, Vol. 2 (D. Osherson, S. Kosslyn, and J. Hollerbach, Eds.), MIT Press, Cambridge, MA.

Biemans, F. (1989). *A reference model for manufacturing and planning and control*, thesis, Franciscus Petrus Maria Biemans, ISBN 90-9002961-3.

Binder, J. R. (1997). Human brain areas identified by fMRI, *Journal of Neurosciences*, *17*(1), pp. 353–362.

Binford, T. (1982). Survey of model based image analysis systems, *International Journal of Robotics Research*, *1*, pp. 18–64.

Blidberg, D. R. (1986). Guidance control architecture for the EAVE vehicle, *IEEE Journal of Oceanic Engineering*, October.

Bloom, H., Furlani, C., and Barbera, A. (1984). Emulation as a design tool in the development of real-time control systems, *Proceedings of the 1984 Winter Simulation Conference*, Dallas, TX, November 28–30.

Bluestone, B., and Harrison, B. (2000). *Growing Prosperity: The Battle for Browth with Equity in the Twenty-first Century*, Houghton Mifflin Co., New York.

Bornstein, J., and Koren, Y. (1991). Histogramic in-motion mapping for mobile robot obstacle avoidance, *IEEE Transactions on Robotics and Automation*, *7*(3), pp. 535–539.

Borst, A. (1990). How do flies land? From behavior to neuronal circuits, *Bioscience*, *40*, pp. 292–299.

Brady, M., and Yuille, A. (1987). An extremum principle for shape from contour, pp. 285–328 in *Vision, Brain, and Cooperative Computation* (M. A. Arbib and A. R. Hanson, Eds.), Bradford Books, imprint of MIT Press, Cambridge, MA.

Brodmann, K. (1914). Physiologie des Gehirns, in. Die Allgemeine Chirurgie der Gehirnkrankheiten, *Neue Deutsch Chirurgie*, Vol. 11, Verlag Ferdinand Enke, Stuttgart, Germany.

Brooks, R. A. (1986). A robust layered control system for a mobile robot, *IEEE Journal of Robotics and Automation, 2*, pp. 14–23.

Brooks, R. A. (1990). Elephants don't play chess, *Robotics and Autonomous Systems*, *6*, pp. 3–15.

Brooks, R. A. (1999). *Cambrian Intelligence: The Early History of the New AI*, MIT Press, Cambridge, MA.

Bryson, A. E., and Ho, Y. C. (1975). *Applied Optimal Control*, 2nd ed., Blaisdell Publishing, Waltham, MA.

Burt, P., and Adelson, E. (1983). The Laplacian pyramid as a compact image code, *IEEE Transactions on Communication*, *31*, pp. 532–540.

Caballero, R., and Jaffe, A. (1993). How high are the giants shoulders: an empirical assessment of knowledge spillovers and creative destruction in a model of economic growth, pp. 15–74 in *NBER Macroeconomics Annual, 1993*, MIT Press, Cambridge, MA.

Camus, T. (1997). Real-time quantized optical flow, *Journal of Real-Time Imaging* (special issue on real-time motion analysis), *3*, pp. 71–86.

Camus, T., Coombs, D., Herman, M., and Hong, T. H. (1999). Real-time single-workstation obstacle avoidance using only wide-field flow divergence, *Journal of Computer Vision Research*, *1*(3).

Capek, K. (1923). *R.U.R. or Rossum's Universal Robots*, first performed at the Garrick Theatre in New York on October 9, 1922.

Carpenter, G., and Grossberg, S. (1988). The ART of adaptive pattern recognition by a self-organizing neural network, *Computer*, *21*, pp. 77–88.

Carpenter, G., and Grossberg, S. (Eds.) (1992). *Neural Networks for Vision and Image Processing*, MIT Press, Cambridge, MA.

Carter, R. (1998). *Mapping the Mind*, University of California Press, Berkeley, CA.

Chang, C., and Lee, R. C. (1973). *Symbolic Logic and Mechanical Theorem Proving*, Academic Press, New York.

Chang, T., Hong, T., Legowik, S., and Nashman-Abrams, M. (1999). Concealment and obstacle detection for autonomous driving, *Proceedings of the International*

Association of Science and Technology: Robotics and Applications, Santa Barbara, CA, November.

Chavez, R., and Meystel, A. (1984). Structure of intelligence for an autonomous vehicle, *Proceedings of the IEEE International Conference on Robotics and Automation*, pp. 584–591.

Chen, C. (1984). *Linear System Theory and Design*, Holt, Rinehart and Winston, New York.

Chomsky, N. (1972). *Language and Mind*, Harcourt Brace Jovanovich, New York.

Chomsky, N. (1988). *Language and Problems of Knowledge: The Managua Lectures*, MIT Press, Cambridge, MA.

Churchland, P. S., and Sejnowski, T. J. (1992). *The Computational Brain*, MIT Press, Cambridge, MA.

Cleveland, B., and Meystel, A. (1990). Predictive planning + fuzzy compensation = intelligent control, *Proceedings of the 5th IEEE International Symposium on Intelligent Control*, Philadelphia, September.

CORBA (1995). *Common Object Request Broker: Architecture and Specification, Revision 2.0*, Object Management Group, Framingham, MA.

Conference on Intelligent Vehicles (1998). *IEEE International Conference on Intelligent Vehicles*, Stuttgart, Germany, October 28–30.

Control Shell. *http://www.rti.com/products/products.html*.

Corbetta, M., Miezin, F., Dobmeyer, S., Shulman, G., and Petersen, S. (1990). Attentional modulation of neural processing of shape, color, and velocity in humans, *Science*, *248*, pp. 1556–1559.

Council of Economic Advisors (1987, 1988, 1989, 1996, 1998, 1999). *Economic Report of the President*, U.S. Government Printing Office, Washington, DC.

Darwin, C. (1859). *Origin of Species* (E. Mayr, Ed.), Harvard University Press, Cambridge, MA, 1964.

Deitel, H., and Deitel, P. (1997). *C++: How to Program*, Prentice Hall, Upper Saddle River, NJ.

Deitel, H., and Deitel, P. (1998). *Java: How to Program*, Prentice Hall, Upper Saddle River, NJ.

Delmia. *http://www.delmia.com*.

Denison, E. F. (1985). *Trends in American Economic Growth, 1929–1982*, Brookings Institution, Washington, DC.

Dennett, D. (1995). *Darwin's Dangerous Idea: Evolution and the Meaning of Life*, Touchstone, imprint of Simon & Shuster, New York.

Dickmanns, E. D., and Graefe, V. (1988). (a) Dynamic monocular machine vision, and (b) Application of dynamic monocular machine vision, *Journal of Machine Vision and Applications*, November, pp. 223–261.

Dickmanns, E. D. (1995). Parallel use of differential and integral representations for realizing efficient mobile robots, *Proceedings of the 7th International Symposium on Robotics Research*, Munich.

Dickmanns, E. D., et al. (1994). The seeing passenger car "*VaMoRs-P*," *International Symposium on Intelligent Vehicles '94*, Paris, October 24–26.

Dickmanns, E. (1992). A General dynamic vision architecture for UGV and UAV, *Journal of Applied Intelligence*, *2*, pp. 251–270.

Dickmanns, E. (1999). An expectation-based, multi-focal, saccadic (EMS) vision system for vehicle guidance, *Proceedings of the 9th International Symposium on Robotics Research (ISRR'99)*, Salt Lake City, UT, October.

Dissanayake, G., Durrant-Whyte, H., and Bailey, T. (2000). A computationally efficient solution to the Simultaneous Localisation and Map Building (SLAM) problem, *Proceedings of IEEE International Conference on Robotics and Automation*, 2, pp. 1009–1014.

Doren, J., and Michie, D. (1966). Experiments with the graph-traverser program, *Proceedings of the Royal Society, A*, pp. 235–259.

Dreyfus, H. (1972). *What Computers Can't Do*, Harper & Row, New York.

Dreyfus, H. (1992). *What Computers Still Can't Do*, MIT Press, Cambridge, MA.

Duda, R., and Hart, P. (1973). *Pattern Classification and Scene Analysis*, Wiley, New York, p. 341.

Dudai, Y. (1989). *The Neurobiology of Memory: Concepts, Finding, and Trends*, Oxford University Press, New York.

Eccles, J., Ito, M., and Szentagothai, J. (1967). *The Cerebellum as a Neuronal Machine*, Springer-Verlag, New York.

Edelman, G. (1999a). *Proceedings of the International Conference on Frontiers of the Mind in the 21st Century*, Library of Congress, Washington DC, June 15.

Edelman, S. (1999b). *Representation and Recognition in Vision*, MIT Press, Cambridge, MA.

Eng, L., Freed, K., Hollister, J., Jobe, C., McGuire, P., Moser, A., Parikh, V., Pratt, M., Waskiewicz, and Yeager, F. (1996). *Computer Integrated Manufacturing (CIM) Application Framework Specification 1.3*, SEMATECH, Inc.

Evans, J., and Krishnamurthy, V. (1989). HELPMATE: a robotic materials transport system, *Robotics and Autonomous Systems*, 5, p. 251.

Everett, B., Laird, R. T., Heath-Pastore, T., Inderieden, R., Grant, K., and Jaffee, D. (1998). Multiple resource host architecture (MRHA) for the mobile detection assessment response system (MDARS), *Technical Note 1710*, Rev. 4, Space and Naval Warfare Systems Center, San Diego, CA.

Expert Systems. *http://www.attar.com*.

Farah, M. J. (1995). The neural bases of mental imagery, pp. 963–975 in *The Cognitive Neurosciences* (M. Gazzaniga, Ed.), Bradford Books, imprint of MIT Press, Cambridge, MA.

Felleman, D. J., and Van Essen, D. C. (1991). Distributed hierarchical processing in primate visual cortex, *Cerebral Cortex*, 1, pp. 1–47.

Fiala, J. C., and Wavering, A. J. (1987). RCS application example: tool changing on a horizontal machining center, *Proceedings of the 2nd International Conference on Robotics and Factories of the Future*, San Diego, CA, July 28–31.

Fikes, R., and Nilsson, N. (1971). STRIPS: a new approach to the application of theorem proving to problem solving, *Artificial Intelligence*, 2, pp. 189–208.

Fikes, R., Hart, P., and Nilsson, N. (1972). Learning and executing generalized robot plans, *Artificial Intelligence*, 3.

Firby, R. (1989). Adaptive execution in complex dynamic worlds, Ph.D. dissertation, Department of Computer Science, Yale University, New Haven, CT.

Fischler, M., and Firschein, O. (1987). *Intelligence: The Eye, the Brain, and the Computer*, Addison-Wesley, Reading, MA, p. 168.

Flanagan, O. (1991). *The Science of the Mind*, MIT Press, Cambridge, MA.

Fodor, J. (1975). *The Language of Thought*, Thomas Y. Crowell, New York.

Fodor, J. (1983). *The Modularity of Mind*, MIT Press, Cambridge, MA.

Fogel, L. (1999). *Intelligence Through Simulated Evolution*, Wiley, New York.

Francis, G., Grossberg, S., and Mingolla, E. (1994). Cortical dynamics of feature binding and reset: control of visual persistence, *Vision Research, 34*, pp. 1089–1104.

Freud, S. (1917). *Introductory Lectures on Psychonalysis*, W.W. Norton, New York.

Fu, K. S. (1969). Learning control systems, in *Advances in Information System Sciences* (J. Tou, Ed.), Plenum Press, New York.

Furlani, C. M., Kent, E. W., Bloom, H. M., and Mclean, C. R. (1983). Automated manufacturing research facility of the National Bureau of Standards, *Proceedings of the Summer Computer Simulation Conference*, Vancouver, British Columbia, Canada, July 11–13.

Galambos, J., Abelson, R., and Black, J. (1986). *Knowledge Structures*, Lawrence Erlbaum, Hillsdale, NJ.

Gat, E. (1992). Integrating planning and reaction in a heterogeneous asynchronous architecture for controlling real-world mobile robots, *Proceedings of the Tenth National Conference on Artificial Intelligence*, *809*, Menlo Park, CA, AAAI Press.

Gazi, V., Moore, M. L., Passino, K. M., Shackleford, W. P., Proctor, F. M., and Albus, J. S. (2001). *The RCS Handbook: Tools for Real Time Control Systems Software Development*, Wiley, New York.

Gazzaniga, M. (Ed.) (1995). *The Cognitive Neurosciences*, Bradford Books, imprint of MIT Press, Cambridge, MA.

Georgeff, M. (1984). A theory of action for multiagent planning, *Proceedings of AAAI'84*.

Georgopoulos, A. P. (1995). Motor cortex and cognitive processing, pp. 507–517 in *The Cognitive Neurosciences* (M. Gazzaniga, Ed.), Bradford Books, imprint of MIT Press, Cambridge, MA.

Georgopoulos, A., Caminiti, R., Kalaska, J., and Massey, J. (1983). Spatial coding of movement: a hypothesis concerning the coding of movement direction by motor cortical populations, *Experimental Brain Research. Suppl. 7*, pp. 327–336.

Gibson, J. (1950). *The Perception of the Visual World*, Houghton Mifflin, Boston.

Gibson, J. (1979). *The Ecological Approach to Visual Perception*, Houghton Mifflin, Boston.

Gibson, J., Olum, P., and Rosenblatt, F. (1955). Parallax and perspective during aircraft landings, *American Journal of Psychology, 68*, pp. 327–385.

Gourley, S. R. (2000). Future combat systems: a revolutionary approach to combat victory, *Army Magazine, 50*(7), pp. 23–26.

Gowdy, J. (1997). SAUSAGES: between planning and action, in *Intelligent Unmanned Ground Vehicles: Autonomous Navigation Research at Carnegie Mellon*, (M. A. Hebert, C. Thorpe, and A. Stentz, Eds.), Kluwer Academic, Boston.

Green, D. M. (1976). *An Introduction to Hearing*, Lawrence Erlbaum, Hillsdale, NJ.

Groover, M. (1980). *Automation, Production Systems, and Computer-Aided Manufacturing*, Prentice Hall, Englewood Cliffs, NJ.

Grossberg, S. (1987). Neural dynamics of surface perception: boundary webs, illuminents, and shape-from-shading, *Computer Vision, Graphics, and Image Processing*, *37*, pp. 116–165.

Grossberg, S. (Ed.) (1989). *Neural Networks and Natural Intelligence*, MIT Press, Cambridge, MA.

Grossberg, S., and Schmajuk, N. A. (1987). Neural dynamics of attentionally modulated pavlovian conditioning: conditioned reinforcement, inhibition, and opponent processing, *Psychobiology*, *15*(3), pp. 195–240.

Grossman, S. P. (1967). *A Textbook of Physiological Psychology*, Wiley, New York.

Grossman, G., and Helpman, E. (1994). Endogenous innovation in the theory of growth, *Journal of Economic Perspectives*, *8*(1), pp. 23–44.

Haldane, and Ross, (Trans.) (1911). *The Philosophical Works of Descartes*, Cambridge University Press, Cambridge.

Hanks, S., Pollack, M., and Cohen, P. (1993). Benchmarks, test beds, controlled experimentation, and design of agent architectures, *Proceedings of the AAAI*, Winter.

Hansen, M. (1998). Video processing technologies and the VFE-200 vision system: Sarnoff pyramid processing technologies and a roadmap for advanced video processing systems, Sarnoff Corporation Internal Presentation, Princeton, NJ, March.

Harrington, J. (1979). *Computer Integrated Manufacturing*, Krieger Publishing, Huntington, NY.

Hart, P., Nilsson, N., and Raphael, B. (1968). A formal basis for the heuristic determination of minimum cost paths, *IEEE Transactions on Systems, Science, and Cybernetics*, *4*(2), pp. 100–107.

Haykin, S. (1994). *Neural Networks: A Comprehensive Foundation*, Macmilian, Englewood Cliffs, NJ.

Haynes, L. S., Barbera, A. J., Albus, J. S., Fitzgerald, M. L., and McCain, H. G. (1984). Application example of the NBS robot control system, *Robotics and Computer Manufacturing*, *1*(1), pp. 81–95.

Hebb, D. O. (1949). *The Organization of Behavior*, Wiley, New York.

Hebb, D. O. (1958). *A Textbook of Psychology*, W. B. Sanders, Philadelphia, PA.

Herman, M., and Albus, J. S. (1988). Overview of the multiple autonomous underwater vehicles (MAUV) project, *Proceedings of the IEEE International Conference on Robotics and Automation*, Philadelphia, April.

Herman, M., Albus, J. S., and Hong, T. H. (1991). Intelligent control for multiple autonomous vehicles, in *Neural Networks for Control* (W. T. Miller and R. Sutton, Eds.), MIT Press, Cambridge, MA.

Hinton, G. (1984). Distributed representations, *Technical Report CMU-CS 84-157*, Department of Computer Science, Carnegie Mellon University, Pittsburgh, PA.

Holland, J. (1975). *Adaption in Natural and Artificial Systems*, University of Michigan Press, Ann Arbor, MI.

Hong, T., Legowik, S., and Nashman, M. (1998). Obstacle detection and mapping system, *NISTIR 6213*, National Institute of Standards and Technology, Gaithersburg, MD, August.

Hopcroft, J., and Ullman, S. (1979). *Introduction to Automata Theory: Languages and Computation*, Addison-Wesley, Reading, MA.

Hopfield, J. (1982). Neural networks and physical systems with emergent collective computational abilities, *Proceedings of the National Academy of Sciences USA*, *79*, April.

Horst, J. A. (1994). Integration of servo control into a large-scale control system design: an example from coal mining, *NISTIR 5446*, National Institute of Standards and Technology, Gaithersburg, MD.

Horst, J. A. (2000). Architecture, design methodology, and component-based tools for a real-time inspection system, *Proceedings of the 3rd IEEE International Symposium on Object-Oriented Real-Time Distrubuted Computing (ISORC 2000)*, Newport Beach, CA, March 15–17.

Horst, J. A., and Barbera, A. J. (1994). Continuous mining machine control using the real-time control system, *NISTIR 5448*, National Institute of Standards and Technology, Gaithersburg, MD.

House, E., and Pansky, B. (1960). *Neuroanatomy*, McGraw-Hill, New York; also, in S. Grossman, 1960, *A Textbook of Physiological Psychology*, Wiley, New York.

Howard, R. (1960). *Dynamic Programming and Markov Processes*, MIT Press, Cambridge, MA.

Howden, W. E. (1968). The sofa problem, *Computer Journal*, *11*(3), pp. 299–301.

Huang, H. M., Quintero, R., and Albus, J. S. (1991). A reference model, design approach, and development illustration toward hierarchical real-time system control for coal mining operations, *Advances in Control and Dynamic Systems*, Vol. 46: *Manufacturing and Automation Systems: Techniques and Technologies*, Part 2 of 5, Edited by C. T. Leondes, Academic Press.

Huang, H. M., Hira, R., and Feldman, P. (1992a). A submarine maneuvering system demonstration using a generic real-time control system (RCS) reference model, *Proceedings of the Summer Computer Simulation Conference '92*, Reno, NV, July 27–29.

Huang, H. M., Horst, J. A., and Quintero, R. (1992b). A motion control algorithm for a continuous mining machine based on a hierarchical real-time control system design methodology, *Journal of Intelligent and Robotic Systems*, *5*, pp. 79–99.

Huang, H. M., Hira, R., and Quintero, R. (1993). A submarine maneuvering system demonstration based on the NIST real-time control system reference model, *Proceedings of the 8th IEEE International Symposium on Intelligent Control*, Chicago, August 24–27.

Huang, H. M., Albus, J. S., Shackleford, W., Scott, H., Kramer, T., Messina, E., and Proctor, F. (2000). An architecting tool for large-scale system control with an application to a manufacturing workstation, 4th International Software Architecture Workshop, in conjunction with the 22nd International Conference on Software Engineering, Limerick, Ireland, June.

Hubel, D. (1988). *Eye, Brain and Vision*, W.H. Freeman, New York.

Hubel, D., and Wiesel, T. (1962). Receptive fields, binocular interaction, and functional architecture in the cat's visual cortex, *Journal of Physiology*, *160*, pp. 106–154.

Hubel, D., and Wiesel, T. (1968). Receptive fields and functional architecture of monkey striate cortex, *Journal of Physiology*, *195*, pp. 215–243.

Hubel, D., and Wiesel, T. (1974). Sequence regularity and geometry of orientation columns in the monkey striate cortex, *Journal of Comparative Neurology*, *158*, pp. 267–293.

Hull, C. L. (1943). *Principles of Behavior*, Appleton-Century, New York.

IMTR (1998). *Integrated Manufacturing Technology Roadmapping Project*, IMTR Project Office, Oak Ridge Centers for Manufacturing Technology, Oak Ridge, TN.

Isik, C., and Meystel, A. (1988). Pilot level of a hierarchical controller for an unmanned mobile robot, *IEEE Journal of Robotics and Automation*, *4*(3), pp. 244–255.

Jacobson, I. (1992). *Object-Oriented Software Engineering*, Addison-Wesley, Reading, MA.

James, W. (1890). *The Principles of Psychology*, Henry Holt, New York.

JAVA. *http://www.javaworld.com* and *http://www.attar.com*.

Jay, M., and Sparks, D. (1984). Auditory receptive fields in primate superior colliculus shift with changes in eye position, *Nature*, *309*, 345–347.

Jay, M., and Sparks, D. (1987). Sensorimotor integration in the primate superior colliculus; II: Coordinates of auditory signals, *Journal of Neurophysiology*, *57*, pp. 35–55.

Jorysz, H. R., and Vernadat, F. B. (1990). CIMOSA, part 1: total enterprise modelling and functional view, *International Journal of Computer Integrated Manufacturing*, *3*(3–4), pp. 144–156.

Jowett, B. (Trans.) (1892). *The Dialogs of Plato*, Random House, New York.

Joy, B. (2000). Why the future doesn't need us, *http://www.wired.com/wired/archive/8.04/joy.html*.

Julliere, M., Marce, L., and Place, H. (1983). A guidance system for a mobile robot, *Proceedings of the 13th International Symposium on Industrial Robots*, Vol. 2, April 17–21.

Kaczynski, T. (1995). The unabomber manifesto, *Washington Post*, September. Full text is available at http://www.soci.niu.edu/~critcrim/uni/uni.txt.

Kalman, R. (1960). A new approach to linear filtering and prediction problems, *Transactions of the ASME Journal of Basic Engineering*, March, pp. 35–45.

Kaufman, L. (1979). *Perception: The World Transformed*, Oxford University Press, New York.

Kelso, L., and Hetter, P. (1967). *Two Factor Theory: The Economics of Reality*, Random House, New York.

Kent, E. (1995). A study of potential applications of automation and robotics technology in construction, maintenance, and operation of highway systems: a final report, *NISTIR 5667, V1-V4*, National Institute of Standards and Technology, Gaithersburg, MD.

Kent, E., and Albus, J. (1984). Servoed world models as interfaces between robot control systems and sensory data, *Robotica*, *2*(1), pp. 17–25.

Khatib, O. (1986). Real-time obstacle avoidance for manipulatiors and mobile robots, *International Journal of Robotics Research*, *5*(1), pp. 90–98.

Khazen, E., and Meystel, A. (1998). Why multiresolutional hierarchies reduce computational complexity of stochastic systems for estimation and control, *Proceedings of the 1998 IEEE International Symposium on Intelligent Control*, a joint conference on the science and technology of intelligent systems, National Institute of Standards and Technology, Gaithersburg, MD, September 14–17, pp. 126–129.

Kickhard, M., and Terveen, L. (1996). *Foundational Issues in Artificial Intelligence and Cognitive Science*, Elsevier, Amsterdam.

Kilmer, R. D., McCain, H. G., Juberts, M., and Legowik, S. A. (1984). Watchdog safety computer design and implementation, *Proceedings of the SME Robots 8 Conference and Exposition*, Detroit, MI, June 4–7.

Kinerva, P. (1988). *Sparse Distributed Memory*, MIT Press, Cambridge, MA.

Knudsen, E., du Lac, S., and Esterly, S. (1987). Computational maps in the brain, *Annual Review of Neuroscience*, *10*, p. 41.

Koenderink, J. J. (1984). The structure of images, *Biological Cybernetics*, *50*.

Koenderink, J., and Van Doorn, A. (1979). The internal representation of solid shape with respect to vision, *Biological Cybernetics*, *32*, pp. 211–216.

Koestler, A. (1967). *The Ghost in the Machine*, Random House, New York.

Koffka, K. (1935). *Principles of Gestalt Psychology*, Harcourt, Brace, New York.

Kohler, W. (1929). *Gestalt Psychology*, Liveright, London.

Kohonen, T. (1977). *Associative Memory: A System-Theoretical Approach*, Springer-Verlag, New York.

Kohonen, T. (1988). *Self-Organization and Associative Memory*, 3rd ed., Springer-Verlag, New York.

Kosslyn, S. (1990). Mental imagery, in *Visual Cognition and Action*, Vol. 2, (D. N. Osherson, S. M. Kosslyn, and J. M. Hollerbach, Eds.), MIT Press, Cambridge, MA.

Kurzweil, R. (1999). *The Age of Spiritual Machines*, Penguin Books, New York, New York.

LabView. *http://www.gngsys.com/cbma/labview.html* and *http://www.ni.com/labview*.

Laird, J., Newell, A., and Rosenbloom, P. (1987). SOAR: an architecture for general intelligence, *Artificial Intelligence*, *33*, pp. 1–64.

Latombe, J. (1991). *Robot Motion Planning*, Kluwer Academic, Boston.

Lea, G. (1975). Chronometric analysis of the method of loci, *Journal of Experimental Psychology: Human Perception and Performance 2*, pp. 95–104.

Leake, S. A., and Kilmer, R. D. (1988). The NBS real-time control system user's reference manual, *NIST Technical Note 1250*, National Institute of Standards and Technology, Gaithersburg, MD.

LeDoux, J. E., and Fellous, J. M. (1998). Emotion and computational neuroscience, in *The Handbook of Brain Theory and Neural Networks* (M. A. Arbib, Ed.), Bradford Books, imprint of MIT Press, Cambridge, MA.

Lenat, D., Guha, R., Pittman, K., Pratt, D., and Shephard, M. (1990). CYC: toward programs with common sense, *Communications of the ACM*, *33*(8), pp. 30–49.

Levine, D. S., and Leven, S. J., Eds. (1992). *Motivation, Emotion, and Goal Direction in Neural Networks*, Lawrence Erlbaum, Hillsdale, NJ.

Lieberman, H., Nardi, B., and Wright, D. (1999). Training agents to recognize text by example, *Proceedings of the ACM Conference on Autonomous Agents*, Seattle, WA.

Lozano-Perez, T. (1981). Automatic planning of manipulator transfer movements, *IEEE Transactions on Systems, Man and Cybernetics*, *11*(10), pp. 581–609.

Lozano-Perez, T., and Wesley, M. (1979). An algorithm for planning collision-free paths among the polyhedral obstacles, *Communications of the ACM*, *22*(10), pp. 560–570.

Lucas, G., Gorman, G., and Pugh, G. (1977). Value-driven decision theory: application to combat simulations, Report DSA-67, Decision Science Applications, Inc., Arlington, VA.

Lumia, R. (1988). CAD-based off-line programming applied to a cleaning and deburring workstation, *Proceedings of the NATO Workshop on CAD-Based Robots*, Il Ciocco, Italy, July.

Lumia, R. (1994). Using NASREM for real-time sensory interactive robot control, *Robotica*, *12*, pp. 127–135.

Lumia, R., Michaloski, J., Russell, R., Wheatley, T., Bake, P., Lee, S., and Steele, R. (1995). Unified telerobotic architecture project (UTAP) standard interface environment (SIE), *NISTIR 5658*, National Institute of Standards and Technology, Gaithersburg, MD, May.

MacLean, P. (1952). Some psychiatric implications of physiological studies on the frontotemporal portion of the limbic system (visceral brain), *Electroencephalography and Clinical Neurophysiology*, *4*, pp. 407–418.

MacLean, P. (1973). *A Triune Concept of the Brain and Behavior*, University of Toronto Press, Toronto, Ontario, Canada.

Maes, P. (1989). The dynamics of action selection, *Proceedings of the 11th International Joint Conference On Artificial Intelligence (IJCAI'89)*, Detroit, MI, pp. 991–997.

Maes, P. (1990). Situated agents can have goals, *Robotics and Autonomous Systems*, *6*, pp. 49–70.

Maes, P. (1994). Agents that reduce work and information overload, *Communications of the ACM*, *37*(7), pp. 31–40.

Maggart, L. E., and Markunas, R. J. (2000). Battlefield dominance through smart technology, *Army Magazine*, *50*(7), pp. 43–46.

Maguire, E., et al. (1997). Recalling routes around London: activation of the right hippocampus in taxi drivers, *Journal of Neuroscience*, *17*, pp. 7103–7110.

Malone, R. (1978). *The Robot Book*, Hartcourt Brace Jovanovich, New York.

Mankiw, G. N. (1992). *Macroeconomics*, Worth Publishers, New York.

Mankiw, G. N. (1995). The growth of nations, *Brookings Papers on Economic Activity*, Vol. 1, Brookings Institution, Washington, DC.

Marlin, T. (1995). *Process Control: Designing Processes and Control Systems for Dynamic Performance*, McGraw-Hill, New York.

Marr, D. (1969). A theory of cerebellar cortex, *Journal of Physiology (London)*, *202*, pp. 437–470.

Marr, D. (1982). *Vision*, W.H. Freeman, San Francisco.

Martin, A. (1998). Organization of semantic knowledge and the origin of words in the brain, in *The Origin and Diversification of Language* (N. Jablonski and L. Aiello, Eds.), *Memoirs of the California Academy of Sciences*, *24*, pp. 69–88.

Martin, A., Haxby, J., Lalonde, F., Wiggs, C., and Ungerleider, L. (1995). Discrete cortical regions associated with knowledge of color and knowledge of action, *Science*, *270*, October 6, pp. 102–105.

Martin, A., Wiggs, C., Ungerleider, L., and Haxby, J. (1996). Neural correlates of catagory-specific knowledge, *Nature*, *379*, February 15, pp. 649–652.

Mataric, M. J. (1992). Integration of representation into goal-driven behavior-based robots, *IEEE Transactions on Robotics and Automation*, *8*(3), pp. 304–312.

MatLab. *http://www.mathworks.com*.

Matthies, L., and Elfes, A. (1987). Sensor integration for robot navigation: combining sonar and stereo range data in a grid-based representation, *Proceedings of the 26th IEEE Decision and Control Conference*, Los Angeles, December 9–11.

Matthies, L., Litwin, T., Owens, K., and Rankin, A. (1996). Performance evaluation of UGV obstacle detection with LADAR and CCD/FLIR stereo systems, *Proceedings of the SPIE 10th Annual AeroSense Symposium, Conference 2738, Navigation and Control*, Technologies for Unmanned Systems, Orlando, FL, April.

Maximov, Y., and Meystel, A. (1992). Optimum design of multiresolutional hierarchical control systems, *Proceedings of the IEEE International Symposium on Intelligent Control*, Glasgow, Scotland, August 11–13, pp. 514–520.

Maxwell, J. (1868). On governors, *Proceedings of the Royal Society (London)*, *16*, pp. 270–283.

McCain, H. G. (1985). Hierarchical controlled, sensory interactive robot in the automated manufacturing research facility, *Proceedings of the IEEE International Conference on Robotics and Automation*, St. Louis, MO, March 25–26.

McCain, H. G., Kilmer, R. D., and Murphy, K. N. (1985). Development of a cleaning and debuffing workstation for the AMRF, *Proceedings of the Deburring and Surface Conditioning '85 Conference*, July.

McCain, H. G., Kilmer, R. D., Szabo, S., and Abrishamian, A. (1986). Hierarchically controlled autonomous robot for heavy payload military field applications, *Proceedings of the International Conference on Intelligent Autonomous Systems*, Amsterdam, December.

McCarthy, J. (1960). Recursive functions of symbolic expressions, *Communications of the Association for Computing Machinery*, *3*.

McCulloch, W., and Pitts, W. (1943). A logical calculus of the ideas immanent in nervous activity, *Bulletin of Mathematical Biophysics*, *5*, pp. 115–133.

McCulloch, W. (1961). What is a number that a man may know it, and a man, that he may know a number? 9th Alfred Korzybski Memorial Lecture, *General Semantics Bulletin*, *26–27*, pp. 115–133, Institute of General Semantics; reprinted in *Embodiments of Mind*, MIT Press, Cambridge, MA, 1965.

McDermott, D. (1982). A temporal logic for reasoning about processes and plans, *Cognitive Science*, *6*, pp. 101–155.

McDermott, D. (1985). Reasoning about plans, in *Formal Theories of the Commonsense World* (J. R. Hobbs, and R. C. Moore, Eds.), Ablex Publishing, Stamford, CT.

Meystel, A. (1982). Intelligent control of a multiactuator system, *Information Control Problems in Manufacturing, Proceedings of the 4th IFAC/IFIP Symposium*, Gaithersburg, MD, October 26–28.

Meystel, A. (1987). Theoretical foundations of planning and navigation for autonomous robots, *International Journal of Intelligent Systems*, *2*, pp. 73–128.

Meystel, A. (1991). *Autonomous Mobile Robots: Vehicles with Cognitive Control*, World Scientific, River Edge, NJ.

Meystel, A. (1994). Multiresolutional system: complexity and reliability, pp. 11–22 in *Intelligent Systems: Safety, Reliability, and Maintainability Issues* (O. Kaynak, G. Honderd, and E. Grant, Eds.), *Computers and Systems Sciences*, Vol. 114, NATO ASI Series F, Springer-Verlag, Berlin.

Meystel, A. (1996). Architectures, representations, and algorithms for intelligent control of robots, pp. 732–788 in *Intelligent Control Systems: Theory and Applications* (M. M. Gupta and N. K. Sinha, Eds.), IEEE Press, New York.

Meystel, A. (1998). Robot path planning, in *The Encyclopedia of Electrical and Electronic Engineers*, Wiley, New York.

Meystel, A. (2000). *Annotated Bibliography of Intelligent Control* (in press).

Meystel, A., and Albus, J. (2002). *Intelligent Systems: Architecture, Design, Control*, Wiley, New York.

Meystel, A., Moskovitz, Y., and Messina, E. (1998). Mission structure for an unmanned vehicle, *Proceedings of the 1998 IEEE International Symposium on Intelligent Control*, a joint conference on the science and technology of intelligent systems, National Institute of Standards and Technology, Gaithersburg, MD, September 14–17, pp. 36–43.

Michaloski, J. (2000). Analysis of module interaction in an OMAC controller, Proceedings of the World Automation Congress Conference (WAC2000), Maui, HI, June 11–16.

Miller, G. A. (1956). The magical number seven, plus or minus two: some limits on our capacity for processing information, *Psychological Review*, *63*, pp. 71–97.

Miller, G. A., and Chomsky, N. (1963). Finitary models of language users, in *Handbook of Mathematical Psychology*, Vol. 2, (R. Luce, R. Bush, and E. Galanter, Eds.), Wiley, New York.

Miller, T., Sutton, R., and Werbos, P. (Eds.) (1990). *Neural Networks for Control*, MIT Press, Cambridge, MA.

Minsky, M. (1975). A framework for representing knowledge, pp. 211–277 in *The Psychology of Computer Vision* (P. Winston, Ed.), McGraw-Hill, New York.

Minsky, M. (1986). *Society of Mind*, Simon & Schuster, New York.

Moravec, H. (1988). Sensor fusion in certainty grids for mobile robots, *AI Magazine*, *9*(2), pp. 61–74.

Moravec, H. (1998). *Mind Children: The Future of Robot and Human Intelligence*, Harvard University Press, Cambridge, MA.

Morganthaler, M., Dickenson, A., and Glass, B. (2000). XUV Demo III multi-vehicle operator control unit, *Proceedings of SPIE Aerosense 2000*, Vol. 4024, *UGV Technology II*, Orlando, FL, April.

Morris, R. G. M. (1981). Spatial location does not require the presence of local cues, *Learning and Motivation*, *12*, pp. 239–260.

Mowrer, O. H. (1960). *Learning Theory and Behavior*, Wiley, New York.

Muller, M. (Trans.) (1881). *Critique of Pure Reason*, London.

Muller, R. U., Kubie, J. L., Bostock, E. M., Taube, J. S., and Quirk, G. J. (1991). Spatial firing correlates of neurons in the hippocampal formation of freely moving rats, pp. 273–295 in *Brain and Space* (J. Paillard, Ed.), Oxford University Press, New York.

Murphy, K. N., and Proctor, F. M. (1990). An advanced deburring and chamfering system, *Proceedings of the IEEE 3rd International Symposium on Robotics and Manufacturing (ISRAM'90)*, Burnaby, British Columbia, Canada, July 18–20.

Murphy, K. N., Norcross, R. J. and Proctor, F. M. (1988). CAD directed robotic deburring, *Proceedings of the 2nd International Symposium on Robotics and Manufacturing Research, Education, and Applications*, Albuquerque, NM, November 16–18.

Murphy, K. N., Juberts, M., Legowik, S., Nashman, M., Schneiderman, H., Scott, H., and Szabo, S. (1993). Ground vehicle control at NIST: from teleoperation to autonomy, *Proceedings of the 7th Annual Space Operations, Applications, and Research Symposium*, Houston, TX, August 3–5.

Nagel, R. N., VanderBrug, G. J., Albus, J. S., and Lowenfeld, E. (1979). Experiments in part acquisition using robot vision, *Proceedings of Autofact II, Robots IV Conference*, Detroit, MI, October 29–November 1.

Narendra, K. (1986). *Adaptive and Learning Systems: Theory and Applications*, Plenum Press, New York.

Newell, A., and Simon, H. (1963). GPS: a program that simulates human thought, in *Computers and Thought* (Feigenbaum, E., and Feldman, J., Eds.), McGraw-Hill, New York.

Newell, A., and Simon, H. (1972). *Human Problem Solving*, Prentice Hall, Englewood Cliffs, NJ.

Newell, A., Shaw, J. C., and Simon, H. A. (1958). Elements of a theory of human problem solving, *Psychological Review*, 65(3), pp. 151–166.

Newell, A., Barnett, J., Forgie, J., Green, C., Klatt, D., Licklider, J., Munson, J., Reddy, D. R., and Woods, W. (1973). *Speech Understanding Systems: Final Report of a Study Group*, North-Holland, Amsterdam and American Elsevier, New York.

Nilsson, N. (1980). *Principles of Artificial Intelligence*, Tioga, Palo Alto, CA.

Norcross, R. J. (1988). Control structure for multi-tasking workstations, *Proceedings of the IEEE International Conference on Robotics and Automation*, Philadelphia, April 24–29, pp. 1133–1135.

Object Management Group (1995). *Common Object Request Broker: Architecture and Specification, Revision 2.0*, OMG, Framingham, MA. See also *http://www.omg.org*.

O'Keefe, J., and Nadel, L. (1978). *The Hippocampus as a Cognitive Map*, Clarendon Press, Oxford.

Ortony, A., Clore, G. L., and Collins, A. (1998). *The Cognitive Structure of Emotions*, Cambridge University Press, New York.

Oskard, D., Hong, T., and Shaffer, C. (1990). Real-time algorithms and data structures for underwater mapping, *Proceedings of SPIE Advances in Intelligent Robotics Systems*, Boston, November 10–11.

Passino, K., and Antsaklis, P. (1989). A system and control theoretic perspective on artificial intelligence planning systems, *Applied Artificial Intelligence*, 3, pp.1–32.

Payton, D. W., (1986). An architecture for reflexive autonomous vehicle control, *Proceedings IEEE Robotics and Automation Conference*, San Francisco.

Peckham, J., and Maryanski, F. (1988). Semantic data models, *ACM Computing Surveys*, 20(3), pp. 153–189.

Peele, T. (1961). *The Neuroanatomic Basis for Clinical Neurology*, 2nd ed., McGraw-Hill, New York.

Penfield, W., and Milner, B. (1958). Memory deficit produced by bilateral lesions in the hippocampal zone, *A. M. A. Arch. Neurol. & Psychiat.*, 79, pp. 475–497.

Penfield, W., and Rasmussen, T. (1950). *The Cerebral Cortex of Man: A Clinical Study of Localization of Function*, Macmillan, New York.

Penrose, R. (1989). *The Emperor's New Mind*, Oxford University Press, New York.

Perret, D. I., Rolls, E. T., and Caan, W. (1982). Visual neurons responsive to faces in the monkey temporal cortex, *Experimental Brain Research, 47*, pp. 329–342.

Perret, D. I., Smith, P. A. J., Potter, D. D., Mistlin, A. J., Head, A. S., Milner, A. D., and Reeves, M. A. (1985). Visual cells in the temporal cortes sensitive to face view and gaze direction, *Proceedings of the Royal Society, B, 223*, pp. 293–317.

Piaget, J. (1952). *The Origins of Intelligence in Children*, International Universities Press, New York.

Piaget, J. (1976). *The Grasp of Consciousness: Action and Concept in the Young Child*, Harvard University Press, Cambridge, MA.

Pinker, S. (1994). *The Language Instinct*, William Morrow, New York.

Pinker, S. (1997). *How the Mind Works*, W.W. Norton, New York.

Polyak, S. (1957). *The Vertebrate Visual System* (H. Kluver, Ed.), University of Chicago Press, Chicago.

Posner, M. I. (Ed.) (1989). *Foundations of Cognitive Science*, MIT Press, Cambridge, MA.

Proctor, F., and Albus, J. S. (1997). Open-architecture controllers, *IEEE Spectrum, 34*(6), pp. 60–64.

Proctor, F., and Michaloski, J. (1993). Enhanced machine controller architecture overview, *NISTIR 5331*, National Institute of Standards and Technology, Gaithersburg, MD, December.

Proctor, F. M., Murphy, K. N., and Norcross, R. J. (1989). Automated robot programming in the cleaning and deburring workstation of the AMRF, *Proceedings of the SME Conference on Deburring and Surface Conditioning '89*, San Diego, CA, February 13–16.

Proctor, F., Michaloski, J., Shackleford, W., and Szabo, S. (1996). Validation of standard interfaces for machine control, *Proceedings of the International Symposium on Robotics and Manufacturing: World Automation Congress '96*, Montpellier, France, May 27–30.

ProEngineer. *http://www.ptc.com/products/proe/index.html.*

Pugh, G. (1977). *The Biological Origin of Human Values*, Basic Books, New York.

Pylyshyn, Z. (1981). The imagery debate: analogue media versus tacit knowledge, *Psychological Review, 87*, pp. 16–45.

Pylyshyn, Z. (1991). The role of cognitive architectures in the theory of cognition, pp. 189–223 in *Architectures for Intelligence* (K. VanLehn, Ed.), Lawrence Erlbaum, Hillsdale, NJ.

Quillian, M. (1968). Semantic memory, pp. 227–270 in *Semantic Information Processing* (M. Minsky, Ed.), MIT Press, Cambridge MA.

Quintero, R., and Barbera, A. J. (1993). A software template approach to building complex large-scale intelligent control systems, *Proceedings of the 8th IEEE International Symposium on Intelligent Control*, Chicago, September 25–27.

Rachlin, H. (1970). *Introduction to Modern Behaviorism*, W.H. Freeman, San Francisco.

Raphael, B. (1968). SIR: semantic information retrieval, in *Semantic Information Processing* (M. Minsky, Ed.), MIT Press, Cambridge, MA.

Raphael, B. (1976). *The Thinking Computer: Mind Inside Matter*, W.H. Freeman, San Francisco.

Rapide. *http://pavg.stanford.edu/rapide.*

Ratliff, F., and Hartline, H. K. (1959). The response of the limulus optic nerve fibers to patterns of illumination on the receptor mosaic, *Journal of General Physiology*, *42*, pp. 1241–1255.

Raviv, D., and Herman, M. (1994). A unified approach to camera fixation and vision-based road following, *IEEE Transactions on Systems, Man and Cybernetics*, *24*(8), pp. 1125–1141.

RCS. *http://www.isd.mel.nist.gov/projects/rcs_lib*.

Richardson, K. (1999). Hyperstructure/brain/cognition, PSYC:PSYCOLOQUY: Refereed Electronic Journal of Peer Discussion in ⟨PSYC@PUCC.PRINCETON.EDU⟩.

Rippey, W. G., and Falco, J. A. (1997). The NIST automated arc welding testbed, *Proceedings of the 7th International Conference on Computer Technology in Welding*, San Francisco, July 8–11.

Riseman, E., and Hanson, A. (1990). A methodology for the development of general knowledge-based vision systems, pp. 285–328 in *Vision, Brain, and Cooperative Computation* (M. A. Arbib and A. R. Hanson, Eds.), Bradford Books, imprint of MIT Press, Cambridge, MA.

Robin, N., and Holyoak, K. J. (1995). Relational complexity and the functions of prefrontal cortex, pp. 987–997 in *The Cognitive Neurosciences* (M. Gassaniga, Ed.), Bradford Books, imprint of MIT Press, Cambridge, MA.

Robinson, J. (1965). A machine-oriented logic based on the resolution principle, *Journal of the Association for Computing Machinery*, *12*(1).

Roeckel, M. W., Rivoir, R. H., and Gibson, R. E. (1999). A behavior based controller architecture and the transition to an industry application, *Proceedings of the 1999 International Symposium on Intelligent Control*, pp. 320–325.

Romer, P. (1990). Endogenous technological change, *Journal of Political Economy*, *Part 2, 98*(5), pp. 71–102.

Romer, P. (1994). The origins of endogenous growth, *Journal of Economic Perspectives*, *8*(1), pp. 3–22.

Rosenblatt, F. (1958). The perceptron: a probabilistic model for information storage and organization in the brain, *Psychological Review*, *65*, pp. 386–408.

Rosenbloom, P. S., Laird, J. E., and Newell, A. (Eds.) (1993). *The Soar Papers: Research on Integrated Intelligence*, MIT Press, Cambridge, MA.

Rosenfeld, A. (1986). Axial representation of shape, *Computer Vision, Graphics and Image Processing*, *33*, pp. 156–173.

Rosenfeld, A., and Kak, A. (1976). *Digital Picture Processing*, Academic Press, New York.

Rosenschein, S. (1981). Plan synthesis: a logical perspective, *Proceedings of the International Joint Conference on Artificial Intelligence*, pp. 331–337.

Ross, W. D. (Ed.) (1928). *Metaphysics, Book 7, The Oxford Aristotle*, Vol. 8, Oxford University Press, Oxford.

Russell, B. (1948). *Human Knowledge: Its Scope and Limits*, Routledge, London.

Sacerdoti, E. (1973). Planning in a hierarchy of abstraction spaces, *SRI Technical Note 78*, Stanford University, Stanford, CA.

Sacerdoti, E. (1975). A structure for plans and behavior, Ph.D. dissertation, *Technical Note 109*, AI Center, SRI International, Menlo Park, CA.

Samuel, A. (1959). Some studies in machine learning using the game of checkers, *IBM Journal of Research and Development, 3*(3).

Samuelson, P., and Nordhaus, W. (1989). *Economics*, 13th ed., McGraw-Hill, New York.

Saridis, G. (1977). *Self-Organizing Control of Stochastic Systems*, Marcel Dekker, New York.

Saridis, G. (1985). Foundations of the theory of intelligent controls, *Proceedings of the IEEE Workshop on Intelligent Control*, Washington, DC.

Saridis, G., and Meystel, A. (Eds.) (1985). *Proceedings of the IEEE Workshop on Intelligent Control*, Troy, NY.

Saridis, G., and Valvanis, K. (1987). On the theory of intelligent controls, *Proceedings of the SPIE Conference on Advances in Intelligent Robotic Systems*, Cambridge, MA, October, pp. 488–95.

Schank, R. C., and Abelson, R. P. (1977). *Scripts, Plans, Goals and Understanding: An Inquiry into Human Knowledge Structures*, Lawrence Erlbaum, Hillsdale, NJ.

Schank, R., and Colby, K. (1973). *Computer Models of Thought and Language*, W.H. Freeman, San Francisco.

Schmajuk, N. (1998). Cognitive maps, pp. 197–199 in *Handbook of Brain Theory and Neural Networks* (M. Arbib, Ed.), Bradford Books, imprint of MIT Press, Cambridge, MA.

Schneiderman, H., and Nashman, M. (1994a). Visual tracking for autonomous driving, *IEEE Transactions on Robotics and Automation, 10*(6), pp. 769–775.

Schneiderman, H., and Nashman, M. (1994b). A discriminating feature tracker for vision-based autonomous driving, *IEEE Transactions on Robotics and Automation*, December.

Schneiderman, H., Wavering, A. J., Nashman, M., and Lumia, R. (1994). Real-time model-based visual tracking, *Proceedings of the Intelligent Robotic Systems '94 Conference*, Grenoble, France, July 11–15.

Schoppers, M. (1987). Universal plans for reactive robots in unpredictable environments, *Proceedings of the IEEE, 77*(1), pp. 81–98.

Scott, H. A. (1996). The inspection workstation-based testbed application for the intelligent systems architecture for manufacturing, *Proceedings of the International Conference on Intelligent Systems: A Semiotic Perspective*, Gaithersburg, MD, October 20–23.

Scott, H., and Strouse, K. (1984). Workstation control in a computer integrated manufacturing system, *Proceedings of Autofact 6*, Anaheim, CA, October.

Searle, J. (1980). Minds, brains, and programs, *Behavioral and Brain Sciences, 3*, pp. 417–424.

Sejnowski, T., and Rosenberg, C. (1987). Parallel networks that learn to pronounce English text, *Complex Systems, 1*, pp. 145–168.

Selby-Bigge, L. (Ed.) (1888). *A Treatise of Human Nature*, Oxford University Press, Oxford.

Selfridge, O. (1959). Pandemonium: a paradigm for learning, in *The Mechanization of Thought Processes*, Her Majesty's Stationary Office, London.

Senehi, M. K., and Kramer, T. R. (1998). A framework for control architectures, *International Journal of Computer Integrated Manufacturing, 11*(4), pp. 347–363.

Senehi, M. K., Kramer, T., Thomas, R., Ray, S., Quintero, R., and Albus, J. (1994a). Hierarchical control architectures from shop level to end effectors, Chapter 2 in *Computer Control of Flexible Manufacturing Systems* (S. Joshi and J. Smith, Eds.), Chapman & Hall, London.

Senehi, M. K., Kramer, T. R., Michaloski, J., Quintero, R., Ray, S. R., Rippey, W. G., and Wallace, S. (1994b). Reference architecture for machine control systems integration: interim report, *NISTIR 5517*, National Institute of Standards and Technology, Gaithersburg, MD.

Shackleford, W. (1999). Real-time control system library: software and documentation, *http://www.isd.mel.nist.gov/projects/rcs_lib/NMLcpp.html*.

Shannon, C. (1950). Automatic chess player, *Scientific American*, *182*(48).

Sheperd, G. (1988). *Neurobiology*, 2nd ed., Oxford University Press, Oxford.

Shoemaker, C., Bornstein, J., Myers, S., and Brendle, B. (1999). Demo III: Department of Defense testbed for unmanned ground mobility, *SPIE Conference on Unmanned Ground Vehicle Technology*, *SPIE Vol. 3693*, Orlando, FL, April.

Simmons, R. (1994). Structured control for autonomous robots, *IEEE Transactions on Robotics and Automation*, *10*(1), pp. 34–43.

Simon, H. (1957). *Models of Man*, Wiley, New York.

Simon, H. (1962). The architecture of complexity, *Proceedings of the American Philosophical Society*, *26*, pp. 467–482.

Simon, H. (1991). Cognitive architectures in a rational analysis: comment, pp. 25–39 in *Architectures for Intelligence* (K. VanLehn, Ed.), Lawrence Erlbaum, Hillsdale, NJ.

Simpson, J. A., Hocken, R. J., and Albus, J. S. (1982). The automated manufacturing research facility of the National Bureau of Standards, *Journal of Manufacturing Systems*, *1*, pp. 17–32.

Skinner, B. F. (1953). *Science and Human Behavior*, Macmillan, New York.

Snippe, H. and Koenderink, J. (1992). Discrimination thresholds for channel-coded systems, *Biological Cybernetics*, *66*, pp. 543–551.

Solow, R. (1970). *Growth Theory: An Exposition*, Oxford University Press, Oxford.

Solow, R. (1994). Perspectives on growth theory, *Journal of Economic Perspectives*, *8*(1), pp. 45–54.

Southall, J. P. C. (Ed.) (1924). *Treatise on Physiological Optics*, Vols. 1–3, translated from the 3rd German edition, Optical Society of America, New York; republished as one volume, Dover, New York, 1962.

Sparks, D. (1986). Translation of sensory signals into commands for the control of saccadic eye movements: role of the primate superior colliculus, *Physiological Reviews*, *66*, pp. 118–171.

Sparks, D., and Groh, J. (1995). The superior colliculus: a window for viewing issues in integrative neuroscience, pp. 565–584 in *The Cognitive Neurosciences* (M. S. Gazzaniga, Ed.), Bradford Books, imprint of MIT Press, Cambridge, MA.

Sparks, D., and Jay, M. (1987). The role of the primate superior colliculus in sensimotor integration, in *Vision, Brain, and Cooperative Computation* (M. A. Arbib and A. R. Hanson, Eds.), MIT Press, Cambridge, MA.

Sparks, D., and Mays, L. (1990). Signal transformation required for the generation of saccadic eye movements, *Annual Review of Neuroscience*, *13*, pp. 309–336.

Sperry, R. W. (1962). Some general aspects of interhemispheric integration, Ch. 3. pp. 43–39 in *Interhemispheric Relations and Cerebral Dominance* (V. B. Montcastle, Ed.), Johns Hopkins Press, Baltimore.

Squire, L., Knowlton, B., et al. (1993). The structure and organization of memory, *Annual Review of Psychology, 44*, pp. 453–495.

Stefik, M. (1981a). Planning with constraints, *Artificial Intelligence, 16*, pp. 111–140.

Stefik, M. (1981b). Planning and meta-planning, *Artificial Intelligence, 16*, pp. 141–170.

STEP (1990). Part 1: Overview and fundamental principles, *ISO TC184/SC4/WG1, Document N494, Version 3*, May.

Stouffer, K., Michaloski, J., Russell, R., and Proctor, F. (1993). ADACS: an automated system for part finishing, *NISTIR 5171*, National Institute of Standards and Technology, Gaithersburg, MD, April; also *Proceedings of the IECON'93 International Conference on Industrial Electronics, Control and Instrumentation*, Maui, Hawaii, November 15–19.

Sutton, R. (1984). Temporal aspects of credit assignment in reinforcement learning, Ph.D. dissertation, University of Massachusetts, Amherst, MA.

Sutton, R., and Barto, A. (1998). *Reinforcement Learning: An Introduction (Adaptive Computation and Machine Learning)*, MIT Press, Cambridge, MA.

Symposia (1988). The slowdown in productivity growth, *Journal of Economic Perspectives, 2*, Fall.

Szabo, S., Scott, H. A., Murphy, K. N., and Legowik, S. A. (1990). Control system architecture for a remotely operated unmanned land vehicle, *Proceedings of the 5th IEEE International Symposium on Intelligent Control*, Philadelphia, September.

Szabo, S., Scott, H. A., Murphy, K. N., Legowik, S. A., and Bostelman, R. V. (1992). High level mobility controller for a remotely operated unmanned land vehicle, *Journal of Intelligent and Robotic Systems, 5*, pp. 63–77.

Tate, A. (1985). A review of knowledge-based planning techniques, in *Expert Systems 85* (M. Merry, Ed.), Cambridge University Press, London.

Thorpe, C., Jochem, T., and Pomerleau, D. (1997). Automated highways and the free agent demonstration, *International Symposium on Robotics Research*, October, *http://www.ri.cmu.edu/people/person_314_pubs.html*.

Thrun, S., Burgard, W., and Fox, D. (2000). A real-time algorithm for mobile robot mapping with applications to multi-robot and 3D mapping, *Proceedings of IEEE International Conference on Robotics and Automation, 1*, pp. 321–328.

Time (1999). Beyond 2000, November 8, p. 65.

Tinbergen, N. (1951). *The Study of Instinct*, Clarendon Press, Oxford.

Toffler, A. (1980). *The Third Wave*, William Morrow, New York.

Tolman, E. D. (1932). Cognitive maps in rats and men, *Psychological Review, 55*, pp. 189–208.

Tomlin, C. D. (1990). *Geographic Information Systems and Cartographic Modeling*, Prentice Hall, Englewood Cliffs, NJ.

Truxel, J. (1955). *Automatic Feedback Control System Synthesis*, McGraw-Hill, New York.

Turing, A. (1950). Computing machinery and intelligence, *Mind*, *59*, pp. 433–460. Reprinted in Feigenbaum, E., and Feldman, J. (Eds.), *Computers and Thought*, McGraw-Hill, New York, 1963.

Ullman, S. (1996). *High-Level Vision: Object Recognition and Visual Cognition*, MIT Press, Cambridge, MA.

UML. *http://www.omg.org/uml*.

U.S. Census Bureau (1999). *Statistical Abstracts of the United States*, 119th ed., Washington, DC.

Van Essen, D. (1985). Functional organization of primate visual cortex, in *Cerebral Cortex*, Vol. 3 (A. Peters and E. Jones, Eds.), Plenum Press, New York.

Van Essen, D., and Deyoe, E. (1995). Concurrent processing in the primate visual cortex, pp. 383–400 in *The Cognitive Neurosciences* (M. Gazzaniga, Ed.), Bradford Books, imprint of MIT Press, Cambridge, MA.

VanderBrug, G. J., Albus, J. S., and Barkmeyer, E. (1979). A vision system for real-time control of robots, *Proceedings of the 9th International Symposium on Industrial Robots*, Washington, DC, March 13–15.

von Holst, E., and Mettelstaedt, H. (1950). Das Reafferenzprinzip, *Naturwissenschafter*, *37*, pp. 464–476.

von Neumann, J., and Morgenstern, O. (1944). *Theory of Games and Economic Behavior*, Princeton University Press, Princeton, NJ.

von Neumann, J. (1966). *Theory of Self-Reproducing Automata* (edited and completed by A. Burks), University of Illinois Press, Urbana, IL.

Vonnegut, K. (1974). *Player Piano*, Mass Market Paperback, amazon.com.

Wandell, B. (1995). *Foundations of Vision*, Sinauer Associates, Sunderland, MA, p. 236.

Warren, W. (1998). The state of flow, in *High Level Motion Processing: Computational, Neurobiological, and Psychophysical Perspectives* (T. Watanabe, Ed.), MIT Press, Cambridge, MA.

Warwick, K. (1996). *An Introduction to Control Systems*, World Scientific, River Edge, NJ.

Watanabe, T. (Ed.) (1998). *High Level Motion Processing: Computational, Neurobiological, and Psychophysical Perspectives*, Bradford Books, imprint of MIT Press, Cambridge, MA.

Watson, J. B. (1913). *Psychological Review*, *20*, pp. 158–167.

Watson, J. B. (1928). *Behaviorism*, London. Reprinted in paperback by W.W. Norton, New York, 1970.

Wavering, A. J., and Fiala, J. C. (1987). Real-time control system of the horizontal workstation robot, *NBSIR 88-3692*, National Institute of Standards and Technology, Gaithersburg, MD, December.

Weiner, N. (1948). *Cybernetics*, MIT Press, Cambridge, MA.

Weiner, N. (1949). *The Extrapolation, Interpolation, and Smoothing of Stationary Time Series*, Wiley, New York.

Werbos, P. (1987). Building and understanding adaptive systems: a statistical/numerical approach to factory automation and brain research, *IEEE Transactions on Systems, Man and Cybernetics*, *17*, pp. 7–19.

Werbos, P. (1994). Neurocontrol and supervised learning: an overview and evaluation, pp. 65–89 in *Handbook of Intelligent Control* (D. White and D. Forge, Eds.), Van Nostrand Reinhold, New York.

Widrow, B. (1995). *Adaptive Inverse Control*, Prentice Hall, Upper Saddle River, NJ.

Wilensky, R. (1983). *Planning and Understanding*, Addison-Wesley, Reading, MA.

Williams, T. J. (1989). A reference model for computer integrated manufacturing from the viewpoint of industrial automation, *International Journal of Computer Integrated Manufacturing*, 2(2), pp. 114–127.

Wilson, W. J. (1996). *When Work Disappears: The World of the New Urban Poor*, Vintage Books, New York.

Winograd, T. (1972). *Understanding Natural Language*, Academic Press, New York.

Winograd, T., and Flores, F. (1986). *Understanding Computers and Cognition*, Addison-Wesley, Reading, MA.

Winston, P. H. (1984). *Artificial Intelligence*, Addison-Wesley, Reading, MA.

Young, M. P., and Yamane, S. (1992). Sparse population coding of faces in inferotemporal cortex, *Science*, 256, pp. 1327–1331.

Zadeh, L. (1965). Fuzzy sets, *Information and Control*, 8, pp. 338–353.

Zadeh, L. A. (1994). Fuzzy logic, neural networks, and soft computing, *Communications of the ACM*, 37(3), pp. 77–84.

Zeki, S., Watson, J. D. G., Lueck, C. J., Friston, K. J., Kennard, C., and Frackowiak, R. S. J. (1991). A direct demonstration of functional specialization in human visual cortex, *Journal of Neuroscience 11*, pp. 641–649.

Zipser, D. (1991). Recurrent network model of the neural mechanism of short-term active memory, *Neural Computation*, 3, pp. 178–192.

INDEX